Lecture Notes in Physics

Volume 846

For further volumes:
http://www.springer.com/series/5304

The Lecture Notes in Physics

The series Lecture Notes in Physics (LNP), founded in 1969, reports new developments in physics research and teaching—quickly and informally, but with a high quality and the explicit aim to summarize and communicate current knowledge in an accessible way. Books published in this series are conceived as bridging material between advanced graduate textbooks and the forefront of research and to serve three purposes:

- to be a compact and modern up-to-date source of reference on a well-defined topic
- to serve as an accessible introduction to the field to postgraduate students and nonspecialist researchers from related areas
- to be a source of advanced teaching material for specialized seminars, courses and schools

Both monographs and multi-author volumes will be considered for publication. Edited volumes should, however, consist of a very limited number of contributions only. Proceedings will not be considered for LNP.

Volumes published in LNP are disseminated both in print and in electronic formats, the electronic archive being available at springerlink.com. The series content is indexed, abstracted and referenced by many abstracting and information services, bibliographic networks, subscription agencies, library networks, and consortia.

Proposals should be sent to a member of the Editorial Board, or directly to the managing editor at Springer:

Christian Caron
Springer Heidelberg
Physics Editorial Department I
Tiergartenstrasse 17
69121 Heidelberg/Germany
christian.caron@springer.com

Éric Gourgoulhon

3+1 Formalism in General Relativity

Bases of Numerical Relativity

Springer

Éric Gourgoulhon
Lab. Univers et Théories (LUTH)
UMR 8102 du CNRS
Observatoire de Paris
Université Paris Diderot
place Jules Janssen 5
92190 Meudon
France

ISSN 0075-8450
ISBN 978-3-642-24524-4
DOI 10.1007/978-3-642-24525-1
Springer Heidelberg New York Dordrecht London

e-ISSN 1616-6361
e-ISBN 978-3-642-24525-1

Library of Congress Control Number: 2011942212

Printed on acid-free paper

Springer is part of Springer Science+Business Media (www.springer.com)

To the memory of
Jean-Alain Marck (1955-2000)

Preface

This monograph originates from lectures given at the General Relativity Trimester at the Institut Henri Poincaré in Paris [1]; at the VII Mexican School on Gravitation and Mathematical Physics in Playa del Carmen (Mexico) [2]; and at the 2008 International Summer School on Computational Methods in Gravitation and Astrophysics held in Pohang (Korea) [3]. It is devoted to the 3+1 formalism of general relativity, which constitutes among other things, the theoretical foundations for numerical relativity. Numerical techniques are not covered here. For a pedagogical introduction to them, we recommend instead the lectures by Choptuik [4] (finite differences) and the review article by Grandclément and Novak [5] (spectral methods), as well as the numerical relativity textbooks by Alcubierre [6], Bona, Palenzuela-Luque and Bona-Casas [7] and Baumgarte and Shapiro [8].

The prerequisites are those of a general relativity course, at the undergraduate or graduate level, like the textbooks by Hartle [9] or Carroll [10], or part I of Wald's book [11], as well as track 1 of the book by Misner, Thorne and Wheeler [12]. The fact that the present text consists of lecture notes implies two things:

- the calculations are rather detailed (the experienced reader might say *too* detailed), with an attempt to made them self-consistent and complete, trying to use as infrequently as possible the famous phrases "as shown in paper XXX" or "see paper XXX for details";
- the bibliographical references do not constitute an extensive survey of the literature on the subject: articles have been cited in so far as they have a direct connection with the main text.

The book starts with a chapter setting the mathematical background, which is differential geometry, at a basic level (Chap. 2). This is followed by two purely geometrical chapters devoted to the study of a single hypersurface embedded in spacetime (Chap. 3) and to the foliation (or slicing) of spacetime by a family of spacelike hypersurfaces (Chap. 4). The presentation is divided in two chapters to distinguish between concepts which are meaningful for a single hypersurface and those that rely on a foliation. The decomposition of the Einstein equation relative to the foliation is given in Chap. 5, giving rise to the Cauchy problem with constraints, which constitutes the core of the 3+1 formalism. The ADM Hamiltonian formulation of general relativity is also introduced in this chapter. Chapter 6 is devoted to the decomposition of the matter and electromagnetic field equations, focusing on the astrophysically relevant cases of a perfect fluid and a perfect conductor (ideal MHD). An important technical chapter occurs then: Chap. 7 introduces some conformal transformation of the 3-metric on each hypersurface and the corresponding rewriting of the 3+1 Einstein equations. As a by-product, we also discuss the Isenberg-Wilson-Mathews (or conformally flat) approximation to general relativity. Chapter 8 details the various global quantities associated with asymptotic flatness (ADM mass, ADM linear momentum and angular momentum) or with some symmetries (Komar mass and Komar angular momentum). In Chap. 9, we study the initial data problem, presenting with some examples two classical methods: the conformal transversetraceless method and the conformal thin-sandwich one. Both methods rely on the conformal decomposition that has been introduced in Chap. 7. The choice of spacetime coordinates within the 3+1 framework is discussed in Chap. 10, starting from the choice of foliation before discussing the choice of the three coordinates in each leaf of the foliation. The major coordinate families used in modern numerical relativity are reviewed. Finally Chap. 11 presents various schemes for the time integration of the 3+1 Einstein equations, putting some emphasis on the most successful scheme to date, the BSSN one. Appendix A is devoted to basic tools of the 3+1 formalism: the conformal Killing operator and the related vector Laplacian, whereas Appendix B provides some computer algebra codes based on the Sage system.

A web page is dedicated to the book, at the URL

http://relativite.obspm.fr/3p1

This page contains the errata, the clickable list of references, the computer algebra codes described in Appendix B and various supplementary material. Readers are invited to use this page to report any error that they may find in the text.

I am deeply indebted to Michał Bejger, Philippe Grandclément, Alexandre Le Tiec, Yuichiro Sekiguchi and Nicolas Vasset for the careful reading of a preliminary version of these notes. I am very grateful to my friends and colleagues Thomas Baumgarte, Michał Bejger, Luc Blanchet, Silvano Bonazzola, Brandon Carter, Isabel Cordero-Carrión, Thibault Damour, Nathalie Deruelle, Guillaume Faye, John Friedman, Philippe Grandclément, José Maria Ibáñez, José Luis Jaramillo, Jean-Pierre Lasota, Jérôme Novak, Jean-Philippe Nicolas, Motoyuki Saijo, Masaru Shibata, Keisuke Taniguchi, Koji Uryu, Nicolas Vasset and Loïc Villain, for the numerous and fruitful discussions that we had about general relativity and the 3+1 formalism. I also warmly thank Robert Beig and Christian Caron for their invitation to publish this text in the Lecture Notes in Physics series.

Meudon, September 2011 Éric Gourgoulhon

References

1. http://www.luth.obspm.fr/IHP06/
2. http://www.smf.mx/~dgfm-smf/EscuelaVII/
3. http://apctp.org/conferences/
4. Choptuik, M.W.: Numerical analysis for numerical relativists, lecture at the VII Mexican school on gravitation and mathematical physics, Playa del Carmen (Mexico), 26 November-1 December 2006 [2]; available at http://laplace.physics.ubc.ca/People/matt/Teaching/06Mexico/
5. Grandclément, P. and Novak, J.: Spectral methods for numerical relativity, Living Rev. Relat. 12, 1 (2009); http://www.livingreviews.org/lrr-2009-1
6. Alcubierre, M.: Introduction to 3+1 Numerical Relativity. Oxford University Press, Oxford (2008)
7. Bona, C., Palenzuela-Luque, C. and Bona-Casas, C.: Elements of numerical relativity and relativistic hydrodynamics: from einstein's equations to astrophysical simulations (2nd edition). Springer, Berlin (2009)
8. Baumgarte, T. W. and Shapiro, S. L.: Numerical relativity. Solving Einstein's equations on the computer, Cambridge University Press, Cambridge (2010)
9. Hartle, J.B.: Gravity: An introduction to Einstein's general relativity, Addison Wesley(Pearson Education), San Fransisco (2003); http://wps.aw.com/aw_hartle_gravity_1/0,6533,512494-,00.html
10. Carroll, S.M.: Spacetime and geometry: an introduction to general relativity, Addison Wesley (Pearson Education), San Fransisco (2004); http://preposterousuniverse.com/spacetimeandgeometry/
11. Wald, R.M.: General relativity, University of Chicago Press, Chicago (1984)
12. Misner, C.W., Thorne, K.S. and Wheeler, J.A.: Gravitation, Freeman, New York (1973)

Contents

Acronyms

ADM	Arnowitt–Deser–Misner
BSSN	Baumgarte-Shapiro-Shibata-Nakamura
CMC	Constant mean curvature
CTS	Conformal thin sandwich
CTT	Conformal transverse traceless
IWM	Isenberg–Wilson–Mathews
MHD	Magnetohydrodynamics
PDE	Partial differential equation
PN	Post-Newtonian
TT	Transverse traceless
XCTS	Extended conformal thin sandwich

Chapter 1
Introduction

The ***3+1 formalism*** is an approach to general relativity that relies on the slicing of the four-dimensional spacetime by three-dimensional surfaces (*hypersurfaces*). These hypersurfaces have to be spacelike, so that the metric induced on them by the *Lorentzian* spacetime metric [signature $(-, +, +, +)$] is *Riemannian* [signature $(+, +, +)$]. From the mathematical point of view, this procedure allows to formulate the problem of resolution of Einstein equations as a *Cauchy problem* with constraints. From the pedestrian point of view, it amounts to a decomposition of spacetime into "space" + "time", so that one manipulates only time-varying tensor fields in some "ordinary" three-dimensional space, where the scalar product is Riemannian. One should stress that this space + time splitting is not some a priori structure of general relativity but relies on the somewhat arbitrary choice of a time coordinate. The 3 + 1 formalism should not be confused with the *1+3 formalism* (cf. e.g. Ref. [1]), where the basic structure is a congruence of one-dimensional curves (mostly timelike curves, i.e. worldlines), instead of a family of three-dimensional surfaces.

The 3+1 formalism originates from studies by Georges Darmois in the 1920s [2], André Lichnerowicz in the 1930–1940s [3–5] and Yvonne Choquet-Bruhat (at that time Yvonne Fourès-Bruhat) in the 1950s [6, 7].[1] Notably, in 1952, Yvonne Choquet-Bruhat was able to show that the Cauchy problem arising from the 3+1 decomposition has locally a unique solution [6]. In the late 1950s and early 1960s, the 3+1 formalism received some considerable impulse, being employed to develop Hamiltonian formulations of general relativity by Dirac [8, 9], and Arnowitt, Deser and Misner [10], referred to by the famous ADM initials. At the same epoch, Wheeler put forward the concept of *geometrodynamics* and coined the names *lapse* and *shift* [11]. In the 1970s, the 3+1 formalism became the basic tool for the nascent numerical relativity. A primordial role has been played by York, who developed a general method to solve the initial data problem [12] and who put the 3+1 equations in the shape used afterwards by the numerical community [13]. In the 1980 and 1990s, numerical

[1] These three persons have some direct affiliation: Georges Darmois was the thesis adviser of André Lichnerowicz, who was himself the thesis adviser of Yvonne Choquet-Bruhat.

É. Gourgoulhon, *3+1 Formalism in General Relativity*, Lecture Notes in Physics 846, DOI: 10.1007/978-3-642-24525-1_1, © Springer-Verlag Berlin Heidelberg 2012

computations increased in complexity, from 1D (spherical symmetry) to 3D (no symmetry at all). In parallel, a lot of studies have been devoted to formulating the 3+1 equations in a form suitable for numerical implementation. The authors who participated to this effort are too numerous to be cited here but it is certainly relevant to mention Takashi Nakamura and his school, who among other things initiated the formulation which would become the popular *BSSN scheme* [14–17]. Needless to say, a strong motivation for the expansion of numerical relativity has been the development of gravitational wave detectors, either ground-based (LIGO, VIRGO, GEO600, TAMA, LCGT) or in space (LISA/NGO project).

Today, most numerical codes for solving Einstein equations are based on the 3+1 formalism. Other approaches are the 2+2 formalism or characteristic formulation, as reviewed by Winicour [18], the conformal field equations by Friedrich [19] as reviewed by Frauendiener [20], or the generalized harmonic decomposition used by Pretorius [21–23] and the Cornell-Caltech group [24, 25] for computing binary black hole mergers.

References

1. Ellis, G.F.R., van Elst, H.: Cosmological models. In: Lachièze-Rey, M. (ed.) Theoretical and Observational Cosmology: Proceedings of the NATO Advanced Study Institute on Theoretical and Observational Cosmology, p. 1. Kluwer Academic, Boston (1999)
2. Darmois, G.: Les équations de la gravitation einsteinienne, Mémorial des Sciences Mathématiques 25. Gauthier-Villars, Paris (1927)
3. Lichnerowicz, A.: Sur certains problèmes globaux relatifs au système des équations d'Einstein. Hermann, Paris (1939). http://www.numdam.org/item?id=THESE_1939__226__1_0
4. Lichnerowicz, A.: L'intégration des équations de la gravitation relativiste et le problème des n corps, J. Math. Pures Appl. 23, 37 (1944); reprinted in Lichnerowicz, A.: Choix d'œuvres mathématiques, p. 4. Hermann, Paris (1982)
5. Lichnerowicz, A.: Sur les équations relativistes de la gravitation. Bulletin de la S.M.F. **80**, 237 (1952). http://www.numdam.org/numdam-bin/item?id=BSMF_1952__80__237_0
6. Fourès-Bruhat, Y. (Choquet-Bruhat, Y.): Théorème d'existence pour certains systèmes d'équations aux dérivées partielles non linéaires. Acta Mathematica **88**, 141 (1952). http://fanfreluche.math.univ-tours.fr
7. Fourès-Bruhat, Y. (Choquet-Bruhat, Y.): Sur l'Intégration des Équations de la Relativité Générale. J. Ration. Mech. Anal. **5** (1956)
8. Dirac, P.A.M.: The theory of gravitation in Hamiltonian form. Proc. Roy. Soc. Lond. A **246**, 333 (1958)
9. Dirac, P.A.M.: Fixation of coordinates in the Hamiltonian theory of gravitation. Phys. Rev. **114**, 924 (1959)
10. Arnowitt, R., Deser, S., Misner, C.W.: The dynamics of general relativity. In: Witten, L. (ed.) Gravitation: An Introduction to Current Research, p. 227. Wiley, New York (1962). http://arxiv.org/abs/gr-qc/0405109
11. Wheeler, J.A.: Geometrodynamics and the issue of the final state. In: DeWitt, C., DeWitt, B.S. (eds.) Relativity, Groups and Topology, p. 316. Gordon and Breach, New York (1964)

12. York, J.W.: Conformally invariant orthogonal decomposition of symmetric tensors on Riemannian manifolds and the initial-value problem of general relativity. J. Math. Phys. **14**, 456 (1973)
13. York, J.W.: Kinematics and dynamics of general relativity. In: Smarr, L.L. (ed.) Sources of gravitational radiation. Cambridge University Press, Cambridge (1979)
14. Nakamura, T., Oohara, K., Kojima, Y.: General relativistic collapse to black holes and gravitational waves from black holes. Prog. Theor. Phys. Suppl. **90**, 1 (1987)
15. Nakamura, T.: 3D numerical relativity. In: Sasaki, M. (ed.) Relativistic Cosmology, Proceedings of the 8th Nishinomiya-Yukawa Memorial Symposium. Universal Academy Press, Tokyo (1994)
16. Shibata, M., Nakamura, T.: Evolution of three-dimensional gravitational waves: harmonic slicing case. Phys. Rev. D **52**, 5428 (1995)
17. Baumgarte, T.W., Shapiro, S.L.: Numerical integration of Einstein's field equations. Phys. Rev. D **59**, 024007 (1999)
18. Winicour, J.: Characteristic evolution and matching. Living Rev. Relativity **12**, 3 (2009). http://www.livingreviews.org/lrr-2009-3
19. Friedrich, H.: Conformal Einstein evolution, in Ref. [26], p. 1
20. Frauendiener, J.: Conformal infinity. Living Rev. Relativity **7**, 1 (2004). http://www.livingreviews.org/lrr-2004-1
21. Pretorius, F.: Numerical relativity using a generalized harmonic decomposition. Class. Quantum Grav. **22**, 425 (2005)
22. Pretorius, F.: Evolution of binary black-hole spacetimes. Phys. Rev. Lett. **95**, 121101 (2005)
23. Pretorius, F.: Simulation of binary black hole spacetimes with a harmonic evolution scheme. Class. Quantum Grav. **23**, S529 (2006)
24. Lindblom, L., Scheel, M.A., Kidder, L.E., Owen, R., Rinne, O.: A new generalized harmonic evolution system. Class. Quantum Grav. **23**, S447 (2006)
25. Scheel, M.A., Boyle, M., Chu, T., Kidder, L.E., Matthews, K.D., Pfeiffer, H.P.: High-accuracy waveforms for binary black hole inspiral, merger, and ringdown. Phys. Rev. D **79**, 024003 (2009)
26. Frauendiener, J., Friedrich, H. (eds.): The Conformal Structure of Space-Times: Geometry, Analysis, Numerics. Lecture Notes in Physics, vol. 604, Springer, Heidelberg (2002)

Chapter 2
Basic Differential Geometry

Abstract This first chapter recapitulates the basic concepts of differential geometry that are used throughout the book. This encompasses differentiable manifolds, tensor fields, affine connections, metric tensors, pseudo-Riemannian manifolds, Levi–Civita connections, curvature tensors and Lie derivatives. The dimension of the manifold and the signature of the metric are kept general so that the results can be subsequently applied either to the whole spacetime or to some submanifold of it.

2.1 Introduction

The mathematical language of general relativity is mostly differential geometry. We recall in this chapter basic definitions and results in this field, which we will use throughout the book. The reader who has some knowledge of general relativity should be familiar with most of them. We recall them here to make the text fairly self-contained and also to provide definitions with sufficient generality, not limited to the dimension 4—the standard spacetime dimension. Indeed we will manipulate manifolds whose dimension differs from 4, such as hypersurfaces (the building blocks of the 3+1 formalism !) or 2-dimensional surfaces. In the same spirit, we do not stick to Lorentzian metrics (such as the spacetime one) but discuss pseudo-Riemannian metrics, which encompass both Lorentzian metrics and Riemannian ones. Accordingly, in this chapter, $\mathscr{M}$ denotes a generic manifold of any dimension and $\boldsymbol{g}$ a pseudo-Riemannian metric on $\mathscr{M}$. In the subsequent chapters, the symbol $\mathscr{M}$ will be restricted to the spacetime manifold and the symbol $\boldsymbol{g}$ to a Lorentzian metric on $\mathscr{M}$.

This chapter is not intended to a be a lecture on differential geometry, but a collection of basic definitions and useful results. In particular, contrary to the other chapters, we state many results without proofs, referring the reader to classical textbooks on the topic [1–6].

É. Gourgoulhon, *3+1 Formalism in General Relativity*, Lecture Notes in Physics 846, DOI: 10.1007/978-3-642-24525-1_2, © Springer-Verlag Berlin Heidelberg 2012

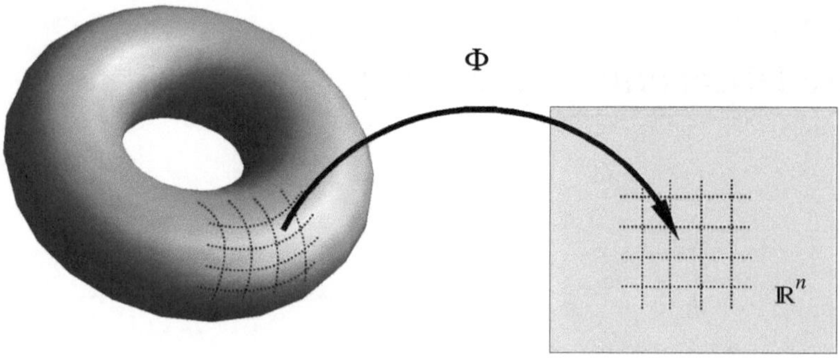

Fig. 2.1 Locally a manifold resembles $\mathbb{R}^n$ ($n = 2$ on the figure), but this is not necessarily the case at the global level

2.2 Differentiable Manifolds

2.2.1 Notion of Manifold

Given an integer $n \geq 1$, a ***manifold of dimension*** n is a topological space $\mathscr{M}$ obeying the following properties:

1. $\mathscr{M}$ is a ***separated space*** (also called ***Hausdorff space***): any two distinct points of $\mathscr{M}$ admit disjoint open neighbourhoods.
2. $\mathscr{M}$ has a ***countable base***[1]: there exists a countable family $(\mathscr{U}_k)_{k\in\mathbb{N}}$ of open sets of $\mathscr{M}$ such that any open set of $\mathscr{M}$ can be written as the union (possibly infinite) of some members of the above family.
3. Around each point of $\mathscr{M}$, there exists a neighbourhood which is homeomorphic to an open subset of $\mathbb{R}^n$.

Property 1 excludes manifolds with "forks" and is very reasonable from a physical point of view: it allows to distinguish between two points even after a small perturbation. Property 2 excludes "too large" manifolds; in particular it permits setting up the theory of integration on manifolds. It also allows for a differentiable manifold of dimension n to be embedded smoothly into the Euclidean space $\mathbb{R}^{2n}$ (Whitney theorem). Property 3 expresses the essence of a manifold: it means that, locally, one can label the points of $\mathscr{M}$ in a continuous way by n real numbers $(x^\alpha)_{\alpha\in\{0,\ldots,n-1\}}$, which are called ***coordinates*** (cf. Fig. 2.1). More precisely, given an open subset $\mathscr{U} \subset \mathscr{M}$, a ***coordinate system*** or ***chart*** on $\mathscr{U}$ is a homeomorphism[2]

$$\begin{aligned} \Phi : \mathscr{U} \subset \mathscr{M} &\longrightarrow \Phi(\mathscr{U}) \subset \mathbb{R}^n \\ p &\longmapsto (x^0, \ldots, x^{n-1}). \end{aligned} \tag{2.1}$$

[1] In the language of topology, one says that $\mathscr{M}$ is a *second-countable space*.

[2] Let us recall that a ***homeomorphism*** between two topological spaces (here $\mathscr{U}$ and $\Phi(\mathscr{U})$) is a one-to-one map Φ such that both Φ and Φ^{-1} are continuous.

Remark 2.1 In relativity, it is customary to label the n coordinates by an index ranging from 0 to $n-1$. Actually, this convention is mostly used when $\mathscr{M}$ is the spacetime manifold ($n=4$ in standard general relativity). The computer-oriented reader will have noticed the similarity with the index ranging of arrays in the C/C++ or Python programming languages.

Remark 2.2 Strictly speaking the definition given above is that of a ***topological manifold***. We are saying *manifold* for short.

Usually, one needs more than one coordinate system to cover $\mathscr{M}$. An ***atlas*** on $\mathscr{M}$ is a finite set of couples $(\mathscr{U}_k, \Phi_k)_{1\le k\le K}$, where K is a non-zero integer, $\mathscr{U}_k$ an open set of $\mathscr{M}$ and Φ_k a chart on $\mathscr{U}_k$, such that the union of all $\mathscr{U}_k$ covers $\mathscr{M}$:

$$\bigcup_{k=1}^{K} \mathscr{U}_k = \mathscr{M}. \tag{2.2}$$

The above definition of a manifold lies at the *topological* level (Remark 2.2), meaning that one has the notion of continuity, but not of differentiability. To provide the latter, one should rely on the differentiable structure of $\mathbb{R}^n$, via the atlases: a ***differentiable manifold*** is a manifold $\mathscr{M}$ equipped with an atlas $(\mathscr{U}_k, \Phi_k)_{1\le k\le K}$ such that for any non-empty intersection $\mathscr{U}_i \cap \mathscr{U}_j$, the mapping

$$\Phi_i \circ \Phi_j^{-1} : \Phi_j(\mathscr{U}_i \cap \mathscr{U}_j) \subset \mathbb{R}^n \longrightarrow \Phi_i(\mathscr{U}_i \cap \mathscr{U}_j) \subset \mathbb{R}^n \tag{2.3}$$

is differentiable (i.e. C^∞). Note that the above mapping is from an open set of $\mathbb{R}^n$ to an open set of $\mathbb{R}^n$, so that the invoked differentiability is nothing but that of $\mathbb{R}^n$. The atlas $(\mathscr{U}_k, \Phi_k)_{1\le k\le K}$ is called a ***differentiable atlas***. In the following, we consider only differentiable manifolds.

Remark 2.3 We are using the word *differentiable* for C^∞, i.e. *smooth*.

Given two differentiable manifolds, $\mathscr{M}$ and $\mathscr{M}'$, of respective dimensions n and n', we say that a map $\phi : \mathscr{M} \to \mathscr{M}'$ is ***differentiable*** iff in some (and hence all) coordinate systems of $\mathscr{M}$ and $\mathscr{M}'$ (belonging to the differentiable atlases of $\mathscr{M}$ and $\mathscr{M}'$), the coordinates of the image $\phi(p)$ are differentiable functions $\mathbb{R}^n \to \mathbb{R}^{n'}$ of the coordinates of p. The map ϕ is said to be a ***diffeomorphism*** iff it is one-to-one and both ϕ and ϕ^{-1} are differentiable. This implies $n = n'$.

Remark 2.4 Strictly speaking a differentiable manifold is a couple $(\mathscr{M}, \mathscr{A})$ where $\mathscr{A}$ is a (maximal) differentiable atlas on $\mathscr{M}$. Indeed a given (topological) manifold $\mathscr{M}$ can have non-equivalent differentiable structures, as shown by Milnor (1956) [7] in the specific case of the unit sphere of dimension 7, $\mathbb{S}^7$: there exist differentiable manifolds, the so-called *exotic spheres*, that are homeomorphic to $\mathbb{S}^7$ but not diffeomorphic to $\mathbb{S}^7$. On the other side, for $n \le 6$, there is a unique differentiable structure for the sphere $\mathbb{S}^n$. Moreover, any manifold of dimension $n \le 3$ admits a unique differentiable structure. Amazingly, in the case of $\mathbb{R}^n$, there exists a unique differentiable structure (the standard one) for any $n \neq 4$, but for $n = 4$ (the space-time case !) there exist uncountably many non-equivalent differentiable structures, the so-called *exotic* $\mathbb{R}^4$ [8].

2.2.2 Vectors on a Manifold

On a manifold, vectors are defined as tangent vectors to a curve. A ***curve*** is a subset $\mathscr{C} \subset \mathscr{M}$ which is the image of a differentiable function $\mathbb{R} \to \mathscr{M}$:

$$\begin{aligned} P : \mathbb{R} &\longrightarrow \mathscr{M} \\ \lambda &\longmapsto p = P(\lambda) \in \mathscr{C}. \end{aligned} \tag{2.4}$$

Hence $\mathscr{C} = \{P(\lambda) | \lambda \in \mathbb{R}\}$. The function P is called a ***parametrization*** of $\mathscr{C}$ and the variable λ is called a ***parameter along*** $\mathscr{C}$. Given a coordinate system (x^α) in a neighbourhood of a point $p \in \mathscr{C}$, the parametrization P is defined by n functions $X^\alpha : \mathbb{R} \to \mathbb{R}$ such that

$$x^\alpha(P(\lambda)) = X^\alpha(\lambda). \tag{2.5}$$

A ***scalar field*** on $\mathscr{M}$ is a function $f : \mathscr{M} \to \mathbb{R}$. In practice, we will always consider differentiable scalar fields. At a point $p = P(\lambda) \in \mathscr{C}$, the ***vector tangent to*** $\mathscr{C}$ associated with the parametrization P is the operator $\boldsymbol{v}$ which maps every scalar field f to the real number

$$\boldsymbol{v}(f) = \left.\frac{\mathrm{d}f}{\mathrm{d}\lambda}\right|_{\mathscr{C}} := \lim_{\varepsilon \to 0} \frac{1}{\varepsilon} \left[f(P(\lambda + \varepsilon)) - f(P(\lambda)) \right]. \tag{2.6}$$

Given a coordinate system (x^α) around some point $p \in \mathscr{M}$, there are n curves $\mathscr{C}_\alpha$ through p associated with (x^α) and called the ***coordinate lines***: for each $\alpha \in \{0, \ldots, n-1\}$, $\mathscr{C}_\alpha$ is defined as the curve through p parametrized by $\lambda = x^\alpha$ and having constant coordinates x^β for all $\beta \neq \alpha$. The vector tangent to $\mathscr{C}_\alpha$ parametrized by x^α is denoted $\boldsymbol{\partial}_\alpha$. Its action on a scalar field f is by definition

$$\boldsymbol{\partial}_\alpha(f) = \left.\frac{\mathrm{d}f}{\mathrm{d}x^\alpha}\right|_{\mathscr{C}_\alpha} = \left.\frac{\mathrm{d}f}{\mathrm{d}x^\alpha}\right|_{\substack{x^\beta = \mathrm{const} \\ \beta \neq \alpha}}.$$

Considering f as a function of the coordinates $(x^0, \ldots, x^{n-1})$ (whereas strictly speaking it is a function of the points on $\mathscr{M}$) we recognize in the last term the partial derivative of f with respect to x^α. Hence

$$\boxed{\boldsymbol{\partial}_\alpha(f) = \frac{\partial f}{\partial x^\alpha}}. \tag{2.7}$$

Similarly, we may rewrite (2.6) as

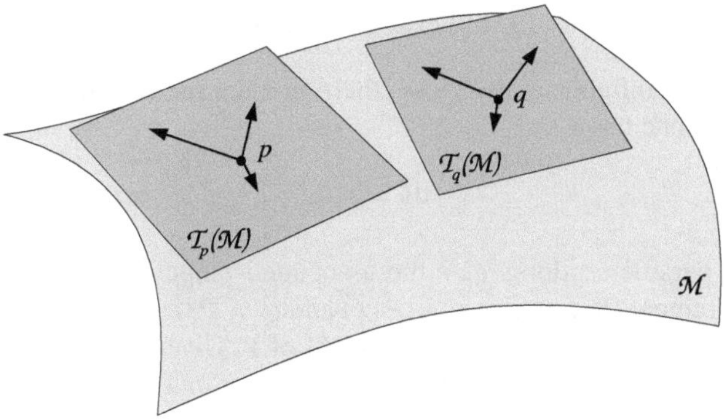

Fig. 2.2 The vectors at two points p and q on the manifold $\mathscr{M}$ belong to two different vector spaces: the tangent spaces $\mathscr{T}_p(\mathscr{M})$ and $\mathscr{T}_q(\mathscr{M})$

$$\begin{aligned} \boldsymbol{v}(f) &= \lim_{\varepsilon \to 0} \frac{1}{\varepsilon}\left[f(X^0(\lambda+\varepsilon), \ldots, X^{n-1}(\lambda+\varepsilon)) - f(X^0(\lambda), \ldots, X^{n-1}(\lambda))\right] \\ &= \frac{\partial f}{\partial x^\alpha}\frac{\mathrm{d}X^\alpha}{\mathrm{d}\lambda} = \boldsymbol{\partial}_\alpha(f)\frac{\mathrm{d}X^\alpha}{\mathrm{d}\lambda}. \end{aligned}$$

In the above equation and throughout the all book, we are using ***Einstein summation convention***: a repeated index implies a summation over all the possible values of this index (here from $\alpha = 0$ to $\alpha = n-1$). The above identity being valid for any scalar field f, we conclude that

$$\boxed{\boldsymbol{v} = v^\alpha \boldsymbol{\partial}_\alpha}, \tag{2.8}$$

with the n real numbers

$$v^\alpha := \frac{\mathrm{d}X^\alpha}{\mathrm{d}\lambda}, \quad 0 \le \alpha \le n-1. \tag{2.9}$$

Since every vector tangent to a curve at p is expressible as (2.8), we conclude that the set of all vectors tangent to a curve at p is a vector space of dimension n and that $(\boldsymbol{\partial}_\alpha)$ constitutes a basis of it. This vector space is called the ***tangent vector space to*** $\mathscr{M}$ ***at*** p and is denoted $\mathscr{T}_p(\mathscr{M})$. The elements of $\mathscr{T}_p(\mathscr{M})$ are simply called ***vectors*** at p. The basis $(\boldsymbol{\partial}_\alpha)$ is called the ***natural basis*** associated with the coordinates (x^α) and the coefficients v^α in (2.8) are called the ***components of the vector*** $\boldsymbol{v}$ ***with respect to the coordinates*** (x^α). The tangent vector space is represented at two different points in Fig. 2.2.

Contrary to what happens for an affine space, one cannot, in general, define a vector connecting two points p and q on a manifold, except if p and q are infinitesimally close to each other. Indeed, in the latter case, we may define the ***infinitesimal displacement vector from*** p ***to*** q as the vector $\mathrm{d}\boldsymbol{\ell} \in \mathscr{T}_p(\mathscr{M})$ whose action on a scalar field f is

$$\mathbf{d}\boldsymbol{\ell}(f) = \mathrm{d}f|_{p\to q} = f(q) - f(p). \tag{2.10}$$

Since p and q are infinitesimally close, there is a unique (piece of) curve $\mathscr{C}$ going from p to q and one has

$$\mathbf{d}\boldsymbol{\ell} = \boldsymbol{v}\mathrm{d}\lambda, \tag{2.11}$$

where λ is a parameter along $\mathscr{C}$, $\boldsymbol{v}$ the associated tangent vector at p and $\mathrm{d}\lambda$ the parameter increment from p to q: $p = P(\lambda)$ and $q = P(\lambda + \mathrm{d}\lambda)$. The relation (2.11) follows immediately from the definition (2.6) of $\boldsymbol{v}$. Given a coordinate system, let (x^α) be the coordinates of p and $(x^\alpha + \mathrm{d}x^\alpha)$ those of q. Then from Eq. (2.10),

$$\mathbf{d}\boldsymbol{\ell}(f) = \mathrm{d}f = \frac{\partial f}{\partial x^\alpha}\mathrm{d}x^\alpha = \mathrm{d}x^\alpha \boldsymbol{\partial}_\alpha(f).$$

The scalar field f being arbitrary, we conclude that

$$\boxed{\mathbf{d}\boldsymbol{\ell} = \mathrm{d}x^\alpha \boldsymbol{\partial}_\alpha}. \tag{2.12}$$

In other words, the components of the infinitesimal displacement vector with respect to the coordinates (x^α) are nothing but the infinitesimal coordinate increments $\mathrm{d}x^\alpha$.

2.2.3 Linear Forms

A fundamental operation on vectors consists in mapping them to real numbers, and this in a linear way. More precisely, at each point $p \in \mathscr{M}$, one defines a ***linear form*** as a mapping[3]

$$\begin{aligned} \boldsymbol{\omega} : \mathscr{T}_p(\mathscr{M}) &\longrightarrow \mathbb{R} \\ \boldsymbol{v} &\longmapsto \langle \boldsymbol{\omega}, \boldsymbol{v} \rangle \end{aligned} \tag{2.13}$$

that is linear: $\langle \boldsymbol{\omega}, \lambda \boldsymbol{v} + \boldsymbol{u} \rangle = \lambda \langle \boldsymbol{\omega}, \boldsymbol{v} \rangle + \langle \boldsymbol{\omega}, \boldsymbol{u} \rangle$ for all $\boldsymbol{u}, \boldsymbol{v} \in \mathscr{T}_p(\mathscr{M})$ and $\lambda \in \mathbb{R}$. The set of all linear forms at p constitutes a n-dimensional vector space, which is called the ***dual space of*** $\mathscr{T}_p(\mathscr{M})$ and denoted by $\mathscr{T}_p^*(\mathscr{M})$. Given the natural basis $(\boldsymbol{\partial}_\alpha)$ of $\mathscr{T}_p(\mathscr{M})$ associated with some coordinates (x^α), there is a unique basis of $\mathscr{T}_p^*(\mathscr{M})$, denoted by $(\mathbf{d}x^\alpha)$, such that

$$\boxed{\langle \mathbf{d}x^\alpha, \boldsymbol{\partial}_\beta \rangle = \delta^\alpha{}_\beta}, \tag{2.14}$$

where $\delta^\alpha{}_\beta$ is the ***Kronecker symbol***: $\delta^\alpha{}_\beta = 1$ if $\alpha = \beta$ and 0 otherwise. The basis $(\mathbf{d}x^\alpha)$ is called the ***dual basis*** of the basis $(\boldsymbol{\partial}_\alpha)$. The notation $(\mathbf{d}x^\alpha)$ stems from the

---

[3] We are using the same bra-ket notation as in quantum mechanics to denote the action of a linear form on a vector.

fact that if we apply the linear form $\mathbf{d}x^\alpha$ to the infinitesimal displacement vector (2.12), we get nothing but the number $\mathrm{d}x^\alpha$:

$$\langle \mathbf{d}x^\alpha, \mathrm{d}\boldsymbol{\ell}\rangle = \langle \mathbf{d}x^\alpha, \mathrm{d}x^\beta \boldsymbol{\partial}_\beta\rangle = \mathrm{d}x^\beta \underbrace{\langle \mathbf{d}x^\alpha, \boldsymbol{\partial}_\beta\rangle}_{\delta^\alpha{}_\beta} = \mathrm{d}x^\alpha. \tag{2.15}$$

Remark 2.5 Do not confuse the linear form $\mathbf{d}x^\alpha$ with the infinitesimal increment $\mathrm{d}x^\alpha$ of the coordinate x^α.

The dual basis can be used to expand any linear form $\boldsymbol{\omega}$, thereby defining its ***components*** ω_α ***with respect to the coordinates*** (x^α):

$$\boldsymbol{\omega} = \omega_\alpha \mathbf{d}x^\alpha. \tag{2.16}$$

In terms of components, the action of a linear form on a vector takes then a very simple form:

$$\boxed{\langle \boldsymbol{\omega}, \boldsymbol{v}\rangle = \omega_\alpha v^\alpha}. \tag{2.17}$$

This follows immediately from (2.16), (2.8) and (2.14).

A field of linear forms, i.e. a (smooth) map which associates to each point $p \in \mathscr{M}$ a member of $\mathscr{T}_p(\mathscr{M})$ is called a ***1-form***. Given a smooth scalar field f on $\mathscr{M}$, there exists a 1-form canonically associated with it, called the ***gradient of*** f and denoted $\boldsymbol{\nabla} f$. At each point $p \in \mathscr{M}$, $\boldsymbol{\nabla} f$ is the unique linear form which, once applied to the infinitesimal displacement vector $\mathrm{d}\boldsymbol{\ell}$ from p to a nearby point q, gives the change in f between points p and q:

$$\mathrm{d}f := f(q) - f(p) = \langle \boldsymbol{\nabla} f, \mathrm{d}\boldsymbol{\ell}\rangle. \tag{2.18}$$

Since $\mathrm{d}f = \partial f/\partial x^\alpha \mathrm{d}x^\alpha$, Eq. (2.15) implies that the components of the gradient with respect to the dual basis are nothing but the partial derivatives of f with respect to the coordinates (x^α):

$$\boxed{\boldsymbol{\nabla} f = \frac{\partial f}{\partial x^\alpha}\mathbf{d}x^\alpha}. \tag{2.19}$$

Remark 2.6 In non-relativistic physics, the gradient is very often considered as a vector field and not as a 1-form. This is so because one associates implicitly a vector $\vec{\omega}$ to any 1-form $\boldsymbol{\omega}$ via the Euclidean scalar product of $\mathbb{R}^3$: $\forall \vec{v} \in \mathbb{R}^3$, $\langle \boldsymbol{\omega}, \vec{v}\rangle = \vec{\omega} \cdot \vec{v}$. Accordingly, formula (2.18) is rewritten as $\mathrm{d}f = \vec{\boldsymbol{\nabla}} f \cdot \mathrm{d}\boldsymbol{\ell}$. But we should keep in mind that, fundamentally, the gradient is a linear form and not a vector.

Remark 2.7 For a fixed value of α, the coordinate x^α can be considered as a scalar field on $\mathscr{M}$. If we apply (2.19) to $f = x^\alpha$, we then get

$$\boldsymbol{\nabla} x^\alpha = \mathbf{d}x^\alpha. \tag{2.20}$$

Hence the dual basis to the natural basis $(\boldsymbol{\partial}_\alpha)$ is formed by the gradients of the coordinates.

Of course the natural bases are not the only possible bases in the vector space $\mathscr{T}_p(\mathscr{M})$. One may use a basis $(\boldsymbol{e}_\alpha)$ that is not related to a coordinate system on $\mathscr{M}$, for instance an orthonormal basis with respect to some metric. There exists then a unique basis $(\boldsymbol{e}^\alpha)$ of the dual space $\mathscr{T}_p^*(\mathscr{M})$ such that[4]

$$\boxed{\langle \boldsymbol{e}^\alpha, \boldsymbol{e}_\beta \rangle = \delta^\alpha{}_\beta}. \tag{2.21}$$

$(\boldsymbol{e}^\alpha)$ is called the ***dual basis*** to $(\boldsymbol{e}_\alpha)$. The relation (2.14) is a special case of (2.21), for which $\boldsymbol{e}_\alpha = \boldsymbol{\partial}_\alpha$ and $\boldsymbol{e}^\alpha = \mathbf{d}x^\alpha$.

2.2.4 Tensors

Tensors are generalizations of both vectors and linear forms. At a point $p \in \mathscr{M}$, a ***tensor of type*** (k, ℓ) with $(k, \ell) \in \mathbb{N}^2$, also called ***tensor k times contravariant and ℓ times covariant***, is a mapping

$$\begin{aligned} \boldsymbol{T} : \underbrace{\mathscr{T}_p^*(\mathscr{M}) \times \cdots \times \mathscr{T}_p^*(\mathscr{M})}_{k \text{ times}} \times \underbrace{\mathscr{T}_p(\mathscr{M}) \times \cdots \times \mathscr{T}_p(\mathscr{M})}_{\ell \text{ times}} &\longrightarrow \mathbb{R} \\ (\boldsymbol{\omega}_1, \ldots, \boldsymbol{\omega}_k, \boldsymbol{v}_1, \ldots, \boldsymbol{v}_\ell) &\longmapsto \boldsymbol{T}(\boldsymbol{\omega}_1, \ldots, \boldsymbol{\omega}_k, \boldsymbol{v}_1, \ldots, \boldsymbol{v}_\ell) \end{aligned} \tag{2.22}$$

that is linear with respect to each of its arguments. The integer $k + \ell$ is called the tensor ***valence***, or sometimes the tensor ***rank*** or ***order***. Let us recall the canonical duality $\mathscr{T}_p^{**}(\mathscr{M}) = \mathscr{T}_p(\mathscr{M})$, which means that every vector $\boldsymbol{v}$ can be considered as a linear form on the space $\mathscr{T}_p^*(\mathscr{M})$, via $\mathscr{T}_p^*(\mathscr{M}) \to \mathbb{R}$, $\boldsymbol{\omega} \mapsto \langle \boldsymbol{\omega}, \boldsymbol{v} \rangle$. Accordingly a vector is a tensor of type $(1, 0)$. A linear form is a tensor of type $(0, 1)$. A tensor of type $(0, 2)$ is called a ***bilinear form***. It maps couples of vectors to real numbers, in a linear way for each vector.

Given a basis $(\boldsymbol{e}_\alpha)$ of $\mathscr{T}_p(\mathscr{M})$ and the corresponding dual basis $(\boldsymbol{e}^\alpha)$ in $\mathscr{T}_p^*(\mathscr{M})$, we can expand any tensor $\boldsymbol{T}$ of type (k, ℓ) as

$$\boxed{\boldsymbol{T} = T^{\alpha_1 \ldots \alpha_k}{}_{\beta_1 \ldots \beta_\ell}\, \boldsymbol{e}_{\alpha_1} \otimes \ldots \otimes \boldsymbol{e}_{\alpha_k} \otimes \boldsymbol{e}^{\beta_1} \otimes \ldots \otimes \boldsymbol{e}^{\beta_\ell}}, \tag{2.23}$$

where the ***tensor product*** $\boldsymbol{e}_{\alpha_1} \otimes \ldots \otimes \boldsymbol{e}_{\alpha_k} \otimes \boldsymbol{e}^{\beta_1} \otimes \ldots \otimes \boldsymbol{e}^{\beta_\ell}$ is the tensor of type (k, ℓ) for which the image of $(\boldsymbol{\omega}_1, \ldots, \boldsymbol{\omega}_k, \boldsymbol{v}_1, \ldots, \boldsymbol{v}_\ell)$ as in (2.22) is the real number

$$\prod_{i=1}^{k} \langle \boldsymbol{\omega}_i, \boldsymbol{e}_{\alpha_i} \rangle \times \prod_{j=1}^{\ell} \langle \boldsymbol{e}^{\beta_j}, \boldsymbol{v}_j \rangle.$$

[4] Notice that, according to the standard usage, the symbol for the vector $\boldsymbol{e}_\alpha$ and that for the linear form $\boldsymbol{e}^\alpha$ differ only by the position of the index α.

Notice that all the products in the above formula are simply products in $\mathbb{R}$. The $n^{k+\ell}$ scalar coefficients $T^{\alpha_1\ldots\alpha_k}{}_{\beta_1\ldots\beta_\ell}$ in (2.23) are called the ***components of the tensor T with respect to the basis*** $(\boldsymbol{e}_\alpha)$. These components are unique and fully characterize the tensor $\boldsymbol{T}$.

Remark 2.8 The notations v^α and ω_α already introduced for the components of a vector $\boldsymbol{v}$ [Eq. (2.8)] or a linear form $\boldsymbol{\omega}$ [Eq. (2.16)] are of course the particular cases $(k, \ell) = (1, 0)$ or $(k, \ell) = (0, 1)$ of (2.23), with, in addition, $\boldsymbol{e}_\alpha = \boldsymbol{\partial}_\alpha$ and $\boldsymbol{e}^\alpha = \mathbf{d}x^\alpha$.

2.2.5 Fields on a Manifold

A ***tensor field*** of type (k, ℓ) is a map which associates to each point $p \in \mathscr{M}$ a tensor of type (k, ℓ) on $\mathscr{T}_p(\mathscr{M})$. By convention, a scalar field is considered as a tensor field of type $(0, 0)$. We shall consider only smooth fields.

Given an integer p, a ***p-form*** is a tensor field of type $(0, p)$, i.e. a field of p-linear forms, that is fully antisymmetric whenever $p \geq 2$. This definition generalizes that of a 1-form given in Sect. 2.2.3.

A ***frame field*** or ***moving frame*** is a n-uplet of vector fields $(\boldsymbol{e}_\alpha)$ such that at each point $p \in \mathscr{M}$, $(\boldsymbol{e}_\alpha(p))$ is a basis of the tangent space $\mathscr{T}_p(\mathscr{M})$. If $n = 4$, a frame field is also called a ***tetrad*** and if $n = 3$, it is called a ***triad***.

Given a vector field $\boldsymbol{v}$ and a scalar field f, the function $\mathscr{M} \to \mathbb{R}$, $p \mapsto \boldsymbol{v}|_p(f)$ clearly defines a scalar field on $\mathscr{M}$, which we denote naturally by $\boldsymbol{v}(f)$. We may then define the ***commutator of two vector fields u*** and ***v*** as the vector field $[\boldsymbol{u}, \boldsymbol{v}]$ whose action on a scalar field f is

$$[\boldsymbol{u}, \boldsymbol{v}](f) := \boldsymbol{u}(\boldsymbol{v}(f)) - \boldsymbol{v}(\boldsymbol{u}(f)). \tag{2.24}$$

With respect to a coordinate system (x^α), it is not difficult, via (2.8), to see that the components of the commutator are

$$\boxed{[\boldsymbol{u}, \boldsymbol{v}]^\alpha = u^\mu \frac{\partial v^\alpha}{\partial x^\mu} - v^\mu \frac{\partial u^\alpha}{\partial x^\mu}}. \tag{2.25}$$

2.3 Pseudo-Riemannian Manifolds

2.3.1 Metric Tensor

A ***pseudo-Riemannian manifold*** is a couple $(\mathscr{M}, \boldsymbol{g})$ where $\mathscr{M}$ is a differentiable manifold and $\boldsymbol{g}$ is a ***metric tensor*** on $\mathscr{M}$, i.e. a tensor field obeying the following properties:

1. $\boldsymbol{g}$ is a tensor field of type (0, 2): at each point $p \in \mathscr{M}$, $\boldsymbol{g}(p)$ is a bilinear form acting on vectors in the tangent space $\mathscr{T}_p(\mathscr{M})$:

$$\begin{aligned} \boldsymbol{g}(p) : \mathscr{T}_p(\mathscr{M}) \times \mathscr{T}_p(\mathscr{M}) &\longrightarrow \mathbb{R} \\ (\boldsymbol{u}, \boldsymbol{v}) \qquad\qquad &\longmapsto \boldsymbol{g}(\boldsymbol{u}, \boldsymbol{v}). \end{aligned} \tag{2.26}$$

2. $\boldsymbol{g}$ is ***symmetric***: $\boldsymbol{g}(\boldsymbol{u}, \boldsymbol{v}) = \boldsymbol{g}(\boldsymbol{v}, \boldsymbol{u})$.
3. $\boldsymbol{g}$ is ***non-degenerate***: at any point $p \in \mathscr{M}$, a vector $\boldsymbol{u}$ such that $\forall \boldsymbol{v} \in \mathscr{T}_p(\mathscr{M})$, $\boldsymbol{g}(\boldsymbol{u}, \boldsymbol{v}) = 0$ is necessarily the null vector.

The properties of being symmetric and non-degenerate are typical of a ***scalar product***. Accordingly, one says that two vectors $\boldsymbol{u}$ and $\boldsymbol{v}$ are ***g-orthogonal*** (or simply ***orthogonal*** if there is no ambiguity) iff $\boldsymbol{g}(\boldsymbol{u}, \boldsymbol{v}) = 0$. Moreover, when there is no ambiguity on the metric (usually the spacetime metric), we are using a dot to denote the scalar product of two vectors taken with $\boldsymbol{g}$:

$$\forall (\boldsymbol{u}, \boldsymbol{v}) \in \mathscr{T}_p(\mathscr{M}) \times \mathscr{T}_p(\mathscr{M}), \quad \boxed{\boldsymbol{u} \cdot \boldsymbol{v} = \boldsymbol{g}(\boldsymbol{u}, \boldsymbol{v})}. \tag{2.27}$$

In a given basis $(\boldsymbol{e}_\alpha)$ of $\mathscr{T}_p(\mathscr{M})$, the components of $\boldsymbol{g}$ is the matrix $(g_{\alpha\beta})$ defined by formula (2.23) with $(k, \ell) = (0, 2)$:

$$\boldsymbol{g} = g_{\alpha\beta} \boldsymbol{e}^\alpha \otimes \boldsymbol{e}^\beta. \tag{2.28}$$

For any couple $(\boldsymbol{u}, \boldsymbol{v})$ of vectors we have then

$$\boldsymbol{g}(\boldsymbol{u}, \boldsymbol{v}) = g_{\alpha\beta} u^\alpha v^\beta. \tag{2.29}$$

In particular, considering the natural basis associated with some coordinate system (x^α), the scalar square of an infinitesimal displacement vector $\mathrm{d}\boldsymbol{\ell}$ [cf. Eq. (2.10)] is

$$\mathrm{d}s^2 := \boldsymbol{g}(\mathrm{d}\boldsymbol{\ell}, \mathrm{d}\boldsymbol{\ell}) = g_{\alpha\beta} \mathrm{d}x^\alpha \mathrm{d}x^\beta. \tag{2.30}$$

This formula, which follows from the value (2.12) of the components of $\mathrm{d}\boldsymbol{\ell}$, is called the expression of the ***line element*** on the pseudo-Riemannian manifold $(\mathscr{M}, \boldsymbol{g})$. It is often used to define the metric tensor in general relativity texts. Note that contrary to what the notation may suggest, $\mathrm{d}s^2$ is not necessarily a positive quantity.

2.3.2 Signature and Orthonormal Bases

An important feature of the metric tensor is its ***signature***: in all bases of $\mathscr{T}_p(\mathscr{M})$ where the components $(g_{\alpha\beta})$ form a diagonal matrix, there is necessarily the same number, s say, of negative components and the same number, $n - s$, of positive components. The independence of s from the choice of the basis where $(g_{\alpha\beta})$ is diagonal is a basic result of linear algebra named ***Sylvester's law of inertia***. One writes:

$$\operatorname{sign} \boldsymbol{g} = (\underbrace{-, \ldots, -}_{s \text{ times}}, \underbrace{+, \ldots, +}_{n-s \text{ times}}). \tag{2.31}$$

If $s = 0$, $\boldsymbol{g}$ is called a ***Riemannian metric*** and $(\mathscr{M}, \boldsymbol{g})$ a ***Riemannian manifold***. In this case, $\boldsymbol{g}$ is ***positive-definite***, which means that

$$\forall \boldsymbol{v} \in \mathscr{T}_p(\mathscr{M}), \quad \boldsymbol{g}(\boldsymbol{v}, \boldsymbol{v}) \geq 0 \tag{2.32}$$

and $\boldsymbol{g}(\boldsymbol{v}, \boldsymbol{v}) = 0$ iff $\boldsymbol{v} = 0$. A standard example of Riemannian metric is of course the scalar product of the Euclidean space $\mathbb{R}^n$.

If $s = 1$, $\boldsymbol{g}$ is called a ***Lorentzian metric*** and $(\mathscr{M}, \boldsymbol{g})$ a ***Lorentzian manifold***. One may then have $\boldsymbol{g}(\boldsymbol{v}, \boldsymbol{v}) < 0$; vectors for which this occurs are called ***timelike***, whereas vectors for which $\boldsymbol{g}(\boldsymbol{v}, \boldsymbol{v}) > 0$ are called ***spacelike***, and those for which $\boldsymbol{g}(\boldsymbol{v}, \boldsymbol{v}) = 0$ are called ***null***. The subset of $\mathscr{T}_p(\mathscr{M})$ formed by all null vectors is termed the ***null cone*** of $\boldsymbol{g}$ at p.

A basis $(\boldsymbol{e}_\alpha)$ of $\mathscr{T}_p(\mathscr{M})$ is called a g-***orthonormal basis*** (or simply ***orthonormal basis*** if there is no ambiguity on the metric) iff [5]

$$\begin{aligned} &\boldsymbol{g}(\boldsymbol{e}_\alpha, \boldsymbol{e}_\alpha) = -1 \quad && \text{for} \quad 0 \leq \alpha \leq s-1 \\ &\boldsymbol{g}(\boldsymbol{e}_\alpha, \boldsymbol{e}_\alpha) = 1 \quad && \text{for} \quad s \leq \alpha \leq n-1 \\ &\boldsymbol{g}(\boldsymbol{e}_\alpha, \boldsymbol{e}_\beta) = 0 \quad && \text{for} \quad \alpha \neq \beta. \end{aligned} \tag{2.33}$$

2.3.3 Metric Duality

Since the bilinear form $\boldsymbol{g}$ is non-degenerate, its matrix $(g_{\alpha\beta})$ in any basis $(\boldsymbol{e}_\alpha)$ is invertible and the inverse is denoted by $(g^{\alpha\beta})$:

$$\boxed{g^{\alpha\mu} g_{\mu\beta} = \delta^\alpha_\beta}. \tag{2.34}$$

The metric $\boldsymbol{g}$ induces an isomorphism between $\mathscr{T}_p(\mathscr{M})$ (vectors) and $\mathscr{T}_p^*(\mathscr{M})$ (linear forms) which, in index notation, corresponds to the lowering or raising of the index by contraction with $g_{\alpha\beta}$ or $g^{\alpha\beta}$. In the present book, an index-free symbol will always denote a tensor with a fixed covariance type (such as a vector, a 1-form, a bilinear form, etc.). We will therefore use a different symbol to denote its image under the metric isomorphism. In particular, we denote by an underbar the isomorphism $\mathscr{T}_p(\mathscr{M}) \to \mathscr{T}_p^*(\mathscr{M})$ and by an arrow the reverse isomorphism $\mathscr{T}_p^*(\mathscr{M}) \to \mathscr{T}_p(\mathscr{M})$:

1. For any vector $\boldsymbol{u}$ in $\mathscr{T}_p(\mathscr{M})$, $\underline{\boldsymbol{u}}$ stands for the unique linear form such that

$$\forall \boldsymbol{v} \in \mathscr{T}_p(\mathscr{M}), \quad \langle \underline{\boldsymbol{u}}, \boldsymbol{v} \rangle = \boldsymbol{g}(\boldsymbol{u}, \boldsymbol{v}). \tag{2.35}$$

[5] No summation on α is implied.

However, we will omit the underbar on the components of $\underline{\boldsymbol{u}}$, since the position of the index allows us to distinguish between vectors and linear forms, following the standard usage: if the components of $\boldsymbol{u}$ in a given basis $(\boldsymbol{e}_\alpha)$ are denoted by u^α, the components of $\underline{\boldsymbol{u}}$ in the dual basis $(\boldsymbol{e}^\alpha)$ are then denoted by u_α and are given by

$$u_\alpha = g_{\alpha\mu} u^\mu. \tag{2.36}$$

2. For any linear form $\boldsymbol{\omega}$ in $\mathscr{T}_p^*(\mathscr{M})$, $\vec{\boldsymbol{\omega}}$ stands for the unique vector of $\mathscr{T}_p(\mathscr{M})$ such that

$$\forall \boldsymbol{v} \in \mathscr{T}_p(\mathscr{M}), \quad \boldsymbol{g}(\vec{\boldsymbol{\omega}}, \boldsymbol{v}) = \langle \boldsymbol{\omega}, \boldsymbol{v} \rangle. \tag{2.37}$$

As for the underbar, we will omit the arrow on the components of $\vec{\boldsymbol{\omega}}$ by denoting them ω^α; they are given by

$$\omega^\alpha = g^{\alpha\mu} \omega_\mu. \tag{2.38}$$

3. We extend the arrow notation to *bilinear* forms on $\mathscr{T}_p(\mathscr{M})$ (type-(0, 2) tensor): for any bilinear form $\boldsymbol{T}$, we denote by $\vec{\boldsymbol{T}}$ the tensor of type (1, 1) such that

$$\forall (\boldsymbol{u}, \boldsymbol{v}) \in \mathscr{T}_p(\mathscr{M}) \times \mathscr{T}_p(\mathscr{M}), \quad \boldsymbol{T}(\boldsymbol{u}, \boldsymbol{v}) = \vec{\boldsymbol{T}}(\underline{\boldsymbol{u}}, \boldsymbol{v}) = \boldsymbol{u} \cdot \vec{\boldsymbol{T}}(\boldsymbol{v}), \tag{2.39}$$

and by $\overset{\twoheadrightarrow}{\boldsymbol{T}}$ the tensor of type (2, 0) defined by

$$\forall (\boldsymbol{u}, \boldsymbol{v}) \in \mathscr{T}_p(\mathscr{M}) \times \mathscr{T}_p(\mathscr{M}), \quad \boldsymbol{T}(\boldsymbol{u}, \boldsymbol{v}) = \overset{\twoheadrightarrow}{\boldsymbol{T}}(\underline{\boldsymbol{u}}, \underline{\boldsymbol{v}}). \tag{2.40}$$

Note that in the second equality of (2.39), we have considered $\vec{\boldsymbol{T}}$ as an endomorphism $\mathscr{T}_p(\mathscr{M}) \to \mathscr{T}_p(\mathscr{M})$, which is always possible for a tensor of type (1, 1). If $T_{\alpha\beta}$ are the components of $\boldsymbol{T}$ in some basis $(\boldsymbol{e}_\alpha)$, the components of $\vec{\boldsymbol{T}}$ and $\overset{\twoheadrightarrow}{\boldsymbol{T}}$ are respectively

$$(\vec{\boldsymbol{T}})^\alpha{}_\beta = T^\alpha{}_\beta = g^{\alpha\mu} T_{\mu\beta} \tag{2.41}$$

$$(\overset{\twoheadrightarrow}{\boldsymbol{T}})^{\alpha\beta} = T^{\alpha\beta} = g^{\alpha\mu} g^{\beta\nu} T_{\mu\nu}. \tag{2.42}$$

Remark 2.9 In mathematical literature, the isomorphism induced by $\boldsymbol{g}$ between $\mathscr{T}_p(\mathscr{M})$ and $\mathscr{T}_p^*(\mathscr{M})$ is called the ***musical isomorphism***, because a flat or a sharp symbol is used instead of, respectively, the underbar and the arrow introduced above:

$$\boldsymbol{u}^\flat = \underline{\boldsymbol{u}} \quad \text{and} \quad \boldsymbol{\omega}^\sharp = \vec{\boldsymbol{\omega}}.$$

2.3.4 Levi–Civita Tensor

Let us assume that $\mathscr{M}$ is an ***orientable manifold***, i.e. that there exists a n-form[6] on $\mathscr{M}$ (n being $\mathscr{M}$'s dimension) that is continuous on $\mathscr{M}$ and nowhere vanishing. Then, given a metric $\boldsymbol{g}$ on $\mathscr{M}$, one can show that there exist only two n-forms $\boldsymbol{\varepsilon}$ such that for any $\boldsymbol{g}$-orthonormal basis $(\boldsymbol{e}_\alpha)$,

$$\boldsymbol{\varepsilon}(\boldsymbol{e}_0, \ldots, \boldsymbol{e}_{n-1}) = \pm 1. \tag{2.43}$$

Picking one of these two n-forms amounts to choosing an ***orientation*** for $\mathscr{M}$. The chosen $\boldsymbol{\varepsilon}$ is then called the ***Levi-Civita tensor*** associated with the metric $\boldsymbol{g}$. Bases for which the right-hand side of (2.43) is $+1$ are called ***right-handed***, and those for which it is -1 are called ***left-handed***. More generally, given a (not necessarily orthonormal) basis $(\boldsymbol{e}_\alpha)$ of $\mathscr{T}_p(\mathscr{M})$, one has necessarily $\boldsymbol{\varepsilon}(\boldsymbol{e}_0, \ldots, \boldsymbol{e}_{n-1}) \neq 0$ and one says that the basis is ***right-handed*** or ***left-handed*** iff $\boldsymbol{\varepsilon}(\boldsymbol{e}_0, \ldots, \boldsymbol{e}_{n-1}) > 0$ or < 0, respectively.

If (x^α) is a coordinate system on $\mathscr{M}$ such that the corresponding natural basis $(\boldsymbol{\partial}_\alpha)$ is right-handed, then the components of $\boldsymbol{\varepsilon}$ with respect to (x^α) are given by

$$\boxed{\varepsilon_{\alpha_1 \ldots \alpha_n} = \sqrt{|g|}\, [\alpha_1, \ldots, \alpha_n]}, \tag{2.44}$$

where g stands for the determinant of the matrix of $\boldsymbol{g}$'s components with respect to the coordinates (x^α):

$$\boxed{g := \det(g_{\alpha\beta})} \tag{2.45}$$

and the symbol $[\alpha_1, \ldots, \alpha_n]$ takes the value 0 if any two indices $(\alpha_1, \ldots, \alpha_n)$ are equal and takes the value 1 or -1 if $(\alpha_1, \ldots, \alpha_n)$ is an even or odd permutation, respectively, of $(0, \ldots, n-1)$.

2.4 Covariant Derivative

2.4.1 Affine Connection on a Manifold

Let us denote by $\mathscr{T}(\mathscr{M})$ the space of smooth vector fields on $\mathscr{M}$.[7] Given a vector field $\boldsymbol{v} \in \mathscr{T}(\mathscr{M})$, it is not possible from the manifold structure alone to define its variation between two neighbouring points p and q. Indeed a formula like $\mathrm{d}\boldsymbol{v} := \boldsymbol{v}(q) - \boldsymbol{v}(p)$ is

---

6 Cf. Sect. 2.2.5 for the definition of a n-form.

7 The experienced reader is warned that $\mathscr{T}(\mathscr{M})$ does not stand for the tangent bundle of $\mathscr{M}$; it rather corresponds to the space of smooth cross-sections of that bundle. No confusion should arise because we shall not use the notion of bundle.

meaningless because the vectors $\boldsymbol{v}(q)$ and $\boldsymbol{v}(p)$ belong to two different vector spaces, $\mathcal{T}_q(\mathcal{M})$ and $\mathcal{T}_p(\mathcal{M})$ respectively (cf. Fig. 2.2). Note that for a scalar field, this problem does not arise [cf. Eq. (2.18)]. The solution is to introduce an extra-structure on the manifold, called an *affine connection* because, by defining the variation of a vector field, it allows one to connect the various tangent spaces on the manifold

An ***affine connection*** on $\mathcal{M}$ is a mapping

$$\begin{aligned} \boldsymbol{\nabla} : \mathcal{T}(\mathcal{M}) \times \mathcal{T}(\mathcal{M}) &\longrightarrow \mathcal{T}(\mathcal{M}) \\ (\boldsymbol{u}, \boldsymbol{v}) &\longmapsto \boldsymbol{\nabla}_{\boldsymbol{u}} \boldsymbol{v} \end{aligned} \tag{2.46}$$

which satisfies the following properties:

1. $\boldsymbol{\nabla}$ is bilinear (considering $\mathcal{T}(\mathcal{M})$ as a vector space over $\mathbb{R}$).
2. For any scalar field f and any pair $(\boldsymbol{u}, \boldsymbol{v})$ of vector fields:

$$\boldsymbol{\nabla}_{f\boldsymbol{u}} \boldsymbol{v} = f \boldsymbol{\nabla}_{\boldsymbol{u}} \boldsymbol{v}. \tag{2.47}$$

3. For any scalar field f and any pair $(\boldsymbol{u}, \boldsymbol{v})$ of vector fields, the following Leibniz rule holds:

$$\boldsymbol{\nabla}_{\boldsymbol{u}}(f\boldsymbol{v}) = \langle \boldsymbol{\nabla} f, \boldsymbol{u} \rangle \boldsymbol{v} + f \boldsymbol{\nabla}_{\boldsymbol{u}} \boldsymbol{v}, \tag{2.48}$$

where $\boldsymbol{\nabla} f$ stands for the gradient of f as defined in Sect. 2.2.3.

The vector $\boldsymbol{\nabla}_{\boldsymbol{u}} \boldsymbol{v}$ is called the ***covariant derivative of*** $\boldsymbol{v}$ ***along*** $\boldsymbol{u}$.

Remark 2.10 Property 2 is not implied by property 1, for f is a scalar field, not a real number. Actually, property 2 ensures that at a given point $p \in \mathcal{M}$, the value of $\boldsymbol{\nabla}_{\boldsymbol{u}} \boldsymbol{v}$ depends only on the vector $\boldsymbol{u}(p) \in \mathcal{T}_p(\mathcal{M})$ and not on the behaviour of $\boldsymbol{u}$ around p; therefore the role of $\boldsymbol{u}$ is only to give the direction of the derivative of $\boldsymbol{v}$.

Given an affine connection, the variation of a vector field $\boldsymbol{v}$ between two neighbouring points p and q, is defined as

$$\mathrm{d}\boldsymbol{v} := \boldsymbol{\nabla}_{\mathrm{d}\boldsymbol{\ell}} \boldsymbol{v}, \tag{2.49}$$

$\mathrm{d}\boldsymbol{\ell}$ being the infinitesimal displacement vector connecting p and q [cf. Eq. (2.10)]. One says that $\boldsymbol{v}$ is ***parallelly transported from*** p ***to*** q ***with respect to the connection*** $\boldsymbol{\nabla}$ iff $\mathrm{d}\boldsymbol{v} = 0$.

Given a frame field $(\boldsymbol{e}_\alpha)$ on $\mathcal{M}$, the ***connection coefficients*** of an affine connection $\boldsymbol{\nabla}$ with respect to $(\boldsymbol{e}_\alpha)$ are the scalar fields $\Gamma^\gamma{}_{\alpha\beta}$ defined by the expansion, at each point $p \in \mathcal{M}$, of the vector $\boldsymbol{\nabla}_{\boldsymbol{e}_\beta} \boldsymbol{e}_\alpha(p)$ onto the basis $(\boldsymbol{e}_\alpha(p))$:

$$\boxed{\boldsymbol{\nabla}_{\boldsymbol{e}_\beta} \boldsymbol{e}_\alpha =: \Gamma^\mu{}_{\alpha\beta} \boldsymbol{e}_\mu}. \tag{2.50}$$

An affine connection is entirely defined by the connection coefficients. In other words, there are as many affine connections on a manifold of dimension n as there are possibilities of choosing n^3 scalar fields $\Gamma^\gamma{}_{\alpha\beta}$.

Given $\boldsymbol{v} \in \mathscr{T}(\mathscr{M})$, one defines a tensor field of type $(1, 1)$, $\boldsymbol{\nabla v}$, called the ***covariant derivative of v with respect to the affine connection*** $\boldsymbol{\nabla}$, by the following action at each point $p \in \mathscr{M}$:

$$\begin{aligned} \boldsymbol{\nabla v}(p) : \mathscr{T}_p^*(\mathscr{M}) \times \mathscr{T}_p(\mathscr{M}) &\longrightarrow \mathbb{R} \\ (\boldsymbol{\omega}, \boldsymbol{u}) &\longmapsto \langle \boldsymbol{\omega}, \boldsymbol{\nabla}_{\tilde{\boldsymbol{u}}} \boldsymbol{v}(p) \rangle, \end{aligned} \tag{2.51}$$

where $\tilde{\boldsymbol{u}}$ is any vector field which performs some extension of $\boldsymbol{u}$ around p: $\tilde{\boldsymbol{u}}(p) = \boldsymbol{u}$. As already noted (cf. Remark 2.10), $\boldsymbol{\nabla}_{\tilde{\boldsymbol{u}}} \boldsymbol{v}(p)$ is independent of the choice of $\tilde{\boldsymbol{u}}$, so that the mapping (2.51) is well defined. By comparing with (2.22), we verify that $\boldsymbol{\nabla v}(p)$ is a tensor of type $(1, 1)$.

One can extend the covariant derivative to all tensor fields by (i) demanding that for a scalar field the action of the affine connection is nothing but the gradient (hence the same notation $\boldsymbol{\nabla} f$) and (ii) using the Leibniz rule. As a result, the covariant derivative of a tensor field $\boldsymbol{T}$ of type (k, ℓ) is a tensor field $\boldsymbol{\nabla T}$ of type $(k, \ell + 1)$. Its components with respect a given field frame $(\boldsymbol{e}_\alpha)$ are denoted

$$\nabla_\gamma T^{\alpha_1 \ldots \alpha_k}{}_{\beta_1 \ldots \beta_\ell} := (\boldsymbol{\nabla T})^{\alpha_1 \ldots \alpha_k}{}_{\beta_1 \ldots \beta_\ell \gamma} \tag{2.52}$$

and are given by

$$\begin{aligned} \nabla_\gamma T^{\alpha_1 \ldots \alpha_k}{}_{\beta_1 \ldots \beta_\ell} = & \, \boldsymbol{e}_\gamma (T^{\alpha_1 \ldots \alpha_k}{}_{\beta_1 \ldots \beta_\ell}) + \sum_{i=1}^{k} \Gamma^{\alpha_i}{}_{\sigma \gamma} \, T^{\alpha_1 \ldots \overset{i}{\overset{\downarrow}{\sigma}} \ldots \alpha_k}{}_{\beta_1 \ldots \beta_\ell} \\ & - \sum_{i=1}^{\ell} \Gamma^{\sigma}{}_{\beta_i \gamma} \, T^{\alpha_1 \ldots \alpha_k}{}_{\beta_1 \ldots \underset{i}{\underset{\uparrow}{\sigma}} \ldots \beta_\ell}, \end{aligned} \tag{2.53}$$

where $\boldsymbol{e}_\gamma (T^{\alpha_1 \ldots \alpha_k}{}_{\beta_1 \ldots \beta_\ell})$ stands for the action of the vector $\boldsymbol{e}_\gamma$ on the scalar field $T^{\alpha_1 \ldots \alpha_k}{}_{\beta_1 \ldots \beta_\ell}$, resulting from the very definition of a vector (cf. Sect. 2.2.2). In particular, if $(\boldsymbol{e}_\alpha)$ is a natural frame associated with some coordinate system (x^α), then $\boldsymbol{e}_\alpha = \boldsymbol{\partial}_\alpha$ and the above formula becomes [cf. (2.7)]

$$\begin{aligned} \nabla_\gamma T^{\alpha_1 \ldots \alpha_k}{}_{\beta_1 \ldots \beta_\ell} = & \, \frac{\partial}{\partial x^\gamma} T^{\alpha_1 \ldots \alpha_k}{}_{\beta_1 \ldots \beta_\ell} + \sum_{i=1}^{k} \Gamma^{\alpha_i}{}_{\sigma \gamma} \, T^{\alpha_1 \ldots \overset{i}{\overset{\downarrow}{\sigma}} \ldots \alpha_k}{}_{\beta_1 \ldots \beta_\ell} \\ & - \sum_{i=1}^{\ell} \Gamma^{\sigma}{}_{\beta_i \gamma} \, T^{\alpha_1 \ldots \alpha_k}{}_{\beta_1 \ldots \underset{i}{\underset{\uparrow}{\sigma}} \ldots \beta_\ell}. \end{aligned} \tag{2.54}$$

Remark 2.11 Notice the position of the index γ in Eq. (2.52): it is the last one on the right-hand side. According to (2.23), $\boldsymbol{\nabla T}$ is then expressed as

$$\boldsymbol{\nabla T} = \nabla_\gamma T^{\alpha_1 \ldots \alpha_k}{}_{\beta_1 \ldots \beta_\ell} \, \boldsymbol{e}_{\alpha_1} \otimes \ldots \otimes \boldsymbol{e}_{\alpha_k} \otimes \boldsymbol{e}^{\beta_1} \otimes \ldots \otimes \boldsymbol{e}^{\beta_\ell} \otimes \boldsymbol{e}^\gamma . \tag{2.55}$$

Because $\boldsymbol{e}^\gamma$ is the *last* 1-form of the tensorial product on the right-hand side, the notation $T^{\alpha_1 \ldots \alpha_k}{}_{\beta_1 \ldots \beta_\ell ; \gamma}$ instead of $\nabla_\gamma T^{\alpha_1 \ldots \alpha_k}{}_{\beta_1 \ldots \beta_\ell}$ would have been more appropriate. The index convention (2.55) agrees with that of MTW [9] [cf. their Eq. (10.17)].

The ***covariant derivative of the tensor field T along a vector v*** is defined by

$$\nabla_{\boldsymbol{v}} \boldsymbol{T} := \nabla \boldsymbol{T}(\underbrace{., \ldots, .}_{k+\ell \text{ slots}}, \boldsymbol{u}). \tag{2.56}$$

The components of $\nabla_{\boldsymbol{v}} \boldsymbol{T}$ are then $v^\mu \nabla_\mu T^{\alpha_1 \ldots \alpha_k}{}_{\beta_1 \ldots \beta_\ell}$. Note that $\nabla_{\boldsymbol{v}} \boldsymbol{T}$ is a tensor field of the same type as $\boldsymbol{T}$. In the particular case of a scalar field f, we will use the notation $\boldsymbol{v} \cdot \nabla f$ for $\nabla_{\boldsymbol{v}} f$:

$$\boldsymbol{v} \cdot \nabla f := \nabla_{\boldsymbol{v}} f = \langle \nabla f, \boldsymbol{v} \rangle = \boldsymbol{v}(f). \tag{2.57}$$

The ***divergence*** with respect to the affine connection ∇ of a tensor field $\boldsymbol{T}$ of type (k, ℓ) with $k \geq 1$ is the tensor field denoted $\nabla \cdot \boldsymbol{T}$ of type $(k-1, \ell)$ and whose components with respect to any frame field are given by

$$(\nabla \cdot \boldsymbol{T})^{\alpha_1 \ldots \alpha_{k-1}}{}_{\beta_1 \ldots \beta_\ell} = \nabla_\mu T^{\alpha_1 \ldots \alpha_{k-1} \mu}{}_{\beta_1 \ldots \beta_\ell}. \tag{2.58}$$

Remark 2.12 For the divergence, the contraction is performed on the *last* upper index of $\boldsymbol{T}$.

2.4.2 Levi–Civita Connection

On a pseudo-Riemannian manifold $(\mathscr{M}, \boldsymbol{g})$ there is a unique affine connection ∇ such that

1. ∇ is ***torsion-free***, i.e. for any scalar field f, $\nabla \nabla f$ is a field of *symmetric* bilinear forms; in components:

$$\nabla_\alpha \nabla_\beta f = \nabla_\beta \nabla_\alpha f. \tag{2.59}$$

2. The covariant derivative of the metric tensor vanishes identically:

$$\boxed{\nabla \boldsymbol{g} = 0}. \tag{2.60}$$

∇ is called the ***Levi–Civita connection associated with g***. In this book, we shall consider only such connections.

With respect to the Levi–Civita connection, the Levi–Civita tensor $\boldsymbol{\varepsilon}$ introduced in Sect. 2.3.4 shares the same property as $\boldsymbol{g}$:

$$\boxed{\nabla \boldsymbol{\varepsilon} = 0}. \tag{2.61}$$

Given a coordinate system (x^α) on $\mathscr{M}$, the connection coefficients of the Levi–Civita connection with respect to the natural basis $(\boldsymbol{\partial}_\alpha)$ are called the ***Christoffel***

symbols; they can be evaluated from the partial derivatives of the metric components with respect to the coordinates:

$$\Gamma^{\gamma}{}_{\alpha\beta} = \frac{1}{2} g^{\gamma\mu} \left(\frac{\partial g_{\mu\beta}}{\partial x^{\alpha}} + \frac{\partial g_{\alpha\mu}}{\partial x^{\beta}} - \frac{\partial g_{\alpha\beta}}{\partial x^{\mu}} \right). \tag{2.62}$$

Note that the Christoffel symbols are symmetric with respect to the lower two indices.

For the Levi–Civita connection, the expression for the divergence of a vector takes a rather simple form in a natural basis associated with some coordinates (x^{α}). Indeed, combining Eqs. (2.58) and (2.54), we get for $\boldsymbol{v} \in \mathcal{T}(\mathcal{M})$,

$$\nabla \cdot \boldsymbol{v} = \nabla_{\mu} v^{\mu} = \frac{\partial v^{\mu}}{\partial x^{\mu}} + \Gamma^{\mu}{}_{\sigma\mu} v^{\sigma}.$$

Now, from (2.62), we have

$$\Gamma^{\mu}{}_{\alpha\mu} = \frac{1}{2} g^{\mu\nu} \frac{\partial g_{\mu\nu}}{\partial x^{\alpha}} = \frac{1}{2} \frac{\partial}{\partial x^{\alpha}} \ln |g| = \frac{1}{\sqrt{|g|}} \frac{\partial}{\partial x^{\alpha}} \sqrt{|g|}, \tag{2.63}$$

where $g := \det(g_{\alpha\beta})$ [Eq. (2.45)]. The last but one equality follows from the general law of variation of the determinant of any invertible matrix A:

$$\boxed{\delta(\ln|\det A|) = \mathrm{tr}(A^{-1} \times \delta A)}, \tag{2.64}$$

where δ denotes any variation (derivative) that fulfills the Leibniz rule, tr stands for the trace and $\times$ for the matrix product. We conclude that

$$\boxed{\nabla \cdot \boldsymbol{v} = \frac{1}{\sqrt{|g|}} \frac{\partial}{\partial x^{\mu}} \left(\sqrt{|g|}\, v^{\mu} \right)}. \tag{2.65}$$

Similarly, for an antisymmetric tensor field of type (2, 0),

$$\nabla_{\mu} A^{\alpha\mu} = \frac{\partial A^{\alpha\mu}}{\partial x^{\mu}} + \underbrace{\Gamma^{\alpha}{}_{\sigma\mu} A^{\sigma\mu}}_{0} + \Gamma^{\mu}{}_{\sigma\mu} A^{\alpha\sigma} = \frac{\partial A^{\alpha\mu}}{\partial x^{\mu}} + \frac{1}{\sqrt{|g|}} \frac{\partial}{\partial x^{\sigma}} \sqrt{|g|}\, A^{\alpha\sigma},$$

where we have used the fact that $\Gamma^{\alpha}_{\sigma\mu}$ is symmetric in (σ, μ), whereas $A^{\sigma\mu}$ is antisymmetric. Hence the simple formula for the divergence of an *antisymmetric* tensor field of (2, 0):

$$\boxed{\nabla_{\mu} A^{\alpha\mu} = \frac{1}{\sqrt{|g|}} \frac{\partial}{\partial x^{\mu}} \left(\sqrt{|g|} A^{\alpha\mu} \right)}. \tag{2.66}$$

2.4.3 Curvature

2.4.3.1 General Definition

The ***Riemann curvature tensor*** of an affine connection ∇ is defined by

$$\begin{array}{llll}\mathbf{Riem}: \mathscr{T}^*(\mathscr{M}) \times \mathscr{T}(\mathscr{M})^3 & \longrightarrow & C^\infty(\mathscr{M}, \mathbb{R}) \\ (\boldsymbol{\omega}, \boldsymbol{w}, \boldsymbol{u}, \boldsymbol{v}) & \longmapsto & \left\langle \boldsymbol{\omega}, \nabla_{\boldsymbol{u}} \nabla_{\boldsymbol{v}} \boldsymbol{w} - \nabla_{\boldsymbol{v}} \nabla_{\boldsymbol{u}} \boldsymbol{w} - \nabla_{[\boldsymbol{u},\boldsymbol{v}]} \boldsymbol{w} \right\rangle, \end{array} \tag{2.67}$$

where $\mathscr{T}^*(\mathscr{M})$ stands for the space of 1-forms on $\mathscr{M}$, $\mathscr{T}(\mathscr{M})$ for the space of vector fields on $\mathscr{M}$ and$C^\infty(\mathscr{M}, \mathbb{R})$ for the space of smooth scalar fields on $\mathscr{M}$. The above formula does define a tensor field on $\mathscr{M}$, i.e. the value of $\mathbf{Riem}(\boldsymbol{\omega}, \boldsymbol{w}, \boldsymbol{u}, \boldsymbol{v})$ at a given point $p \in \mathscr{M}$ depends only upon the values of the fields $\boldsymbol{\omega}$, $\boldsymbol{w}$, $\boldsymbol{u}$ and $\boldsymbol{v}$ at p and not upon their behaviours away from p, as the gradients in Eq. (2.67) might suggest. We denote the components of this tensor in a given basis $(\boldsymbol{e}_\alpha)$, not by $\mathrm{Riem}^\gamma{}_{\delta\alpha\beta}$, but by $R^\gamma{}_{\delta\alpha\beta}$. The definition (2.67) leads then to the following expression, named the ***Ricci identity***:

$$\forall \boldsymbol{w} \in \mathscr{T}(\mathscr{M}), \quad \left(\nabla_\alpha \nabla_\beta - \nabla_\beta \nabla_\alpha\right) w^\gamma = R^\gamma_{\ \mu\alpha\beta} w^\mu. \tag{2.68}$$

Remark 2.13 In view of this identity, one may say that the Riemann tensor measures the lack of commutativity of two successive covariant derivatives of a vector field. On the opposite, for a scalar field and a torsion-free connection, two successive covariant derivatives always commute [cf. Eq. (2.59)].

In a coordinate basis, the components of the Riemann tensor are given in terms of the connection coefficients by

$$\boxed{R^\alpha{}_{\beta\mu\nu} = \frac{\partial \Gamma^\alpha{}_{\beta\nu}}{\partial x^\mu} - \frac{\partial \Gamma^\alpha{}_{\beta\mu}}{\partial x^\nu} + \Gamma^\alpha{}_{\sigma\mu} \Gamma^\sigma{}_{\beta\nu} - \Gamma^\alpha{}_{\sigma\nu} \Gamma^\sigma{}_{\beta\mu}}. \tag{2.69}$$

From the definition (2.67), the Riemann tensor is clearly antisymmetric with respect to its last two arguments $(\boldsymbol{u}, \boldsymbol{v})$:

$$\mathbf{Riem}(., ., \boldsymbol{u}, \boldsymbol{v}) = -\mathbf{Riem}(., ., \boldsymbol{v}, \boldsymbol{u}). \tag{2.70}$$

In addition, it satisfies the cyclic property

$$\mathbf{Riem}(., \boldsymbol{u}, \boldsymbol{v}, \boldsymbol{w}) + \mathbf{Riem}(., \boldsymbol{w}, \boldsymbol{u}, \boldsymbol{v}) + \mathbf{Riem}(., \boldsymbol{v}, \boldsymbol{w}, \boldsymbol{u}) = 0. \tag{2.71}$$

The covariant derivatives of the Riemann tensor obeys the ***Bianchi identity***

$$\boxed{\nabla_\rho R^\alpha_{\ \beta\mu\nu} + \nabla_\mu R^\alpha_{\ \beta\nu\rho} + \nabla_\nu R^\alpha_{\ \beta\rho\mu} = 0}. \tag{2.72}$$

2.4.3.2 Case of a Pseudo-Riemannian Manifold

The Riemann tensor of the Levi–Civita connection obeys the additional antisymmetry:

$$\mathbf{Riem}(\boldsymbol{\omega}, \boldsymbol{w}, ., .) = -\mathbf{Riem}(\underline{\boldsymbol{w}}, \vec{\boldsymbol{\omega}}, ., .). \tag{2.73}$$

Combined with (2.70) and (2.71), this implies the symmetry property

$$\mathbf{Riem}(\boldsymbol{\omega}, \boldsymbol{w}, \boldsymbol{u}, \boldsymbol{v}) = \mathbf{Riem}(\underline{\boldsymbol{u}}, \boldsymbol{v}, \vec{\boldsymbol{\omega}}, \boldsymbol{w}). \tag{2.74}$$

A pseudo-Riemannian manifold $(\mathscr{M}, \boldsymbol{g})$ with a vanishing Riemann tensor is called a ***flat manifold***; in this case, $\boldsymbol{g}$ is said to be a ***flat metric***. If in addition, it has a Riemannian signature, $\boldsymbol{g}$ is called an ***Euclidean metric***.

2.4.3.3 Ricci Tensor

The ***Ricci tensor*** of the affine connection $\boldsymbol{\nabla}$ is the field of bilinear forms $\boldsymbol{R}$ defined by

$$\begin{array}{rcl} \boldsymbol{R} : \mathscr{T}(\mathscr{M}) \times \mathscr{T}(\mathscr{M}) & \longrightarrow & C^\infty(\mathscr{M}, \mathbb{R}) \\ (\boldsymbol{u}, \boldsymbol{v}) & \longmapsto & \mathbf{Riem}(\boldsymbol{e}^\mu, \boldsymbol{u}, \boldsymbol{e}_\mu, \boldsymbol{v}). \end{array} \tag{2.75}$$

This definition is independent of the choice of the basis $(\boldsymbol{e}_\alpha)$ and its dual counterpart $(\boldsymbol{e}^\alpha)$. In terms of components:

$$\boxed{R_{\alpha\beta} := R^\mu{}_{\alpha\mu\beta}}. \tag{2.76}$$

Remark 2.14 Following the standard usage, we denote the components of the Riemann and Ricci tensors by the same letter R, the number of indices allowing us to distinguish between the two tensors. On the other hand, we are using different symbols, **Riem** and $\boldsymbol{R}$, when employing the 'intrinsic' notation.

For the Levi–Civita connection associated with the metric $\boldsymbol{g}$, the property (2.74) implies that the Ricci tensor is symmetric:

$$\boldsymbol{R}(\boldsymbol{u}, \boldsymbol{v}) = \boldsymbol{R}(\boldsymbol{v}, \boldsymbol{u}). \tag{2.77}$$

In addition, one defines the ***Ricci scalar*** (also called ***scalar curvature***) as the trace of the Ricci tensor with respect to the metric $\boldsymbol{g}$:

$$R := g^{\mu\nu} R_{\mu\nu}. \tag{2.78}$$

The Bianchi identity (2.72) implies the divergence-free property

$$\boxed{\boldsymbol{\nabla} \cdot \vec{\boldsymbol{G}} = 0}, \tag{2.79}$$

where $\vec{G}$ in the type-(1, 1) tensor associated by metric duality [cf. (2.39)] to the ***Einstein tensor***:

$$\boxed{\boldsymbol{G} := \boldsymbol{R} - \frac{1}{2} R \boldsymbol{g}}. \tag{2.80}$$

Equation (2.79) is called the ***contracted Bianchi identity***.

2.4.4 Weyl Tensor

Let $(\mathscr{M}, \boldsymbol{g})$ be a pseudo-Riemannian manifold of dimension n.

For $n = 1$, the Riemann tensor vanishes identically, i.e. $(\mathscr{M}, \boldsymbol{g})$ is necessarily flat. The reader who has in mind a curved line in the Euclidean plane $\mathbb{R}^2$ might be surprised by the above statement. This is because the Riemann tensor represents the *intrinsic* curvature of a manifold. For a line, the curvature that is not vanishing is the *extrinsic* curvature, i.e. the curvature resulting from the embedding of the line in $\mathbb{R}^2$. We shall discuss in more details the concepts of intrinsic and extrinsic curvatures in Chap. 3.

For $n = 2$, the Riemann tensor is entirely determined by the knowledge of the Ricci scalar R, according to the formula:

$$R^{\gamma}{}_{\delta\alpha\beta} = R \left(\delta^{\gamma}{}_{\alpha} g_{\delta\beta} - \delta^{\gamma}{}_{\beta} g_{\delta\alpha} \right) \qquad (n = 2). \tag{2.81}$$

For $n = 3$, the Riemann tensor is entirely determined by the knowledge of the Ricci tensor, according to

$$\begin{aligned} R^{\gamma}{}_{\delta\alpha\beta} =& R^{\gamma}{}_{\alpha} g_{\delta\beta} - R^{\gamma}{}_{\beta} g_{\delta\alpha} + \delta^{\gamma}{}_{\alpha} R_{\delta\beta} - \delta^{\gamma}{}_{\beta} R_{\delta\alpha} \\ &+ \frac{R}{2} \left(\delta^{\gamma}{}_{\beta} g_{\delta\alpha} - \delta^{\gamma}{}_{\alpha} g_{\delta\beta} \right) \qquad (n = 3). \end{aligned} \tag{2.82}$$

For $n \geq 4$, the Riemann tensor can be split into (i) a "trace-trace" part, represented by the Ricci scalar R [Eq. (2.78)], (ii) a "trace" part, represented by the Ricci tensor $\boldsymbol{R}$ [Eq. (2.76)], and (iii) a "traceless" part, which is constituted by the ***Weyl conformal curvature tensor***, $\boldsymbol{C}$:

$$\begin{aligned} R^{\gamma}{}_{\delta\alpha\beta} =& C^{\gamma}{}_{\delta\alpha\beta} + \frac{1}{n-2} \left(R^{\gamma}_{\alpha} g_{\delta\beta} - R^{\gamma}_{\beta} g_{\delta\alpha} + \delta^{\gamma}_{\alpha} R_{\delta\beta} - \delta^{\gamma}_{\beta} R_{\delta\alpha} \right) \\ &+ \frac{1}{(n-1)(n-2)} R \left(\delta^{\gamma}_{\beta} g_{\delta\alpha} - \delta^{\gamma}_{\alpha} g_{\delta\beta} \right). \end{aligned} \tag{2.83}$$

The above relation may be taken as the definition of $\boldsymbol{C}$. It implies that $\boldsymbol{C}$ is traceless: $C^{\mu}{}_{\alpha\mu\beta} = 0$. The other possible traces are zero thanks to the symmetry properties of the Riemann tensor.

Remark 2.15 The decomposition (2.83) is also meaningful for $n = 3$, but it then implies that the Weyl tensor vanishes identically [compare with (2.82)].

2.5 Lie Derivative

As discussed in Sect. 2.4.1, the notion of a derivative of a vector field on a manifold $\mathscr{M}$ requires the introduction of some extra-structure on $\mathscr{M}$. In Sect. 2.4.1, this extra-structure was an affine connection and in Sect. 2.4.2 a metric $\boldsymbol{g}$ (which provides naturally an affine connection: the Levi–Civita one). Another possible extra-structure is a "reference" vector field, with respect to which the derivative is to be defined. This is the concept of the *Lie derivative*, which we discuss here.

2.5.1 Lie Derivative of a Vector Field

Consider a vector field $\boldsymbol{u}$ on $\mathscr{M}$, called hereafter the ***flow***. Let $\boldsymbol{v}$ be another vector field on $\mathscr{M}$, the variation of which is to be studied. We can use the flow $\boldsymbol{u}$ to transport the vector $\boldsymbol{v}$ from one point p to a neighbouring one q and then define rigorously the variation of $\boldsymbol{v}$ as the difference between the actual value of $\boldsymbol{v}$ at q and the transported value via $\boldsymbol{u}$. More precisely the definition of the Lie derivative of $\boldsymbol{v}$ with respect to $\boldsymbol{u}$ is as follows (see Fig. 2.3). We first define the image $\Phi_\varepsilon(p)$ of the point p by the transport by an infinitesimal "distance" ε along the field lines of $\boldsymbol{u}$ as $\Phi_\varepsilon(p) = q$, where q is the point close to p such that the infinitesimal displacement vector from p to q is $\overrightarrow{pq} = \varepsilon \boldsymbol{u}(p)$ (cf. Sect. 2.2.2). Besides, if we multiply the vector $\boldsymbol{v}(p)$ by some infinitesimal parameter λ, it becomes an infinitesimal vector at p. Then there exists a unique point p' close to p such that $\lambda \boldsymbol{v}(p) = \overrightarrow{pp'}$. We may transport the point p' to a point q' along the field lines of $\boldsymbol{u}$ by the same "distance" ε as that used to transport p to q: $q' = \Phi_\varepsilon(p')$ (see Fig. 2.3). $\overrightarrow{qq'}$ is then an infinitesimal vector at q and we define the transport by the distance ε of the vector $\boldsymbol{v}(p)$ along the field lines of $\boldsymbol{u}$ according to

$$\Phi_\varepsilon(\boldsymbol{v}(p)) := \frac{1}{\lambda}\overrightarrow{qq'}. \tag{2.84}$$

$\Phi_\varepsilon(\boldsymbol{v}(p))$ is a vector in $\mathscr{T}_q(\mathscr{M})$. We may then subtract it from the actual value of the field $\boldsymbol{v}$ at q and define the ***Lie derivative*** of $\boldsymbol{v}$ along $\boldsymbol{u}$ by

$$\mathscr{L}_{\boldsymbol{u}} \boldsymbol{v} := \lim_{\varepsilon \to 0} \frac{1}{\varepsilon} \left[\boldsymbol{v}(q) - \Phi_\varepsilon(\boldsymbol{v}(p))\right]. \tag{2.85}$$

Let us consider a coordinate system (x^α) adapted to the field $\boldsymbol{u}$ in the sense that $\boldsymbol{u} = \boldsymbol{\partial}_0$, where $\boldsymbol{\partial}_0$ is the first vector of the natural basis associated with the coordinates (x^α). We have, from the definitions of points q, p' and q',

$$\begin{aligned}
x^\alpha(q) &= x^\alpha(p) + \varepsilon \delta^\alpha{}_0 \\
x^\alpha(p') &= x^\alpha(p) + \lambda v^\alpha(p) \\
x^\alpha(q') &= x^\alpha(p') + \varepsilon \delta^\alpha{}_0,
\end{aligned}$$

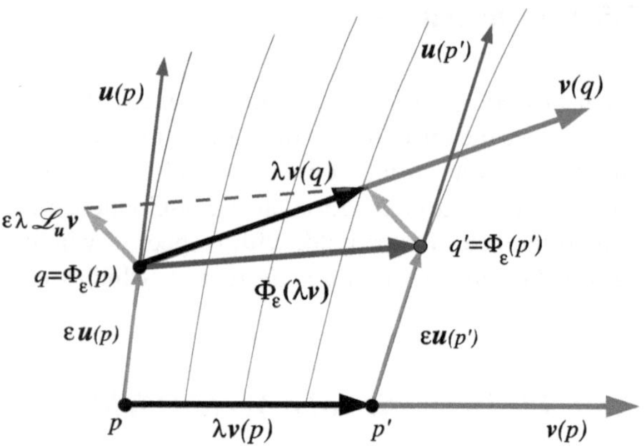

Fig. 2.3 Geometrical construction of the Lie derivative of a vector field: given a small parameter λ, each extremity of the arrow $\lambda \boldsymbol{v}$ is dragged by some small parameter ε along $\boldsymbol{u}$, to form the vector denoted by $\Phi_\varepsilon(\lambda \boldsymbol{v})$. The latter is then compared with the actual value of $\lambda \boldsymbol{v}$ at the point q, the difference (divided by $\lambda\varepsilon$) defining the Lie derivative $\mathscr{L}_{\boldsymbol{u}} \boldsymbol{v}$

so that

$$(qq')^\alpha = x^\alpha(p') - x^\alpha(p) = \lambda v^\alpha(p).$$

Accordingly, (2.84) and (2.85) result in

$$\begin{aligned}(\mathscr{L}_{\boldsymbol{u}} \boldsymbol{v})^\alpha &= \lim_{\varepsilon\to 0} \frac{1}{\varepsilon}\left[v^\alpha(q) - v^\alpha(p)\right] \\ &= \lim_{\varepsilon\to 0} \frac{1}{\varepsilon}\left[v^\alpha(x^0+\varepsilon,\ x^1,\ \ldots,\ x^{n-1}) - v^\alpha(x^0,\ x^1,\ \ldots,\ x^{n-1})\right].\end{aligned}$$

Hence, in adapted coordinates, the Lie derivative is simply obtained by taking the partial derivative of the vector components with respect to x^0:

$$\mathscr{L}_{\boldsymbol{u}} v^\alpha = \frac{\partial v^\alpha}{\partial x^0}, \tag{2.86}$$

where we have used the standard notation for the components of a Lie derivative: $\mathscr{L}_{\boldsymbol{u}} v^\alpha := (\mathscr{L}_{\boldsymbol{u}} \boldsymbol{v})^\alpha$. Besides, using the fact that the components of $\boldsymbol{u}$ are $u^\alpha = (1, 0, \ldots, 0)$ in the adapted coordinate system, we notice that the components of the commutator of $\boldsymbol{u}$ and $\boldsymbol{v}$, as given by (2.25), are

$$[\boldsymbol{u}, \boldsymbol{v}]^\alpha = \frac{\partial v^\alpha}{\partial x^0}.$$

This is exactly (2.86): $[\boldsymbol{u}, \boldsymbol{v}]^\alpha = \mathscr{L}_{\boldsymbol{u}} v^\alpha$. We conclude that the Lie derivative of a vector with respect to another one is actually nothing but the commutator of these two vectors:

$$\boxed{\mathscr{L}_{\boldsymbol{u}} \boldsymbol{v} = [\boldsymbol{u}, \boldsymbol{v}]}. \tag{2.87}$$

Thanks to formula (2.25), we may then express the components of the Lie derivative in an arbitrary coordinate system:

$$\boxed{\mathcal{L}_{\boldsymbol{u}} v^\alpha = u^\mu \frac{\partial v^\alpha}{\partial x^\mu} - v^\mu \frac{\partial u^\alpha}{\partial x^\mu}}. \tag{2.88}$$

Thanks to the symmetry property of the Christoffel symbols, the partial derivatives in Eq. (2.88) can be replaced by the Levi–Civita connection $\boldsymbol{\nabla}$ associated with some metric $\boldsymbol{g}$, yielding

$$\mathcal{L}_{\boldsymbol{u}} v^\alpha = u^\mu \nabla_\mu v^\alpha - v^\mu \nabla_\mu u^\alpha. \tag{2.89}$$

2.5.2 Generalization to Any Tensor Field

The Lie derivative is extended to any tensor field by (i) demanding that for a scalar field f, $\mathcal{L}_{\boldsymbol{u}} f = \langle \boldsymbol{\nabla} f, \boldsymbol{u}\rangle$ and (ii) using the Leibniz rule. As a result, the Lie derivative $\mathcal{L}_{\boldsymbol{u}} \boldsymbol{T}$ of a tensor field $\boldsymbol{T}$ of type (k, ℓ) is a tensor field of the same type, the components of which with respect to a given coordinate system (x^α) are

$$\mathcal{L}_{\boldsymbol{u}} T^{\alpha_1\ldots\alpha_k}{}_{\beta_1\ldots\beta_\ell} = u^\mu \frac{\partial}{\partial x^\mu} T^{\alpha_1\ldots\alpha_k}{}_{\beta_1\ldots\beta_\ell} - \sum_{i=1}^{k} T^{\alpha_1\ldots \overset{i}{\downarrow}{\sigma} \ldots\alpha_k}{}_{\beta_1\ldots\beta_\ell} \frac{\partial u^{\alpha_i}}{\partial x^\sigma}$$
$$+ \sum_{i=1}^{\ell} T^{\alpha_1\ldots\alpha_k}{}_{\beta_1\ldots \underset{\uparrow i}{\sigma} \ldots\beta_\ell} \frac{\partial u^\sigma}{\partial x^{\beta_i}}. \tag{2.90}$$

In particular, for a 1-form,

$$\mathcal{L}_{\boldsymbol{u}} \omega_\alpha = u^\mu \frac{\partial \omega_\alpha}{\partial x^\mu} + \omega_\mu \frac{\partial u^\mu}{\partial x^\alpha}. \tag{2.91}$$

As for the vector case [Eq. (2.88)], the partial derivatives in Eq. (2.90) can be replaced by the covariant derivative $\boldsymbol{\nabla}$ (or any other connection without torsion), yielding

$$\mathcal{L}_{\boldsymbol{u}} T^{\alpha_1\ldots\alpha_k}{}_{\beta_1\ldots\beta_\ell} = u^\mu \nabla_\mu T^{\alpha_1\ldots\alpha_k}{}_{\beta_1\ldots\beta_\ell} - \sum_{i=1}^{k} T^{\alpha_1\ldots \overset{i}{\downarrow}{\sigma} \ldots\alpha_k}{}_{\beta_1\ldots\beta_\ell} \nabla_\sigma u^{\alpha_i}$$
$$+ \sum_{i=1}^{\ell} T^{\alpha_1\ldots\alpha_k}{}_{\beta_1\ldots \underset{\uparrow i}{\sigma} \ldots\beta_\ell} \nabla_{\beta_i} u^\sigma. \tag{2.92}$$

Remark 2.16 Both the covariant derivative (affine connection) and the Lie derivative act on any kind of tensor field. For the specific class of tensor fields composed of p-forms (cf. Sect. 2.2.5), there exists a third type of derivative, which does not require any extra-structure on $\mathcal{M}$: the *exterior derivative* $\mathbf{d}$. For a 0-form (scalar field), $\mathbf{d}$ coincides with the gradient, hence the notation $\mathbf{d}x^\alpha$ used to denote the gradient of coordinates [cf. (2.20)]. We shall not use the exterior derivative in this book and so will not discus it further (see the classical textbooks [3, 6, 9] or Ref. [10] for an elementary introduction).

References

1. Berger, M.: A Panoramic View of Riemannian Geometry. Springer, Berlin (2003)
2. Choquet-Bruhat, Y.: General Relativity and Einstein's Equations. Oxford University Press, New York (2009)
3. Choquet-Bruhat, Y., De Witt-Moretten, C., Dillard-Bleick, M.: Analysis, Manifolds and Physics. North-Holland, Amsterdam (1977)
4. Eschrig, H.: Topology and Geometry for Physics. Springer, Berlin (2011)
5. Straumann, N.: General Relavity, with Applications to Astrophysics. Springer, Berlin (2004)
6. Wald, R.M.: General Relativity. University of Chicago Press, Chicago (1984)
7. Milnor, J.W.: On manifolds homeomorphic to the 7-sphere. Ann. Math. **64**, 399 (1956)
8. Taubes, C.H.: Gauge theory on asymptotically periodic 4-manifolds. J. Differ. Geom. **25**, 363 (1987)
9. Misner, C.W., Thorne, K.S., Wheeler, J.A.: Gravitation. Freeman, New York (1973)
10. Gourgoulhon, E.: An introduction to relativistic hydrodynamics. In: Rieutord, M., Dubrulle, B. (eds.) Stellar Fluid Dynamics and Numerical Simulations: From the Sun to Neutron Stars. EAS Publications Series 21, p. 43. EDP Sciences, Les Ulis (2006) available at http://arxiv.org/abs/gr-qc/0603009

Chapter 3
Geometry of Hypersurfaces

Abstract This chapter is devoted to hypersurfaces, which are at the basis of the 3+1 formalism for general relativity. After introducing the general notion of hypersurface embedded in spacetime, we focus on spacelike hypersurfaces, which are those involved in the 3+1 formalism. We present the first and second fundamental forms, giving rise to the notions of intrinsic and extrinsic curvatures. Finally, we derive the Gauss–Codazzi equations relating the intrinsic and extrinsic curvatures of an hypersurface to the curvature of the ambient spacetime. All results in this chapter are valid for any spacetime endowed with a Lorentzian metric, whether the latter is or not a solution of Einstein equation.

3.1 Introduction

The basic geometrical settings on which the 3+1 formalism is built is a foliation of spacetime by a one-parameter family of hypersurfaces. Before considering the whole family of hypersurfaces in the next chapter, it is natural to start by examining the properties of a single hypersurface.

Elementary presentations of hypersurfaces are given in numerous textbooks. To mention a few in the physics literature, let us quote Chap. 3 of Poisson's book [1], Appendix D of Carroll's one [2] and Appendix A of Straumann's one [3]. The presentation performed hereafter is relatively self-contained and requires only some elementary knowledge of differential geometry, at the level of an introductory course in general relativity (e.g. [4–6]).

3.2 Framework and Notations

We consider a ***spacetime*** $(\mathscr{M}, \boldsymbol{g})$, i.e. a differentiable manifold $\mathscr{M}$ of dimension 4 endowed with a metric $\boldsymbol{g}$ of signature $(-, +, +, +)$. $(\mathscr{M}, \boldsymbol{g})$ is a Lorentzian manifold (cf. Sect. 2.3.2).

É. Gourgoulhon, *3+1 Formalism in General Relativity*, Lecture Notes in Physics 846,
DOI: 10.1007/978-3-642-24525-1_3, © Springer-Verlag Berlin Heidelberg 2012

We assume that $(\mathscr{M}, \boldsymbol{g})$ is ***time orientable***, that is, it is possible to divide *continuously over* $\mathscr{M}$ each light cone of the metric $\boldsymbol{g}$ in two parts, ***past*** and ***future*** [5, 7]. We denote by ∇ the Levi–Civita connection associated with $\boldsymbol{g}$ (cf. Sect. 2.4.2) and shall call it the ***spacetime connection*** to distinguish it from other connections introduced in the text. The Riemann tensor, Ricci tensor and Ricci scalar of the metric $\boldsymbol{g}$ (cf. Sect. 2.4.2) are denoted with a superscript '4', i.e. respectively ${}^4\mathbf{Riem}$, ${}^4\boldsymbol{R}$ and ${}^4R$, to distinguish them from their analogs on 3-dimensional hypersurfaces.

When dealing with indices, we adopt the following conventions: all Greek indices run in $\{0, 1, 2, 3\}$. We will use letters from the beginning of the alphabet $(\alpha, \beta, \gamma, \ldots)$ for free indices, and letters starting from μ $(\mu, \nu, \rho, \ldots)$ as dumb indices for contraction (in this way the tensorial degree (valence) of any equation is immediately apparent). Lower case Latin indices starting from the letter i $(i, j, k, \ldots)$ run in $\{1, 2, 3\}$, while those starting from the beginning of the alphabet $(a, b, c, \ldots)$ run in $\{2, 3\}$ only.

3.3 Hypersurface Embedded in Spacetime

3.3.1 Definition

A ***hypersurface*** of $\mathscr{M}$ is the image Σ of a 3-dimensional manifold $\hat{\Sigma}$ by an embedding $\Phi : \hat{\Sigma} \to \mathscr{M}$ (Fig. 3.1):

$$\Sigma = \Phi(\hat{\Sigma}). \tag{3.1}$$

Let us recall that ***embedding*** means that $\Phi : \hat{\Sigma} \to \Sigma$ is a homeomorphism, i.e. a one-to-one mapping such that both Φ and Φ^{-1} are continuous. The one-to-one character guarantees that Σ does not "intersect itself". A hypersurface can be defined locally as the set of points for which a scalar field on $\mathscr{M}$ is constant. Denoting the scalar field by t and setting the constant to zero, we get

$$\forall p \in \mathscr{M}, \quad p \in \Sigma \iff t(p) = 0. \tag{3.2}$$

For instance, let us assume that Σ is a connected submanifold of $\mathscr{M}$ with topology $\mathbb{R}^3$. Then we may introduce locally a coordinate system of $\mathscr{M}$, $(x^\alpha) = (t, x, y, z)$, such that t spans $\mathbb{R}$ and (x,y,z) are Cartesian coordinates spanning $\mathbb{R}^3$. Σ is then defined by the coordinate condition $t = 0$ [Eq. (3.2)] and an explicit form of the mapping Φ can be obtained by considering $(x^i) = (x, y, z)$ as coordinates on the 3-manifold $\hat{\Sigma}$:

$$\begin{aligned} \Phi : \hat{\Sigma} &\longrightarrow \mathscr{M} \\ (x, y, z) &\longmapsto (0, x, y, z). \end{aligned} \tag{3.3}$$

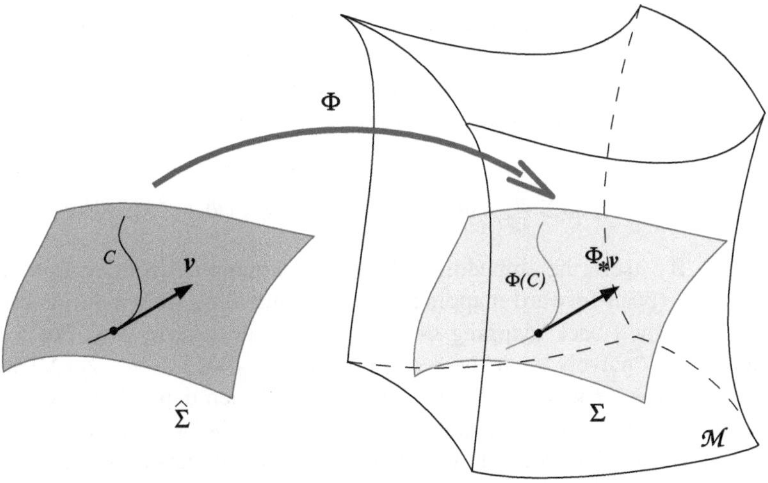

Fig. 3.1 Embedding Φ of the three-dimensional manifold $\hat{\Sigma}$ into the four-dimensional manifold $\mathscr{M}$, defining the hypersurface $\Sigma = \Phi(\hat{\Sigma})$. The push-forward $\Phi_* \boldsymbol{v}$ of a vector $\boldsymbol{v}$ tangent to some curve C in $\hat{\Sigma}$ is a vector tangent to $\Phi(C)$ in $\mathscr{M}$

The embedding Φ "carries along" curves in $\hat{\Sigma}$ to curves in $\mathscr{M}$. Consequently it also "carries along" vectors on $\hat{\Sigma}$ to vectors on $\mathscr{M}$ (cf. Fig. 3.1). In other words, it defines a mapping between $\mathscr{T}_p(\hat{\Sigma})$ and $\mathscr{T}_p(\mathscr{M})$. This mapping is denoted by Φ_* and is called the ***push-forward mapping***; thanks to the adapted coordinate system $(x^\alpha) = (t, x, y, z)$, it can be made explicit as follows

$$\begin{aligned} \Phi_* : \mathscr{T}_p(\hat{\Sigma}) &\longrightarrow \mathscr{T}_p(\mathscr{M}) \\ \boldsymbol{v} = (v^x, v^y, v^z) &\longmapsto \Phi_* \boldsymbol{v} = (0, v^x, v^y, v^z), \end{aligned} \tag{3.4}$$

where $v^i = (v^x, v^y, v^z)$ denotes the components of the vector $\boldsymbol{v}$ with respect to the natural basis $(\partial/\partial x^i)$ of $\mathscr{T}_p(\Sigma)$ associated with the coordinates (x^i).

Conversely, the embedding Φ induces a mapping, called the ***pull-back mapping*** and denoted Φ^*, between the linear forms on $\mathscr{T}_p(\mathscr{M})$ and those on $\mathscr{T}_p(\hat{\Sigma})$ as follows

$$\begin{aligned} \Phi^* : \mathscr{T}_p^*(\mathscr{M}) &\longrightarrow \mathscr{T}_p^*(\hat{\Sigma}) \\ \boldsymbol{\omega} &\longmapsto \Phi^* \boldsymbol{\omega} : \mathscr{T}_p(\hat{\Sigma}) \rightarrow \mathbb{R} \\ &\qquad\qquad \boldsymbol{v} \mapsto \langle \boldsymbol{\omega}, \Phi_* \boldsymbol{v} \rangle . \end{aligned} \tag{3.5}$$

Taking into account (3.4), the pull-back mapping can be made explicit:

$$\begin{aligned} \Phi^* : \quad \mathscr{T}_p^*(\mathscr{M}) &\longrightarrow \mathscr{T}_p^*(\hat{\Sigma}) \\ \boldsymbol{\omega} = (\omega_t, \omega_x, \omega_y, \omega_z) &\longmapsto \Phi^* \boldsymbol{\omega} = (\omega_x, \omega_y, \omega_z), \end{aligned} \tag{3.6}$$

where ω_α denotes the components of the 1-form $\boldsymbol{\omega}$ with respect to the basis $(\mathbf{d}x^\alpha)$ associated with the coordinates (x^α) (cf. Sect. 2.2.3).

In what follows, we identify $\hat{\Sigma}$ and $\Sigma = \Phi(\hat{\Sigma})$. In particular, we identify any vector on $\hat{\Sigma}$ with its push-forward image in $\mathscr{M}$, writing simply $\boldsymbol{v}$ instead of $\Phi_* \boldsymbol{v}$.

The pull-back operation can be extended to the multi-linear forms on $\mathscr{T}_p(\mathscr{M})$ in an obvious way: if $\boldsymbol{T}$ is a n-linear form on $\mathscr{T}_p(\mathscr{M})$, $\Phi^* \boldsymbol{T}$ is the n-linear form on $\mathscr{T}_p(\Sigma)$ defined by

$$\forall (\boldsymbol{v}_1, \ldots, \boldsymbol{v}_n) \in \mathscr{T}_p(\Sigma)^n, \quad \Phi^* \boldsymbol{T}(\boldsymbol{v}_1, \ldots, \boldsymbol{v}_n) = \boldsymbol{T}(\Phi_* \boldsymbol{v}_1, \ldots, \Phi_* \boldsymbol{v}_n). \tag{3.7}$$

Remark 3.1 By itself, the embedding Φ induces a mapping from vectors on Σ to vectors on $\mathscr{M}$ (push-forward mapping Φ_*) and a mapping from 1-forms on $\mathscr{M}$ to 1-forms on Σ (pull-back mapping Φ^*), but not in the reverse way. For instance, one may define "naively" a reverse mapping $F : \mathscr{T}_p(\mathscr{M}) \longrightarrow \mathscr{T}_p(\Sigma)$ by $\boldsymbol{v} = (v^t, v^x, v^y, v^z) \longmapsto F\boldsymbol{v} = (v^x, v^y, v^z)$, but it would then depend on the choice of coordinates (t, x, y, z), which is not the case of the push-forward mapping defined by Eq. (3.4). As we shall see below, if Σ is a spacelike hypersurface, a coordinate-independent reverse mapping is provided by the *orthogonal projector* (with respect to the ambient metric $\boldsymbol{g}$) onto Σ.

A very important case of pull-back operation is that of the bilinear form $\boldsymbol{g}$ (i.e. the spacetime metric), which defines the ***induced metric on*** Σ:

$$\boxed{\boldsymbol{\gamma} := \Phi^* \boldsymbol{g}} \tag{3.8}$$

$\boldsymbol{\gamma}$ is also called the ***first fundamental form of*** Σ. We shall also use the short-hand name ***3-metric*** to design it. Notice that

$$\forall (\boldsymbol{u}, \boldsymbol{v}) \in \mathscr{T}_p(\Sigma) \times \mathscr{T}_p(\Sigma), \quad \boldsymbol{u} \cdot \boldsymbol{v} = \boldsymbol{g}(\boldsymbol{u}, \boldsymbol{v}) = \boldsymbol{\gamma}(\boldsymbol{u}, \boldsymbol{v}). \tag{3.9}$$

In terms of the coordinate system[1] $(x^i) = (x, y, z)$ of Σ, the components of $\boldsymbol{\gamma}$ are deduced from (3.6):

$$\boxed{\gamma_{ij} = g_{ij}}\,. \tag{3.10}$$

The hypersurface Σ is said to be (cf. Sect. 2.3.2)

- ***spacelike*** iff the metric $\boldsymbol{\gamma}$ is Riemannian, i.e. has signature $(+, +, +)$;
- ***timelike*** iff the metric $\boldsymbol{\gamma}$ is Lorentzian, i.e. has signature $(-, +, +)$;
- ***null*** iff the metric $\boldsymbol{\gamma}$ is degenerate, i.e. has signature $(0, +, +)$.

3.3.2 Normal Vector

Given a scalar field t on $\mathscr{M}$ such that the hypersurface Σ is defined as a level surface of t [cf. Eq. (3.2)], the gradient 1-form $\boldsymbol{\nabla} t$ is normal to Σ, in the sense that for every

---

[1] Let us recall that by convention Latin indices run in $\{1, 2, 3\}$.

vector $\boldsymbol{v}$ tangent to Σ, $\langle \boldsymbol{\nabla} t, \boldsymbol{v} \rangle = 0$. The metric dual to $\boldsymbol{\nabla} t$, i.e. the vector $\vec{\boldsymbol{\nabla}} t$ (the component of which are $\nabla^\alpha t = g^{\alpha\mu} \nabla_\mu t$) is a vector normal to Σ and satisfies to the following properties

- $\vec{\boldsymbol{\nabla}} t$ is timelike iff Σ is spacelike;
- $\vec{\boldsymbol{\nabla}} t$ is spacelike iff Σ is timelike;
- $\vec{\boldsymbol{\nabla}} t$ is null iff Σ is null.

The vector $\vec{\boldsymbol{\nabla}} t$ defines the unique direction normal to Σ. In other words, any other vector $\boldsymbol{v}$ normal to Σ must be collinear to $\vec{\boldsymbol{\nabla}} t$: $\boldsymbol{v} = \lambda \vec{\boldsymbol{\nabla}} t$. Notice a characteristic property of null hypersurfaces: a vector normal to them is also tangent to them. This is because null vectors are orthogonal to themselves.

In the case where Σ is not null, we can re-normalize $\vec{\boldsymbol{\nabla}} t$ to make it a unit vector, by setting

$$\boldsymbol{n} := \left(\pm \vec{\boldsymbol{\nabla}} t \cdot \vec{\boldsymbol{\nabla}} t \right)^{-1/2} \vec{\boldsymbol{\nabla}} t, \tag{3.11}$$

with the sign + for a timelike hypersurface and the sign − for a spacelike one. The vector $\boldsymbol{n}$ is by construction a unit vector:

$$\boldsymbol{n} \cdot \boldsymbol{n} = -1 \quad \text{if } \Sigma \text{ is spacelike}, \tag{3.12}$$

$$\boldsymbol{n} \cdot \boldsymbol{n} = 1 \quad \text{if } \Sigma \text{ is timelike}. \tag{3.13}$$

$\boldsymbol{n}$ is one of the two unit vectors normal to Σ, the other one being $\boldsymbol{n}' = -\boldsymbol{n}$.

Remark 3.2 In the case where Σ is a null hypersurface, such a construction is not possible since $\vec{\boldsymbol{\nabla}} t \cdot \vec{\boldsymbol{\nabla}} t = 0$. Therefore there is no natural way to pick a privileged normal vector in this case. Actually, given a null normal $\boldsymbol{n}$, any vector $\boldsymbol{n}' = \lambda \boldsymbol{n}$, with $\lambda \in \mathbb{R}^*$, is a perfectly valid alternative to $\boldsymbol{n}$.

3.3.3 Intrinsic Curvature

If Σ is a spacelike or timelike hypersurface, then the induced metric $\boldsymbol{\gamma}$ is not degenerate. This implies that there is a unique connection (or covariant derivative) $\boldsymbol{D}$ on the manifold Σ that is torsion-free and satisfies

$$\boxed{\boldsymbol{D}\boldsymbol{\gamma} = 0}. \tag{3.14}$$

$\boldsymbol{D}$ is the Levi–Civita connection associated with the metric $\boldsymbol{\gamma}$ (cf. Sect. 2.4.2). The Riemann tensor associated with this connection represents what can be called the ***intrinsic curvature*** of $(\Sigma, \boldsymbol{\gamma})$. We shall denote it by **Riem** (without any superscript '4', cf. Sect. 3.2), and its components by the letter R, as $R^k{}_{lij}$. **Riem** measures the non-commutativity of two successive covariant derivatives $\boldsymbol{D}$, as expressed by the Ricci identity (2.68):

$$\forall \boldsymbol{v} \in \mathscr{T}(\Sigma),\ (D_i D_j - D_j D_i)v^k = R^k{}_{lij} v^l. \tag{3.15}$$

The corresponding Ricci tensor is denoted $\mathbf{R}$: $R_{ij} = R^k{}_{ikj}$ [Eq. (2.76)] and the Ricci scalar (scalar curvature) is denoted R: $R = \gamma^{ij} R_{ij}$ [Eq. (2.78) with $\boldsymbol{g} \to \boldsymbol{\gamma}$]. R is also called the ***Gaussian curvature*** of $(\Sigma, \boldsymbol{\gamma})$.

Let us recall that in dimension 3, the Riemann tensor can be fully deduced from the Ricci tensor, according to formula (2.82). In other words, the Weyl tensor identically vanishes in dimension 3 (cf. Sect. 2.4.4).

3.3.4 Extrinsic Curvature

Beside the intrinsic curvature discussed above, there is another type of "curvature" regarding hypersurfaces, namely that related to the "bending" of Σ in $\mathscr{M}$. This "bending" corresponds to the change of direction of the normal $\boldsymbol{n}$ as one moves on Σ. More precisely, one defines the ***Weingarten map*** (sometimes called the ***shape operator***) as the endomorphism of $\mathscr{T}_p(\Sigma)$ which associates with each vector tangent to Σ the variation of the normal along that vector, the variation being evaluated via the spacetime connection $\boldsymbol{\nabla}$:

$$\boxed{\begin{aligned} \boldsymbol{\chi} : \mathscr{T}_p(\Sigma) &\longrightarrow \mathscr{T}_p(\Sigma) \\ \boldsymbol{v} &\longmapsto \boldsymbol{\nabla}_{\boldsymbol{v}} \boldsymbol{n} \end{aligned}} \tag{3.16}$$

This application is well defined (i.e. its image is in $\mathscr{T}_p(\Sigma)$) since the constant character of $\boldsymbol{n} \cdot \boldsymbol{n}$ implies

$$\boldsymbol{n} \cdot \boldsymbol{\chi}(\boldsymbol{v}) = \boldsymbol{n} \cdot \boldsymbol{\nabla}_{\boldsymbol{v}} \boldsymbol{n} = \frac{1}{2} \boldsymbol{\nabla}_{\boldsymbol{v}} (\boldsymbol{n} \cdot \boldsymbol{n}) = 0,$$

which shows that $\boldsymbol{\chi}(\boldsymbol{v}) \in \mathscr{T}_p(\Sigma)$. If Σ is not a null hypersurface, the Weingarten map is uniquely defined (modulo the choice $+\boldsymbol{n}$ or $-\boldsymbol{n}$ for the unit normal), whereas if Σ is null, the definition of $\boldsymbol{\chi}$ depends upon the choice of the null normal $\boldsymbol{n}$.

The fundamental property of the Weingarten map is to be *self-adjoint* with respect to the induced metric $\boldsymbol{\gamma}$:

$$\boxed{\forall (\boldsymbol{u}, \boldsymbol{v}) \in \mathscr{T}_p(\Sigma) \times \mathscr{T}_p(\Sigma), \quad \boldsymbol{u} \cdot \boldsymbol{\chi}(\boldsymbol{v}) = \boldsymbol{\chi}(\boldsymbol{u}) \cdot \boldsymbol{v}}, \tag{3.17}$$

where the dot means the scalar product with respect to $\boldsymbol{\gamma}$ [considering $\boldsymbol{u}$ and $\boldsymbol{v}$ as vectors of $\mathscr{T}_p(\Sigma)$] or $\boldsymbol{g}$ [considering $\boldsymbol{u}$ and $\boldsymbol{v}$ as vectors of $\mathscr{T}_p(\mathscr{M})$]. It is not difficult to prove (3.17): it suffices to use the fact that the normal vector $\boldsymbol{n}$ is colinear to the gradient of the scalar field t defining Σ via Eq. (3.2):

$$n = \alpha \vec{\nabla} t,$$

where $\alpha := (-\vec{\nabla} t \cdot \vec{\nabla} t)^{-1/2}$ if Σ is spacelike, $\alpha := (\vec{\nabla} t \cdot \vec{\nabla} t)^{-1/2}$ if Σ is timelike and α is some non-vanishing field if Σ is null [cf. Eq. (3.11)]. Then

$$\begin{aligned}
\boldsymbol{u} \cdot \boldsymbol{\chi}(\boldsymbol{v}) - \boldsymbol{\chi}(\boldsymbol{u}) \cdot \boldsymbol{v} &= \boldsymbol{u} \cdot \nabla_{\boldsymbol{v}} \boldsymbol{n} - \boldsymbol{v} \cdot \nabla_{\boldsymbol{u}} \boldsymbol{n} \\
&= u^\mu v^\nu \nabla_\nu n_\mu - v^\mu u^\nu \nabla_\nu n_\mu \\
&= u^\mu v^\nu (\nabla_\nu n_\mu - \nabla_\mu n_\nu) \\
&= u^\mu v^\nu (\nabla_\nu \alpha \nabla_\mu t + \alpha \nabla_\nu \nabla_\mu t - \nabla_\mu \alpha \nabla_\nu t - \alpha \nabla_\mu \nabla_\nu t) \\
&= \underbrace{u^\mu \nabla_\mu t}_{0}\, v^\nu \nabla_\nu \alpha - u^\mu \nabla_\mu \alpha \underbrace{v^\nu \nabla_\nu t}_{0} + \alpha u^\mu v^\nu \underbrace{(\nabla_\nu \nabla_\mu t - \nabla_\mu \nabla_\nu t)}_{0} \\
&= 0,
\end{aligned}$$

where the vanishing of the last term results from the torsion-free property (2.59) and we have used the fact that $\boldsymbol{u}$ and $\boldsymbol{v}$ are tangent to Σ to set to zero the terms $u^\mu \nabla_\mu t$ and $v^\nu \nabla_\nu t$. We have thus proved (3.17).

The eigenvalues of the Weingarten map, which are all real numbers since $\boldsymbol{\chi}$ is self-adjoint, are called the ***principal curvatures*** of the hypersurface Σ and the corresponding eigenvectors define the so-called ***principal directions*** of Σ. The ***mean curvature*** of the hypersurface Σ is the arithmetic mean of the principal curvatures:

$$H := \frac{1}{3}(\kappa_1 + \kappa_2 + \kappa_3) \tag{3.18}$$

where the κ_i are the three eigenvalues of $\boldsymbol{\chi}$.

Remark 3.3 The curvatures defined above are not to be confused with the Gaussian curvature introduced in Sect. 3.3.3. The latter is an *intrinsic* quantity, independent of the way the manifold $(\Sigma, \boldsymbol{\gamma})$ is embedded in $(\mathscr{M}, \boldsymbol{g})$. On the contrary the principal curvatures and mean curvature depend on the embedding. For this reason, they are qualified of *extrinsic*.

The self-adjointness of $\boldsymbol{\chi}$ implies that the bilinear form defined on Σ's tangent space by

$$\boxed{\begin{aligned} \boldsymbol{K} : \mathscr{T}_p(\Sigma) \times \mathscr{T}_p(\Sigma) &\longrightarrow \mathbb{R} \\ (\boldsymbol{u}, \boldsymbol{v}) &\longmapsto -\boldsymbol{u} \cdot \boldsymbol{\chi}(\boldsymbol{v}) \end{aligned}} \tag{3.19}$$

is symmetric. It is called the ***second fundamental form*** of the hypersurface Σ. It is also called the ***extrinsic curvature tensor of Σ*** (cf. the remark above regarding the qualifier 'extrinsic'). $\boldsymbol{K}$ contains the same information as the Weingarten map.

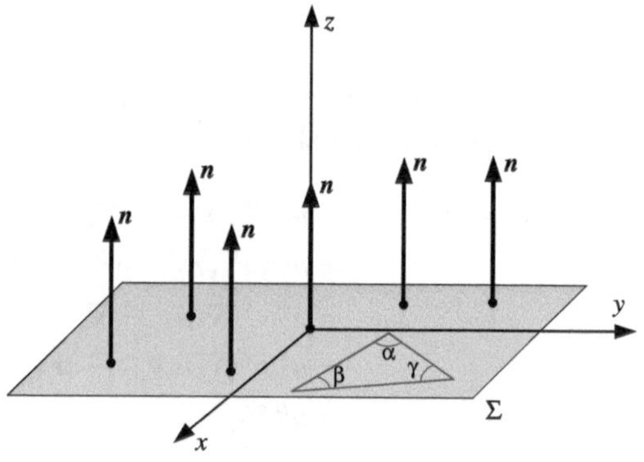

Fig. 3.2 Plane Σ as a hypersurface of the Euclidean space $\mathbb{R}^3$. Notice that the unit normal vector $\boldsymbol{n}$ stays constant along Σ; this implies that the extrinsic curvature of Σ vanishes identically. Besides, the sum of angles of any triangle lying in Σ is $\alpha+\beta+\gamma=\pi$, which shows that the *intrinsic* curvature of $(\Sigma, \boldsymbol{\gamma})$ vanishes as well

Remark 3.4 The minus sign in the definition (3.19) is chosen so that $\boldsymbol{K}$ agrees with the convention used in the numerical relativity community, as well as in the MTW book [4]. Some other authors (e.g. Carroll [2], Poisson [1], Wald [5]) choose the opposite convention.

If we make explicit the value of $\boldsymbol{\chi}$ in the definition (3.19), we get

$$\forall (\boldsymbol{u}, \boldsymbol{v}) \in \mathscr{T}_p(\Sigma) \times \mathscr{T}_p(\Sigma), \quad \boxed{\boldsymbol{K}(\boldsymbol{u}, \boldsymbol{v}) = -\boldsymbol{u} \cdot \boldsymbol{\nabla}_{\boldsymbol{v}} \boldsymbol{n}}. \tag{3.20}$$

We shall denote by K the trace of the bilinear form $\boldsymbol{K}$ with respect to the metric $\boldsymbol{\gamma}$; it is the opposite of the trace of the endomorphism $\boldsymbol{\chi}$ and is equal to -3 times the mean curvature of Σ:

$$K := \gamma^{ij} K_{ij} = -3H. \tag{3.21}$$

3.3.5 Examples: Surfaces Embedded in the Euclidean Space $\mathbb{R}^3$

Let us illustrate the previous definitions with some hypersurfaces of a space which we are very familiar with, namely $\mathbb{R}^3$ endowed with the standard Euclidean metric. In this case, the dimension is reduced by one unit with respect to the spacetime $\mathscr{M}$ and the ambient metric $\boldsymbol{g}$ is Riemannian instead of Lorentzian (cf. Sect. 2.3.2). The hypersurfaces are two-dimensional submanifolds of $\mathbb{R}^3$, namely they are *surfaces* by the ordinary meaning of this word.

In this section, and in this section only, we change our index convention to take into account that the base manifold is of dimension 3 and not 4: until the next section, the Greek indices run in $\{1, 2, 3\}$ and the Latin indices run in $\{1, 2\}$.

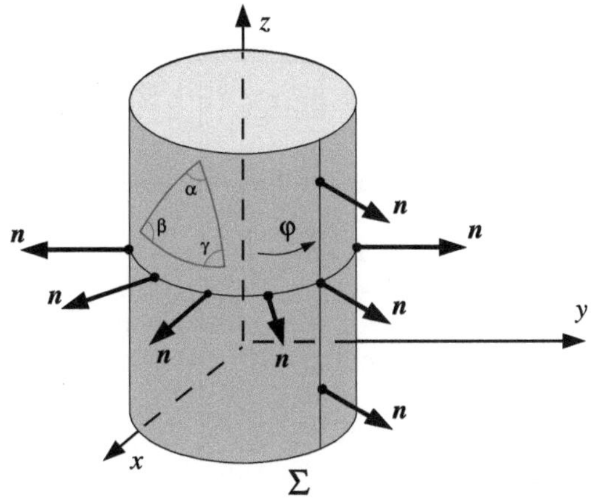

Fig. 3.3 Cylinder Σ as a hypersurface of the Euclidean space $\mathbb{R}^3$. Notice that the unit normal vector $\boldsymbol{n}$ stays constant when z varies at fixed φ, whereas its direction changes as φ varies at fixed z. Consequently the extrinsic curvature of Σ vanishes in the z direction, but is non zero in the φ direction. Besides, the sum of angles of any triangle lying in Σ is $\alpha + \beta + \gamma = \pi$, which shows that the *intrinsic* curvature of $(\Sigma, \boldsymbol{\gamma})$ is identically zero

Example 3.1 **A plane in $\mathbb{R}^3$**
Let us take for Σ the simplest surface one may think of: a plane (cf. Fig. 3.2). Let us consider Cartesian coordinates $(X^\alpha) = (x, y, z)$ on $\mathbb{R}^3$, such that Σ is the $z = 0$ plane. The scalar function t defining Σ according to Eq. (3.2) is then simply $t = z$. $(x^i) = (x, y)$ constitutes a coordinate system on Σ and the metric $\boldsymbol{\gamma}$ induced by $\boldsymbol{g}$ on Σ has the components $\gamma_{ij} = \mathrm{diag}(1, 1)$ with respect to these coordinates. It is obvious that this metric is flat: **Riem** $= 0$. The unit normal $\boldsymbol{n}$ has components $n^\alpha = (0, 0, 1)$ with respect to the coordinates (X^α). The components of the gradient $\boldsymbol{\nabla n}$ being simply given by the partial derivatives $\nabla_\beta n^\alpha = \partial n^\alpha / \partial X^\beta$ [the Christoffel symbols vanishes for the coordinates (X^α)], we get immediately $\boldsymbol{\nabla n} = 0$. Consequently, the Weingarten map and the extrinsic curvature vanish identically:

$$\boldsymbol{\chi} = 0 \quad \text{and} \quad \boldsymbol{K} = 0. \tag{3.22}$$

Example 3.2 **A cylinder in $\mathbb{R}^3$**
Let us at present consider for Σ the cylinder defined by the equation $t := \rho - a = 0$, where $\rho := \sqrt{x^2 + y^2}$ and a is a positive constant—the radius of the cylinder (cf. Fig. 3.3). Let us introduce the cylindrical coordinates $(x^\alpha) = (\rho, \varphi, z)$, such that $\varphi \in [0, 2\pi)$, $x = \rho \cos\varphi$ and $y = \rho \sin\varphi$. Then $(x^i) = (\varphi, z)$ constitutes a coordinate system on Σ. The components of the induced

metric in this coordinate system are given by

$$\gamma_{ij} \mathrm{d}x^i \mathrm{d}x^j = a^2 \mathrm{d}\varphi^2 + \mathrm{d}z^2. \tag{3.23}$$

It appears that this metric is flat, as for the plane considered above:

$$\mathbf{Riem} = 0. \tag{3.24}$$

Indeed, the change of coordinate $\eta := a\varphi$ (remember a is a constant!) transforms the metric components into

$$\gamma_{i'j'} \mathrm{d}x^{i'} \mathrm{d}x^{j'} = \mathrm{d}\eta^2 + \mathrm{d}z^2, \tag{3.25}$$

which exhibits the standard Cartesian shape of the flat metric.

To evaluate the extrinsic curvature of Σ, let us consider the unit normal $\boldsymbol{n}$ to Σ. Its components with respect to the Cartesian coordinates $(X^\alpha) = (x, y, z)$ are

$$n^\alpha = \left(\frac{x}{a}, \frac{y}{a}, 0\right). \tag{3.26}$$

a being constant, it is immediate to compute $\nabla_\beta n^\alpha = \partial n^\alpha / \partial X^\beta$:

$$\nabla_\beta n^\alpha = \mathrm{diag}\left(a^{-1}, a^{-1}, 0\right). \tag{3.27}$$

From Eq. (3.20), the components of the extrinsic curvature $\boldsymbol{K}$ with respect to the basis $(x^i) = (\varphi, z)$ are

$$K_{ij} = \boldsymbol{K}(\boldsymbol{\partial}_i, \boldsymbol{\partial}_j) = -\nabla_\beta n_\alpha (\partial_i)^\alpha (\partial_j)^\beta, \tag{3.28}$$

where $(\boldsymbol{\partial}_i) = (\boldsymbol{\partial}_\varphi, \boldsymbol{\partial}_z) = (\partial/\partial\varphi, \partial/\partial z)$ denotes the natural basis associated with the coordinates (φ, z) and $(\partial_i)^\alpha$ the components of the vector $\boldsymbol{\partial}_i$ with respect to the natural basis $(\boldsymbol{\partial}_\alpha) = (\boldsymbol{\partial}_x, \boldsymbol{\partial}_y, \boldsymbol{\partial}_z)$ associated with the Cartesian coordinates $(X^\alpha) = (x, y, z)$. Specifically, since $\boldsymbol{\partial}_\varphi = -y\boldsymbol{\partial}_x + x\boldsymbol{\partial}_y$, one has $(\partial_\varphi)^\alpha = (-y, x, 0)$ and $(\partial_z)^\alpha = (0, 0, 1)$. From Eqs. (3.27) and (3.28), we then obtain

$$K_{ij} = \begin{pmatrix} K_{\varphi\varphi} & K_{\varphi z} \\ K_{z\varphi} & K_{zz} \end{pmatrix} = \begin{pmatrix} -a & 0 \\ 0 & 0 \end{pmatrix}. \tag{3.29}$$

From Eq. (3.23), $\gamma^{ij} = \mathrm{diag}(a^{-2}, 1)$, so that the trace of $\boldsymbol{K}$ is

$$K = -\frac{1}{a}. \tag{3.30}$$

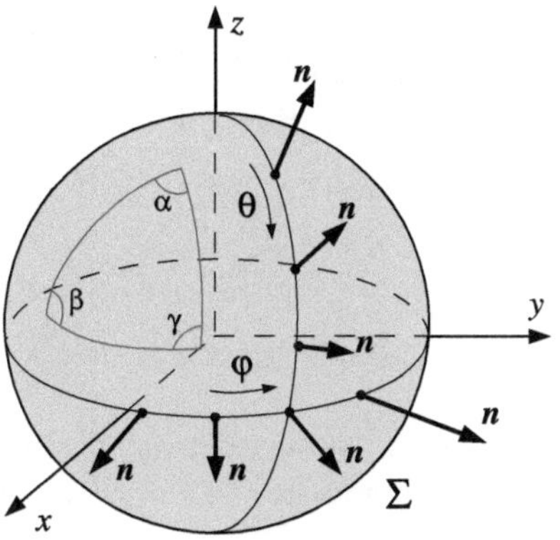

Fig. 3.4 Sphere Σ as a hypersurface of the Euclidean space $\mathbb{R}^3$. Notice that the unit normal vector $\boldsymbol{n}$ changes its direction when displaced on Σ. This shows that the extrinsic curvature of Σ does not vanish. Moreover all directions being equivalent at the surface of the sphere, $\boldsymbol{K}$ is necessarily proportional to the induced metric $\boldsymbol{\gamma}$, as found by the explicit calculation leading to Eq. (3.36). Besides, the sum of angles of any triangle lying in Σ is $\alpha + \beta + \gamma > \pi$, which shows that the *intrinsic* curvature of $(\Sigma, \boldsymbol{\gamma})$ does not vanish either

Example 3.3 **A sphere in $\mathbb{R}^3$**

Our final simple example is constituted by the sphere of radius a (cf. Fig. 3.4), the equation of which is $t := r - a = 0$, with $r = \sqrt{x^2 + y^2 + z^2}$. Introducing the spherical coordinates $(x^\alpha) = (r, \theta, \varphi)$ such that $x = r\sin\theta\cos\varphi$, $y = r\sin\theta\sin\varphi$ and $z = r\cos\theta$, $(x^i) = (\theta, \varphi)$ constitutes a coordinate system on Σ. The components of the induced metric $\boldsymbol{\gamma}$ in this coordinate system are given by

$$\gamma_{ij}\mathrm{d}x^i\mathrm{d}x^j = a^2\left(\mathrm{d}\theta^2 + \sin^2\theta\mathrm{d}\varphi^2\right). \tag{3.31}$$

Contrary to the previous two examples, this metric is not flat: the Ricci scalar, Ricci tensor and Riemann tensor of $(\Sigma, \boldsymbol{\gamma})$ are respectively (cf. Appendix B for the computation)

$$R = \frac{2}{a^2}, \quad R_{ij} = \frac{1}{a^2}\gamma_{ij}, \quad R^i{}_{jkl} = \frac{1}{a^2}\left(\delta^i{}_k\gamma_{jl} - \delta^i{}_l\gamma_{jk}\right). \tag{3.32}$$

The non vanishing of the Riemann tensor is reflected by the well-known property that the sum of angles of any triangle drawn at the surface of a sphere is larger than π (cf. Fig. 3.4).

The unit vector $\boldsymbol{n}$ normal to Σ (and oriented towards the exterior of the sphere) has the following components with respect to the coordinates $(X^\alpha) = (x, y, z)$:

$$n^\alpha = \left(\frac{x}{a}, \frac{y}{a}, \frac{z}{a}\right). \tag{3.33}$$

Within the Cartesian coordinates $(X^\alpha) = (x, y, z)$, we have $\nabla_\beta n^\alpha = \partial n^\alpha / \partial X^\beta$, hence $\nabla_\beta n^\alpha = a^{-1} \delta^\alpha_\beta$ and

$$\nabla_\beta n_\alpha = \frac{1}{a} g_{\alpha\beta}. \tag{3.34}$$

From Eq. (3.20) the components of the extrinsic curvature tensor in the basis $(\boldsymbol{\partial}_\theta, \boldsymbol{\partial}_\varphi)$ associated with the coordinates $(x^i) = (\theta, \varphi)$ on Σ are given by

$$K_{ij} = \boldsymbol{K}(\boldsymbol{\partial}_i, \boldsymbol{\partial}_j) = -\nabla_\beta n_\alpha (\partial_i)^\alpha (\partial_j)^\beta = -\frac{1}{a} g_{\alpha\beta} (\partial_i)^\alpha (\partial_j)^\beta = -\frac{1}{a} \boldsymbol{g}(\boldsymbol{\partial}_i, \boldsymbol{\partial}_j).$$

Since $\boldsymbol{\partial}_i$ is tangent to Σ, $\boldsymbol{g}(\boldsymbol{\partial}_i, \boldsymbol{\partial}_j) = \boldsymbol{\gamma}(\boldsymbol{\partial}_i, \boldsymbol{\partial}_j)$ from the very definition of $\boldsymbol{\gamma}$ [Eq. (3.9)]. Hence we conclude that

$$\boldsymbol{K} = -\frac{1}{a} \boldsymbol{\gamma}. \tag{3.35}$$

Explicitly

$$K_{ij} = \begin{pmatrix} K_{\theta\theta} & K_{\theta\varphi} \\ K_{\varphi\theta} & K_{\varphi\varphi} \end{pmatrix} = \begin{pmatrix} -a & 0 \\ 0 & -a \sin^2\theta \end{pmatrix}. \tag{3.36}$$

The trace of $\boldsymbol{K}$ with respect to $\boldsymbol{\gamma}$ is then

$$K = -\frac{2}{a}. \tag{3.37}$$

With these examples, we have encountered hypersurfaces with intrinsic and extrinsic curvature both vanishing (the plane), the intrinsic curvature vanishing but not the extrinsic one (the cylinder), and with both curvatures non vanishing (the sphere). As we shall see in Sect. 3.5, the extrinsic curvature is not fully independent from the intrinsic one: they are related by the Gauss equation.

3.3.6 An Example in Minkowski Spacetime: The Hyperbolic Space $\mathbb{H}^3$

The examples presented above are not directly connected to general relativity for the ambient manifold $(\mathscr{M}, \boldsymbol{g})$ is Riemannian (Euclidean space) and not Lorentzian. Here we provide a more relevant example, where $(\mathscr{M}, \boldsymbol{g})$ is the simplest spacetime that one may think of: Minkowski spacetime. A trivial example of hypersurface of

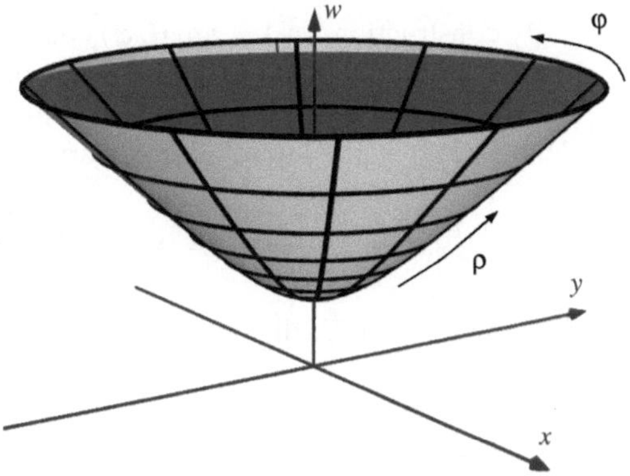

Fig. 3.5 Hyperbolic 3-space $\mathbb{H}^3$ as a hypersurface Σ embedded in Minkowski space. The dimension along z is suppressed

this spacetime is of course a hyperplane, both the intrinsic and extrinsic curvature of which are zero, as for the plane in Euclidean space (Example 3.1). A more instructive example is provided by a hyperbolic 3-space.

Example 3.4 **Hyperbolic 3-space in Minkowski spacetime** $(\mathscr{M}, \boldsymbol{g})$ being Minkowski spacetime, let $(X^\alpha) = (w, x, y, z)$ be Minkowskian coordinates, i.e. coordinates such that $g_{\alpha\beta} = \mathrm{diag}(-1, 1, 1, 1)$, and $b > 0$ some constant having the dimension of a length. Let us consider the hypersurface Σ defined by the equation

$$\Sigma : t := w - \sqrt{b^2 + x^2 + y^2 + z^2} = 0. \tag{3.38}$$

Σ is the upper sheet of a two-sheeted 3-dimensional hyperboloid. Indeed Eq. (3.38) implies

$$x^2 + y^2 + z^2 - w^2 = -b^2. \tag{3.39}$$

b being constant, we recognize the equation of the two-sheeted 3-dimensional hyperboloid oriented along the w axis, with summits $(b, 0, 0, 0)$ and $(-b, 0, 0, 0)$. Σ represents only the upper sheet of this hyperboloid (cf. Fig. 3.5), i.e. the sheet with $w > 0$, the lower sheet would correspond to $w + \sqrt{b^2 + x^2 + y^2 + z^2} = 0$ instead of (3.38). In view of (3.39) and the identity $\sinh^2 \rho - \cosh^2 \rho = -1$, it is natural to introduce a coordinate ρ such that $b \sinh \rho = \sqrt{x^2 + y^2 + z^2}$ and $b \cosh \rho = w$. Let us supplement ρ with two other coordinates (θ, φ) such that

$$\rho \in [0, +\infty), \quad \theta \in [0, \pi], \quad \varphi \in [0, 2\pi), \tag{3.40}$$

$$\Sigma : \begin{cases} x = b\sinh \rho \sin\theta \cos\varphi \\ y = b\sinh \rho \sin\theta \sin\varphi \\ z = b\sinh \rho \cos \theta \\ w = b\cosh\rho. \end{cases} \tag{3.41}$$

By construction, $(x^i) = (\rho, \theta, \varphi)$ constitutes a coordinate system on Σ and (3.41) is nothing but the parametric equation of Σ as an hypersurface of $\mathscr{M}$. The metric $\boldsymbol{\gamma}$ induced on Σ by the Minkowski metric $\boldsymbol{g}$ is obtained via the identity (3.9) applied to two elementary displacement vectors tangent to Σ:

$$\gamma_{ij}\mathrm{d}x^i\mathrm{d}y^j = \left(g_{\mu\nu}\mathrm{d}X^\mu\mathrm{d}X^\nu\right)\big|_\Sigma = \left(-\mathrm{d}w^2 + \mathrm{d}x^2 + \mathrm{d}y^2 + \mathrm{d}z^2\right)\Big|_\Sigma .$$

Noticing that, on Σ, $\mathrm{d}w^2 = b^2\sinh^2\rho\mathrm{d}\rho^2$ and

$$\begin{aligned}\mathrm{d}x^2 + \mathrm{d}y^2 + \mathrm{d}z^2 &= \mathrm{d}r^2 + r^2(\mathrm{d}\theta^2 + \sin^2\theta\mathrm{d}\varphi^2)\\ &= b^2\cosh^2\rho\mathrm{d}\rho^2 + b^2\sinh^2\rho(\mathrm{d}\theta^2 + \sin^2\theta\mathrm{d}\varphi^2), \quad (3.42)\end{aligned}$$

with $r := \sqrt{x^2 + y^2 + z^2} = b\sinh\rho$, we obtain

$$\gamma_{ij}\mathrm{d}x^i\mathrm{d}y^j = b^2\left[\mathrm{d}\rho^2 + \sinh^2\rho(\mathrm{d}\theta^2 + \sin^2\theta\mathrm{d}\varphi^2)\right]. \quad (3.43)$$

It is clear from the above expression that the metric $\boldsymbol{\gamma}$ is Riemannian (i.e. definite positive). In other words, Σ is a spacelike hypersurface.

The scalar curvature, Ricci tensor and Riemann tensor of $(\Sigma, \boldsymbol{\gamma})$ are respectively (cf. Appendix B for the computation)

$$R = -\frac{6}{b^2}, \quad R_{ij} = -\frac{2}{b^2}\gamma_{ij}, \quad R^i{}_{jkl} = -\frac{1}{b^2}\left(\delta^i{}_k\gamma_{jl} - \delta^i{}_l\gamma_{jk}\right). \quad (3.44)$$

Note that, contrary to the sphere (Example 3.3), the scalar curvature is negative. Note also that the Riemann tensor has the same simple structure than that of the sphere [Eq. (3.32)]. Actually this structure is common to all ***maximally symmetric spaces*** (cf. e.g. Sect. 3.9 of Ref. [2]), i.e. the spaces having the maximum number of independent continuous symmetries, which is $N = n(n+1)/2$, n being the dimension of the space. A maximally symmetric space has necessarily a constant curvature ($R = -6/b^2$ here). In dimension $n=3$, there are only three types of maximally symmetric spaces: (i) the hyperplane $\mathbb{R}^3$ ($R = 0$), (ii) the hypersphere $\mathbb{S}^3$ ($R > 0$) and (iii) the ***hyperbolic 3-space*** $\mathbb{H}^3$ ($R < 0$). The latter is precisely the case in which we are here and we may say that $(\Sigma, \boldsymbol{\gamma})$ is a concrete realization of $\mathbb{H}^3$.

From Eq. (3.38), the gradient of the scalar field t defining Σ is

$$\nabla_\alpha t = \left(1, -\frac{x}{w}, -\frac{y}{w}, -\frac{z}{w}\right) = -\frac{1}{w}(-w, x, y, z).$$

Note that the above expression gives the components with respect to the Minkowskian coordinates $(X^\alpha) = (w, x, y, z)$ and is valid on Σ only, for we have used Eq. (3.38) to let appear w. From ∇t, we get the unit normal via Eq. (3.11):

$$n^\alpha = \frac{1}{b}(w, x, y, z). \tag{3.45}$$

From this expression, we have immediately $\boldsymbol{n} \cdot \boldsymbol{n} = b^{-2}(-w^2 + x^2 + y^2 + z^2)$. Therefore, in view of (3.39), we check that $\boldsymbol{n} \cdot \boldsymbol{n} = -1$ on Σ. Note however that if we use expression (3.45) to extend $\boldsymbol{n}$ outside Σ, we can no longer guarantee that $\boldsymbol{n}$ is a unit vector. Within the Minkowskian coordinates $(X^\alpha) = (w, x, y, z)$, we have $\nabla_\beta n^\alpha = \partial n^\alpha / \partial X^\beta$, so that we deduce immediately from the above expression that $\nabla_\beta n^\alpha = b^{-1} \delta^\alpha_{\ \beta}$. Hence

$$\nabla_\beta n_\alpha = \frac{1}{b} g_{\alpha\beta}. \tag{3.46}$$

From Eq. (3.20), we have, for any couple of vectors $(\boldsymbol{u}, \boldsymbol{v})$ tangent to Σ,

$$\boldsymbol{K}(\boldsymbol{u}, \boldsymbol{v}) = -\boldsymbol{u} \cdot \nabla_{\boldsymbol{v}} \boldsymbol{n} = -\frac{1}{b} \boldsymbol{g}(\boldsymbol{u}, \boldsymbol{v}) = -\frac{1}{b} \boldsymbol{\gamma}(\boldsymbol{u}, \boldsymbol{v}),$$

where the last equality follows from (3.9). Hence we conclude that

$$\boldsymbol{K} = -\frac{1}{b} \boldsymbol{\gamma}. \tag{3.47}$$

In particular, the trace of $\boldsymbol{K}$, $K = \gamma^{ij} K_{ij}$, is

$$K = -\frac{3}{b}. \tag{3.48}$$

Note that it is constant.

3.4 Spacelike Hypersurfaces

From now on, we focus on spacelike hypersurfaces, i.e. hypersurfaces Σ such that the induced metric $\boldsymbol{\gamma}$ is definite positive (Riemannian), or equivalently such that the unit normal vector $\boldsymbol{n}$ is timelike (cf. Sects. 3.3.1 and Sects. 3.3.2). Indeed these are the hypersurfaces involved in the 3+1 formalism.

3.4.1 The Orthogonal Projector

At each point $p \in \Sigma$, the space of all spacetime vectors can be orthogonally decomposed as

$$\boxed{\mathcal{T}_p(\mathcal{M}) = \mathcal{T}_p(\Sigma) \oplus \mathrm{span}(\boldsymbol{n})}, \tag{3.49}$$

where span($\boldsymbol{n}$) stands for the 1-dimensional subspace of $\mathcal{T}_p(\mathcal{M})$ generated by the vector $\boldsymbol{n}$.

Remark 3.5 The orthogonal decomposition (3.49) holds for spacelike and timelike hypersurfaces, but not for the null ones. Indeed for any normal $\boldsymbol{n}$ to a null hypersurface Σ, $\mathrm{span}(\boldsymbol{n}) \subset \mathcal{T}_p(\Sigma)$.

The ***orthogonal projector onto*** Σ is the operator $\vec{\boldsymbol{\gamma}}$ associated with the decomposition (3.49) according to

$$\boxed{\begin{array}{llll} \vec{\boldsymbol{\gamma}} : & \mathcal{T}_p(\mathcal{M}) & \longrightarrow & \mathcal{T}_p(\Sigma) \\ & \boldsymbol{v} & \longmapsto & \boldsymbol{v} + (\boldsymbol{n} \cdot \boldsymbol{v})\boldsymbol{n}. \end{array}} \tag{3.50}$$

In particular, as a direct consequence of $\boldsymbol{n} \cdot \boldsymbol{n} = -1$, $\vec{\boldsymbol{\gamma}}$ satisfies

$$\vec{\boldsymbol{\gamma}}(\boldsymbol{n}) = 0. \tag{3.51}$$

Besides, it reduces to the identity operator for any vector tangent to Σ :

$$\forall \boldsymbol{v} \in \mathcal{T}_p(\Sigma), \quad \vec{\boldsymbol{\gamma}}(\boldsymbol{v}) = \boldsymbol{v}. \tag{3.52}$$

According to Eq. (3.50), the components of $\vec{\boldsymbol{\gamma}}$ with respect to any basis $(\boldsymbol{e}_\alpha)$ of $\mathcal{T}_p(\mathcal{M})$ are

$$\boxed{\gamma^\alpha{}_\beta = \delta^\alpha{}_\beta + n^\alpha n_\beta.} \tag{3.53}$$

We have noticed in Sect. 3.3.1 that the embedding Φ of Σ in $\mathcal{M}$ induces a mapping $\mathcal{T}_p(\Sigma) \to \mathcal{T}_p(\mathcal{M})$ (push-forward) and a mapping $\mathcal{T}_p^*(\mathcal{M}) \to \mathcal{T}_p^*(\Sigma)$ (pull-back), but does not provide any mapping in the reverse ways, i.e. from $\mathcal{T}_p(\mathcal{M})$ to $\mathcal{T}_p(\Sigma)$ and from $\mathcal{T}_p^*(\Sigma)$ to $\mathcal{T}_p^*(\mathcal{M})$. The orthogonal projector naturally provides these reverse mappings: from its very definition, it is a mapping $\mathcal{T}_p(\mathcal{M}) \to \mathcal{T}_p(\Sigma)$ and we can construct from it a mapping $\vec{\boldsymbol{\gamma}}^*_{\mathcal{M}} : \mathcal{T}_p^*(\Sigma) \to \mathcal{T}_p^*(\mathcal{M})$ by setting, for any linear form $\boldsymbol{\omega} \in \mathcal{T}_p^*(\Sigma)$,

$$\begin{array}{llll} \vec{\boldsymbol{\gamma}}^*_{\mathcal{M}}\boldsymbol{\omega} : & \mathcal{T}_p(\mathcal{M}) & \longrightarrow & \mathbb{R} \\ & \boldsymbol{v} & \longmapsto & \langle \boldsymbol{\omega}, \vec{\boldsymbol{\gamma}}(\boldsymbol{v}) \rangle. \end{array} \tag{3.54}$$

This clearly defines a linear form belonging to $\mathcal{T}_p^*(\mathcal{M})$. Obviously, we can extend the operation $\vec{\boldsymbol{\gamma}}^*_{\mathcal{M}}$ to any multilinear form $\boldsymbol{A}$ acting on $\mathcal{T}_p(\Sigma)$, by setting

$$\begin{aligned} \vec{\gamma}^*_{\mathcal{M}} \boldsymbol{A} : \quad \mathcal{T}_p(\mathcal{M})^n &\longrightarrow \mathbb{R} \\ (\boldsymbol{v}_1, \ldots, \boldsymbol{v}_n) &\longmapsto \boldsymbol{A}\left(\vec{\gamma}(\boldsymbol{v}_1), \ldots, \vec{\gamma}(\boldsymbol{v}_n)\right). \end{aligned} \tag{3.55}$$

Let us apply this definition to the bilinear form on Σ constituted by the induced metric $\boldsymbol{\gamma}$: $\vec{\gamma}^*_{\mathcal{M}}\boldsymbol{\gamma}$ is then a bilinear form on $\mathcal{M}$, which coincides with $\boldsymbol{\gamma}$ if its two arguments are vectors tangent to Σ and which gives zero if any of its argument is a vector orthogonal to Σ, i.e. parallel to $\boldsymbol{n}$. Since it constitutes an "extension" of $\boldsymbol{\gamma}$ to all vectors in $\mathcal{T}_p(\mathcal{M})$, we shall denote it by the same symbol:

$$\boxed{\boldsymbol{\gamma} := \vec{\gamma}^*_{\mathcal{M}}\boldsymbol{\gamma}}. \tag{3.56}$$

This extended $\boldsymbol{\gamma}$ can be expressed in terms of the metric tensor $\boldsymbol{g}$ and the linear form $\underline{\boldsymbol{n}}$ dual to the normal vector $\boldsymbol{n}$ according to

$$\boxed{\boldsymbol{\gamma} = \boldsymbol{g} + \underline{\boldsymbol{n}} \otimes \underline{\boldsymbol{n}}}. \tag{3.57}$$

In components:

$$\gamma_{\alpha\beta} = g_{\alpha\beta} + n_\alpha n_\beta. \tag{3.58}$$

Indeed, if $\boldsymbol{v}$ and $\boldsymbol{u}$ are vectors both tangent to Σ, $\boldsymbol{\gamma}(\boldsymbol{u},\boldsymbol{v}) = \boldsymbol{g}(\boldsymbol{u},\boldsymbol{v}) + \langle\underline{\boldsymbol{n}},\boldsymbol{u}\rangle\langle\underline{\boldsymbol{n}},\boldsymbol{v}\rangle = \boldsymbol{g}(\boldsymbol{u},\boldsymbol{v}) + 0 = \boldsymbol{g}(\boldsymbol{u},\boldsymbol{v})$, and if $\boldsymbol{u} = \lambda\boldsymbol{n}$, then, for any $\boldsymbol{v} \in \mathcal{T}_p(\mathcal{M})$, $\boldsymbol{\gamma}(\boldsymbol{u},\boldsymbol{v}) = \lambda\boldsymbol{g}(\boldsymbol{n},\boldsymbol{v}) + \lambda\langle\underline{\boldsymbol{n}},\boldsymbol{n}\rangle\langle\underline{\boldsymbol{n}},\boldsymbol{v}\rangle = \lambda[\boldsymbol{g}(\boldsymbol{n},\boldsymbol{v}) - \langle\underline{\boldsymbol{n}},\boldsymbol{v}\rangle] = 0$. This establishes Eq. (3.57).

Remark 3.6 Comparing Eq. (3.58) with Eq. (3.53) justifies the notation $\vec{\gamma}$ employed for the orthogonal projector onto Σ, according to the convention set in Sect. 2.3.3 [see Eq. (2.39)]: $\vec{\gamma}$ is nothing but the "extended" induced metric $\boldsymbol{\gamma}$ with the first index raised by the metric $\boldsymbol{g}$.

Similarly, we may use the $\vec{\gamma}^*_{\mathcal{M}}$ operation to extend the extrinsic curvature tensor $\boldsymbol{K}$, defined a priori as a bilinear form on Σ [Eq. (3.19)], to a bilinear form on $\mathcal{M}$, and we shall use the same symbol to denote this extension:

$$\boxed{\boldsymbol{K} := \vec{\gamma}^*_{\mathcal{M}}\boldsymbol{K}}. \tag{3.59}$$

Remark 3.7 In this book, we will very often use such a "four-dimensional point of view", i.e. we shall treat tensor fields defined on Σ as if they were defined on $\mathcal{M}$. For covariant tensors (multilinear forms), if not mentioned explicitly, the four-dimensional extension is performed via the $\vec{\gamma}^*_{\mathcal{M}}$ operator, as above for $\boldsymbol{\gamma}$ and $\boldsymbol{K}$. For contravariant tensors, the identification is provided by the push-forward mapping Φ_* discussed in Sect. 3.3.1. This four-dimensional point of view has been advocated by Carter [8–10] and results in easier manipulations of tensors defined in Σ, by treating them as ordinary tensors on $\mathcal{M}$. In particular this avoids the introduction of special coordinate systems and complicated notations.

In addition to the extension of three dimensional tensors to four dimensional ones, we use the orthogonal projector $\vec{\gamma}$ to define an "orthogonal projection operation" for *all tensors on* $\mathscr{M}$ in the following way. Given a tensor $\boldsymbol{T}$ of type $\binom{p}{q}$ on $\mathscr{M}$, we denote by $\vec{\gamma}^*\boldsymbol{T}$ another tensor on $\mathscr{M}$, of the same type and whose components in any basis $(\boldsymbol{e}_\alpha)$ of $\mathscr{T}_p(\mathscr{M})$ are deduced from those of $\boldsymbol{T}$ by contracting with $\gamma^\alpha_{\ \beta}$ on all indices:

$$(\vec{\gamma}^*\boldsymbol{T})^{\alpha_1\ldots\alpha_p}{}_{\beta_1\ldots\beta_q} = \gamma^{\alpha_1}{}_{\mu_1}\cdots\gamma^{\alpha_p}{}_{\mu_p}\gamma^{\nu_1}{}_{\beta_1}\cdots\gamma^{\nu_q}{}_{\beta_q}T^{\mu_1\ldots\mu_p}{}_{\nu_1\ldots\nu_q}. \tag{3.60}$$

Notice that for any multilinear form $\boldsymbol{A}$ on Σ, $\vec{\gamma}^*(\vec{\gamma}^*_{\mathscr{M}}\boldsymbol{A}) = \vec{\gamma}^*_{\mathscr{M}}\boldsymbol{A}$, for a vector $\boldsymbol{v} \in \mathscr{T}_p(\mathscr{M})$, $\vec{\gamma}^*\boldsymbol{v} = \vec{\gamma}(\boldsymbol{v})$, for a linear form $\boldsymbol{\omega} \in \mathscr{T}_p^*(\mathscr{M})$, $\vec{\gamma}^*\boldsymbol{\omega} = \boldsymbol{\omega}\circ\vec{\gamma}$, and for any tensor $\boldsymbol{T}$, $\vec{\gamma}^*\boldsymbol{T}$ is ***tangent to*** Σ, in the sense that $\vec{\gamma}^*\boldsymbol{T}$ results in zero if one of its arguments is $\boldsymbol{n}$ or $\underline{\boldsymbol{n}}$.

3.4.2 Relation Between *K* and $\nabla \boldsymbol{n}$

A priori the unit vector $\boldsymbol{n}$ normal to Σ is defined only at points belonging to Σ. Let us consider some extension of $\boldsymbol{n}$ in an open neighbourhood of Σ. If Σ is a level surface of some scalar field t, such a natural extension is provided by the gradient of t, according to Eq. (3.11). Then the tensor fields $\nabla\boldsymbol{n}$ and $\nabla\underline{\boldsymbol{n}}$ are well defined quantities. In particular, we can introduce the vector

$$\boldsymbol{a} := \nabla_{\boldsymbol{n}}\boldsymbol{n}. \tag{3.61}$$

Since $\boldsymbol{n}$ is a timelike unit vector, it can be regarded as the 4-velocity of some observer, and $\boldsymbol{a}$ is then the corresponding 4-acceleration. $\boldsymbol{a}$ is orthogonal to $\boldsymbol{n}$ and hence tangent to Σ, since $\boldsymbol{n}\cdot\boldsymbol{a} = \boldsymbol{n}\cdot\nabla_{\boldsymbol{n}}\boldsymbol{n} = 1/2\nabla_{\boldsymbol{n}}(\boldsymbol{n}\cdot\boldsymbol{n}) = 1/2\nabla_{\boldsymbol{n}}(-1) = 0$.

Let us make explicit the definition of the tensor $\boldsymbol{K}$ extended to $\mathscr{M}$ by Eq. (3.59). From the definition of the operator $\vec{\gamma}^*_{\mathscr{M}}$ [Eq. (3.55)] and the original definition of $\boldsymbol{K}$ [Eq. (3.20)], we have

$$\begin{aligned}
\forall(\boldsymbol{u},\boldsymbol{v}) \in \mathscr{T}_p(\mathscr{M})^2, \quad \boldsymbol{K}(\boldsymbol{u},\boldsymbol{v}) &= \boldsymbol{K}(\vec{\gamma}(\boldsymbol{u}),\vec{\gamma}(\boldsymbol{v})) = -\vec{\gamma}(\boldsymbol{u})\cdot\nabla_{\vec{\gamma}(\boldsymbol{v})}\boldsymbol{n}\\
&= -\vec{\gamma}(\boldsymbol{u})\cdot\nabla_{\boldsymbol{v}+(\boldsymbol{n}\cdot\boldsymbol{v})\boldsymbol{n}}\boldsymbol{n}\\
&= -[\boldsymbol{u}+(\boldsymbol{n}\cdot\boldsymbol{u})\boldsymbol{n}]\cdot[\nabla_{\boldsymbol{v}}\boldsymbol{n}+(\boldsymbol{n}\cdot\boldsymbol{v})\nabla_{\boldsymbol{n}}\boldsymbol{n}]\\
&= -\boldsymbol{u}\cdot\nabla_{\boldsymbol{v}}\boldsymbol{n} - (\boldsymbol{n}\cdot\boldsymbol{v})\boldsymbol{u}\cdot\underbrace{\nabla_{\boldsymbol{n}}\boldsymbol{n}}_{\boldsymbol{a}} - (\boldsymbol{n}\cdot\boldsymbol{u})\underbrace{\boldsymbol{n}\cdot\nabla_{\boldsymbol{v}}\boldsymbol{n}}_{0}\\
&\quad - (\boldsymbol{n}\cdot\boldsymbol{u})(\boldsymbol{n}\cdot\boldsymbol{v})\underbrace{\boldsymbol{n}\cdot\nabla_{\boldsymbol{n}}\boldsymbol{n}}_{0}\\
&= -\boldsymbol{u}\cdot\nabla_{\boldsymbol{v}}\boldsymbol{n} - (\boldsymbol{a}\cdot\boldsymbol{u})(\boldsymbol{n}\cdot\boldsymbol{v}),\\
&= -\nabla\underline{\boldsymbol{n}}(\boldsymbol{u},\boldsymbol{v}) - \langle\underline{\boldsymbol{a}},\boldsymbol{u}\rangle\langle\underline{\boldsymbol{n}},\boldsymbol{v}\rangle,
\end{aligned} \tag{3.62}$$

where we have used the fact that $\boldsymbol{n}\cdot\boldsymbol{n} = -1$ to set $\boldsymbol{n}\cdot\nabla_{\boldsymbol{x}}\boldsymbol{n} = 0$ for any vector $\boldsymbol{x}$. Since Eq. (3.62) is valid for any pair of vectors $(\boldsymbol{u},\boldsymbol{v})$ in $\mathscr{T}_p(\mathscr{M})$, we conclude that

$$\boxed{\nabla \underline{n} = -\boldsymbol{K} - \underline{\boldsymbol{a}} \otimes \underline{\boldsymbol{n}}}. \tag{3.63}$$

In components:

$$\boxed{\nabla_\beta n_\alpha = -K_{\alpha\beta} - a_\alpha n_\beta}. \tag{3.64}$$

Notice that Eq. (3.63) implies that the (extended) extrinsic curvature tensor is nothing but the gradient of the 1-form $\underline{\boldsymbol{n}}$ to which the projector operator $\vec{\boldsymbol{\gamma}}^*$ is applied:

$$\boxed{\boldsymbol{K} = -\vec{\boldsymbol{\gamma}}^* \nabla \underline{\boldsymbol{n}}}. \tag{3.65}$$

Remark 3.8 Whereas the bilinear form $\nabla \underline{\boldsymbol{n}}$ is a priori not symmetric, its projected part, $-\boldsymbol{K}$, is a symmetric bilinear form.

Taking the trace of Eq. (3.63) with respect to the metric $\boldsymbol{g}$ (i.e. contracting Eq. (3.64) with $g^{\alpha\beta}$) yields a simple relation between the divergence of the vector $\boldsymbol{n}$ and the trace of the extrinsic curvature tensor:

$$\boxed{K = -\nabla \cdot \boldsymbol{n}}. \tag{3.66}$$

3.4.3 Links Between the ∇ and D Connections

Given a tensor field $\boldsymbol{T}$ on Σ, its covariant derivative $\boldsymbol{DT}$ with respect to the Levi–Civita connection $\boldsymbol{D}$ of the metric $\boldsymbol{\gamma}$ (cf. Sect. 3.3.3) is expressible in terms of the covariant derivative $\nabla \boldsymbol{T}$ with respect to the spacetime connection ∇ according to the formula

$$\boxed{\boldsymbol{DT} = \vec{\boldsymbol{\gamma}}^* \nabla \boldsymbol{T}}, \tag{3.67}$$

the component version of which is [cf. Eq. (3.60)]:

$$\boxed{D_\rho T^{\alpha_1 \ldots \alpha_p}{}_{\beta_1 \ldots \beta_q} = \gamma^{\alpha_1}{}_{\mu_1} \cdots \gamma^{\alpha_p}{}_{\mu_p} \gamma^{\nu_1}{}_{\beta_1} \cdots \gamma^{\nu_q}{}_{\beta_q} \gamma^{\sigma}{}_{\rho} \nabla_\sigma T^{\mu_1 \ldots \mu_p}{}_{\nu_1 \ldots \nu_q}}. \tag{3.68}$$

Before proceeding to the demonstration of this formula, some comments are appropriate: first of all, the $\boldsymbol{T}$ in the right-hand side of Eq. (3.67) should be the four-dimensional extension $\vec{\boldsymbol{\gamma}}^*_{\mathscr{M}} \boldsymbol{T}$ provided by Eq. (3.55), for ∇ acts on tensors on $\mathscr{M}$. Following the remark made above, we write $\boldsymbol{T}$ instead of $\vec{\boldsymbol{\gamma}}^*_{\mathscr{M}} \boldsymbol{T}$. Similarly the right-hand side should write $\vec{\boldsymbol{\gamma}}^*_{\mathscr{M}} \boldsymbol{DT}$, so that Eq. (3.67) is a equality between tensors on $\mathscr{M}$. Therefore the rigorous version of Eq. (3.67) is

$$\vec{\boldsymbol{\gamma}}^*_{\mathscr{M}} \boldsymbol{DT} = \vec{\boldsymbol{\gamma}}^* [\nabla (\vec{\boldsymbol{\gamma}}^*_{\mathscr{M}} \boldsymbol{T})]. \tag{3.69}$$

Besides, even if $\boldsymbol{T} := \vec{\boldsymbol{\gamma}}^*_{\mathscr{M}}\boldsymbol{T}$ is a four-dimensional tensor, its support (domain of definition) remains the hypersurface Σ. In order to define the covariant derivative $\boldsymbol{\nabla T}$, the support must be an open set of $\mathscr{M}$, which Σ is not. Accordingly, one must first construct some extension $\boldsymbol{T}'$ of $\boldsymbol{T}$ in an open neighbourhood of Σ in $\mathscr{M}$ and then compute $\boldsymbol{\nabla T}'$. The key point is that thanks to the operator $\vec{\boldsymbol{\gamma}}^*$ acting on $\boldsymbol{\nabla T}'$, the result does not depend of the choice of the extension $\boldsymbol{T}'$, provided that $\boldsymbol{T}' = \boldsymbol{T}$ at every point in Σ.

The demonstration of formula (3.67) takes two steps. First, one can show easily that $\vec{\boldsymbol{\gamma}}^*\boldsymbol{\nabla}$ (or more precisely the pull-back of $\vec{\boldsymbol{\gamma}}^*\boldsymbol{\nabla}\vec{\boldsymbol{\gamma}}^*_{\mathscr{M}}$) is a torsion-free connection on Σ, for it satisfies all the defining properties of a connection (linearity, reduction to the gradient for a scalar field, commutation with contractions and Leibniz' rule) and its torsion vanishes. Secondly, this connection vanishes when applied to the metric tensor $\boldsymbol{\gamma}$: indeed, using Eqs. (3.60) and (3.58),

$$\begin{aligned}
\left(\vec{\boldsymbol{\gamma}}^*\boldsymbol{\nabla\gamma}\right)_{\alpha\beta\gamma} &= \gamma^\mu{}_\alpha\gamma^\nu{}_\beta\gamma^\rho{}_\gamma\nabla_\rho\gamma_{\mu\nu} \\
&= \gamma^\mu{}_\alpha\gamma^\nu{}_\beta\gamma^\rho{}_\gamma(\underbrace{\nabla_\rho g_{\mu\nu}}_{0} + \nabla_\rho n_\mu n_\nu + n_\mu\nabla_\rho n_\nu) \\
&= \gamma^\rho{}_\gamma(\underbrace{\gamma^\mu{}_\alpha\,\gamma^\nu{}_\beta n_\nu}_{0}\nabla_\rho n_\mu + \underbrace{\gamma^\mu{}_\alpha n_\mu}_{0}\nabla_\rho n_\nu) \\
&= 0.
\end{aligned}$$

Invoking the uniqueness of the torsion-free connection associated with a given non-degenerate metric (the Levi–Civita connection), we conclude that necessarily $\vec{\boldsymbol{\gamma}}^*\boldsymbol{\nabla} = \boldsymbol{D}$.

One can deduce from Eq. (3.67) an interesting formula about the derivative of a vector field $\boldsymbol{v}$ along another vector field $\boldsymbol{u}$, when both vectors are tangent to Σ. Indeed, from Eq. (3.67),

$$\begin{aligned}
(\boldsymbol{D_u v})^\alpha &= u^\sigma D_\sigma v^\alpha = \underbrace{u^\sigma\gamma^\nu{}_\sigma}_{u^\nu}\gamma^\alpha{}_\mu\nabla_\nu v^\mu = u^\nu\left(\delta^\alpha{}_\mu + n^\alpha n_\mu\right)\nabla_\nu v^\mu \\
&= u^\nu\nabla_\nu v^\alpha + n^\alpha u^\nu\underbrace{n_\mu\nabla_\nu v^\mu}_{-v^\mu\nabla_\nu n_\mu} = u^\nu\nabla_\nu v^\alpha - n^\alpha u^\nu v^\mu\nabla_\nu n_\mu,
\end{aligned}$$

where we have used $n_\mu v^\mu = 0$ ($\boldsymbol{v}$ being tangent to Σ) to write $n_\mu\nabla_\nu v^\mu = -v^\mu\nabla_\nu n_\mu$. Now, from Eq. (3.20) and the symmetry of $\boldsymbol{K}$, $u^\nu v^\mu\nabla_\nu n_\mu = -\boldsymbol{K}(\boldsymbol{v},\boldsymbol{u}) = -\boldsymbol{K}(\boldsymbol{u},\boldsymbol{v})$, so that the above formula becomes

$$\forall(\boldsymbol{u},\boldsymbol{v}) \in \mathscr{T}(\Sigma)\times\mathscr{T}(\Sigma), \quad \boxed{\boldsymbol{D_u v} = \boldsymbol{\nabla_u v} + \boldsymbol{K}(\boldsymbol{u},\boldsymbol{v})\boldsymbol{n}}. \tag{3.70}$$

This equation provides another interpretation of the extrinsic curvature tensor $\boldsymbol{K}$: $\boldsymbol{K}$ measures the deviation of the derivative of any vector of Σ along another vector of Σ, taken with the intrinsic connection $\boldsymbol{D}$ of Σ from the derivative taken with the spacetime connection $\boldsymbol{\nabla}$. Notice from Eq. (3.70) that this deviation is always in the direction of the normal vector $\boldsymbol{n}$.

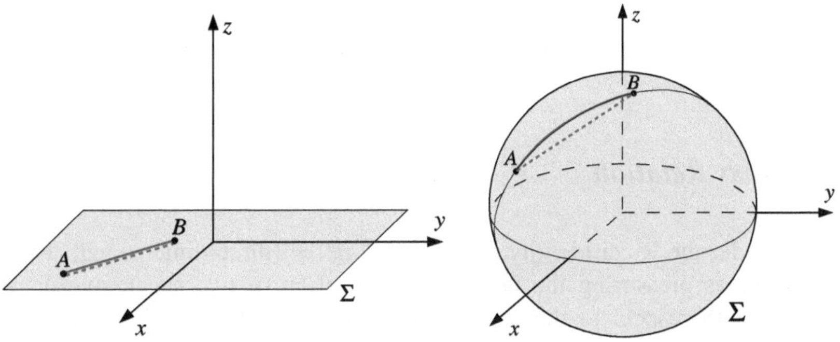

Fig. 3.6 In the Euclidean space $\mathbb{R}^3$, the plane Σ is a totally geodesic hypersurface, for the geodesic between two points A and B within $(\Sigma, \boldsymbol{\gamma})$ (solid line) coincides with the geodesic in the ambient space (dashed line). On the contrary, for the sphere, the two geodesics are distinct, whatever the position of points A and B

Consider a geodesic curve $\mathscr{L}$ in $(\Sigma, \boldsymbol{\gamma})$ and the tangent vector $\boldsymbol{u}$ associated with some affine parametrization of $\mathscr{L}$. Then $\boldsymbol{D_u u} = 0$ and Eq. (3.70) leads to $\boldsymbol{\nabla_u u} = -\boldsymbol{K}(\boldsymbol{u}, \boldsymbol{u})\boldsymbol{n}$. If $\mathscr{L}$ were a geodesic of $(\mathscr{M}, \boldsymbol{g})$, one should have $\boldsymbol{\nabla_u u} = \kappa \boldsymbol{u}$, for some non-affinity parameter κ. Since $\boldsymbol{u}$ is never parallel to $\boldsymbol{n}$, we conclude that the extrinsic curvature tensor $\boldsymbol{K}$ measures the failure of a geodesic of $(\Sigma, \boldsymbol{\gamma})$ to be a geodesic of $(\mathscr{M}, \boldsymbol{g})$. Only in the case where $\boldsymbol{K}$ vanishes, the two notions of geodesics coincide. For this reason, hypersurfaces for which $\boldsymbol{K} = 0$ are called ***totally geodesic hypersurfaces***.

Example 3.5 The plane in the Euclidean space $\mathbb{R}^3$ discussed as Example 3.1 in Sect. 3.3.5 is a totally geodesic hypersurface: $\boldsymbol{K} = 0$. This is obvious since the geodesics of the plane are straight lines, which are also geodesics of $\mathbb{R}^3$ (cf. Fig. 3.6). A counter-example is provided by the sphere embedded in $\mathbb{R}^3$ (Example 3.3 in Sect. 3.3.5): given two points A and B, the geodesic curve with respect to $(\Sigma, \boldsymbol{\gamma})$ joining them is a portion of a sphere's great circle, whereas from the point of view of $\mathbb{R}^3$, the geodesic from A to B is a straight line (cf. Fig. 3.6). The same things holds with the cylinder (Example 3.2): the geodesics on the cylinder are helices, which differ from the straight lines of $\mathbb{R}^3$ (except for the special case where points A and B are vertically aligned: the helix degenerates into a straight line).

3.5 Gauss–Codazzi Relations

We derive here equations that will constitute the basis of the 3+1 formalism for general relativity. They are decompositions of the spacetime Riemann tensor, ${}^4\mathbf{Riem}$ [Eq. (2.67) with $\mathbf{Riem} \to {}^4\mathbf{Riem}$], in terms of quantities relative to the spacelike hypersurface Σ, namely the Riemann tensor associated with the induced metric $\boldsymbol{\gamma}$, **Riem** [Eq. (3.15)] and the extrinsic curvature tensor of Σ, $\boldsymbol{K}$.

3.5.1 Gauss Relation

Let us consider the Ricci identity (3.15) defining the (three-dimensional) Riemann tensor **Riem** as measuring the lack of commutation of two successive covariant derivatives with respect to the connection $\boldsymbol{D}$ associated with Σ's metric $\boldsymbol{\gamma}$. The 4-dimensional version of this identity is

$$D_\alpha D_\beta v^\gamma - D_\beta D_\alpha v^\gamma = R^\gamma{}_{\mu\alpha\beta} v^\mu, \tag{3.71}$$

where $\boldsymbol{v}$ is a generic vector field tangent to Σ. Let us use formula (3.68) which relates the $\boldsymbol{D}$-derivative to the $\boldsymbol{\nabla}$-derivative, to write

$$D_\alpha D_\beta v^\gamma = D_\alpha (D_\beta v^\gamma) = \gamma^\mu{}_\alpha \gamma^\nu{}_\beta \gamma^\gamma{}_\rho \nabla_\mu (D_\nu v^\rho).$$

Using again formula (3.68) to express $D_\nu v^\rho$ yields

$$D_\alpha D_\beta v^\gamma = \gamma^\mu{}_\alpha \gamma^\nu{}_\beta \gamma^\gamma{}_\rho \nabla_\mu \left(\gamma^\sigma{}_\nu \gamma^\rho{}_\lambda \nabla_\sigma v^\lambda\right).$$

Let us expand this formula by making use of Eq. (3.53) to replace $\nabla_\mu \gamma^\sigma_\nu$ by $\nabla_\mu n^\sigma n_\nu + n^\sigma \nabla_\mu n_\nu$. Since $\gamma^\nu_\beta n_\nu = 0$, we get

$$\begin{aligned} D_\alpha D_\beta v^\gamma &= \gamma^\mu{}_\alpha \gamma^\nu{}_\beta \gamma^\gamma{}_\rho \Big(n^\sigma \nabla_\mu n_\nu \gamma^\rho{}_\lambda \nabla_\sigma v^\lambda + \gamma^\sigma{}_\nu \nabla_\mu n^\rho \underbrace{n_\lambda \nabla_\sigma v^\lambda}_{-v^\lambda \nabla_\sigma n_\lambda} + \gamma^\sigma{}_\nu \gamma^\rho{}_\lambda \nabla_\mu \nabla_\sigma v^\lambda \Big) \\ &= \gamma^\mu{}_\alpha \gamma^\nu{}_\beta \gamma^\gamma{}_\lambda \nabla_\mu n_\nu n^\sigma \nabla_\sigma v^\lambda - \gamma^\mu{}_\alpha \gamma^\sigma{}_\beta \gamma^\gamma{}_\rho v^\lambda \nabla_\mu n^\rho \nabla_\sigma n_\lambda + \gamma^\mu{}_\alpha \gamma^\sigma{}_\beta \gamma^\gamma{}_\lambda \nabla_\mu \nabla_\sigma v^\lambda \\ &= -K_{\alpha\beta} \gamma^\gamma{}_\lambda n^\sigma \nabla_\sigma v^\lambda - K^\gamma{}_\alpha K_{\beta\lambda} v^\lambda + \gamma^\mu{}_\alpha \gamma^\sigma{}_\beta \gamma^\gamma{}_\lambda \nabla_\mu \nabla_\sigma v^\lambda, \end{aligned} \tag{3.72}$$

where we have used the idempotence of the projection operator $\vec{\gamma}$, i.e. $\gamma^\gamma_\rho \gamma^\rho_\lambda = \gamma^\gamma{}_\lambda$ to get the second line and $\gamma^\mu{}_\alpha \gamma^\nu{}_\beta \nabla_\mu n_\nu = -K_{\beta\alpha}$ [Eq. (3.65)] to get the third one. When we permute the indices α and β and subtract from Eq. (3.72) to form $D_\alpha D_\beta v^\gamma - D_\beta D_\gamma v^\gamma$, the first term vanishes since $K_{\alpha\beta}$ is symmetric in (α, β). There remains

$$D_\alpha D_\beta v^\gamma - D_\beta D_\gamma v^\gamma = \left(K_{\alpha\mu} K^\gamma_\beta - K_{\beta\mu} K^\gamma_\alpha\right) v^\mu + \gamma^\rho_\alpha \gamma^\sigma{}_\beta \gamma^\gamma{}_\lambda \left(\nabla_\rho \nabla_\sigma v^\lambda - \nabla_\sigma \nabla_\rho v^\lambda\right).$$

Now the Ricci identity (2.68) for the connection $\boldsymbol{\nabla}$ gives $\nabla_\rho \nabla_\sigma v^\lambda - \nabla_\sigma \nabla_\rho v^\lambda = {}^4R^\lambda_{\mu\rho\sigma} v^\mu$. Therefore

$$D_\alpha D_\beta v^\gamma - D_\beta D_\gamma v^\gamma = \left(K_{\alpha\mu} K^\gamma{}_\beta - K_{\beta\mu} K^\gamma{}_\alpha\right) v^\mu + \gamma^\rho{}_\alpha \gamma^\sigma{}_\beta \gamma^\gamma_\lambda {}^4R^\lambda{}_{\mu\rho\sigma} v^\mu.$$

Substituting this relation for the left-hand side of Eq. (3.71) results in

$$\left(K_{\alpha\mu}K^{\gamma}{}_{\beta} - K_{\beta\mu}K^{\gamma}{}_{\alpha}\right)v^{\mu} + \gamma^{\rho}{}_{\alpha}\gamma^{\sigma}{}_{\beta}\gamma^{\gamma}{}_{\lambda}{}^{4}R^{\lambda}{}_{\mu\rho\sigma}v^{\mu} = R^{\gamma}{}_{\mu\alpha\beta}v^{\mu},$$

or equivalently, since $v^{\mu} = \gamma^{\mu}_{\sigma}v^{\sigma}$,

$$\gamma^{\mu}{}_{\alpha}\gamma^{\nu}{}_{\beta}\gamma^{\gamma}{}_{\rho}\gamma^{\sigma}{}_{\lambda}{}^{4}R^{\rho}{}_{\sigma\mu\nu}v^{\lambda} = R^{\gamma}{}_{\lambda\alpha\beta}v^{\lambda} + \left(K^{\gamma}{}_{\alpha}K_{\lambda\beta} - K^{\gamma}{}_{\beta}K_{\alpha\lambda}\right)v^{\lambda}.$$

In this identity, $\boldsymbol{v}$ can be replaced by any vector of $\mathscr{T}(\mathscr{M})$ without changing the results, thanks to the presence of the projector operator $\vec{\boldsymbol{\gamma}}$ and to the fact that both $\boldsymbol{K}$ and **Riem** are tangent to Σ. Therefore we conclude that

$$\boxed{\gamma^{\mu}{}_{\alpha}\gamma^{\nu}{}_{\beta}\gamma^{\gamma}{}_{\rho}\gamma^{\sigma}{}_{\delta}{}^{4}R^{\rho}{}_{\sigma\mu\nu} = R^{\gamma}{}_{\delta\alpha\beta} + K^{\gamma}{}_{\alpha}K_{\delta\beta} - K^{\gamma}{}_{\beta}K_{\alpha\delta}}. \tag{3.73}$$

This is the ***Gauss relation***.

If we contract the Gauss relation on the indices γ and α and use $\gamma^{\mu}{}_{\alpha}\gamma^{\alpha}{}_{\rho} = \gamma^{\mu}{}_{\rho} = \delta^{\mu}{}_{\rho} + n^{\mu}n_{\rho}$, we obtain an expression that lets appear the Ricci tensors ${}^{4}\boldsymbol{R}$ and $\boldsymbol{R}$ associated with $\boldsymbol{g}$ and $\boldsymbol{\gamma}$ respectively:

$$\boxed{\gamma^{\mu}{}_{\alpha}\gamma^{\nu}{}_{\beta}{}^{4}R_{\mu\nu} + \gamma_{\alpha\mu}n^{\nu}\gamma^{\rho}{}_{\beta}n^{\sigma}{}^{4}R^{\mu}{}_{\nu\rho\sigma} = R_{\alpha\beta} + KK_{\alpha\beta} - K_{\alpha\mu}K^{\mu}{}_{\beta}}. \tag{3.74}$$

This equation is naturally called the ***contracted Gauss relation***. Let us take its trace with respect to $\boldsymbol{\gamma}$, taking into account that $K^{\mu}_{\mu} = K^{i}_{i} = K$, $K_{\mu\nu}K^{\mu\nu} = K_{ij}K^{ij}$ and

$$\gamma^{\alpha\beta}\gamma_{\alpha\mu}n^{\nu}\gamma^{\rho}{}_{\beta}n^{\sigma}{}^{4}R^{\mu}{}_{\nu\rho\sigma} = \gamma^{\rho}{}_{\mu}n^{\nu}n^{\sigma}{}^{4}R^{\mu}{}_{\nu\rho\sigma} = \underbrace{{}^{4}R^{\mu}{}_{\nu\mu\sigma}}_{{}^{4}R_{\nu\sigma}}n^{\nu}n^{\sigma} + \underbrace{{}^{4}R^{\mu}{}_{\nu\rho\sigma}n^{\rho}n_{\mu}n^{\nu}n^{\sigma}}_{0}$$
$$= {}^{4}R_{\mu\nu}n^{\mu}n^{\nu}.$$

We obtain

$$\boxed{{}^{4}R + 2\,{}^{4}R_{\mu\nu}n^{\mu}n^{\nu} = R + K^{2} - K_{ij}K^{ij}}. \tag{3.75}$$

This equation is called the ***scalar Gauss relation*** . It constitutes a generalization of Gauss' famous ***Theorema Egregium*** (*remarkable theorem*) [11, 12]. It relates the intrinsic curvature of Σ, represented by the Ricci scalar R, to its extrinsic curvature, represented by $K^{2} - K_{ij}K^{ij}$. Actually, the original version of Gauss' theorem was for two-dimensional surfaces embedded in the Euclidean space $\mathbb{R}^{3}$. Since the curvature of the latter is zero, the left-hand side of Eq. (3.75) vanishes identically in this case. Moreover, the metric $\boldsymbol{g}$ of the Euclidean space $\mathbb{R}^{3}$ is Riemannian, not Lorentzian. Consequently the term $K^{2} - K_{ij}K^{ij}$ has the opposite sign, so that Eq. (3.75) becomes

$$R - K^{2} + K_{ij}K^{ij} = 0 \quad (\boldsymbol{g}\text{ Euclidean}). \tag{3.76}$$

This change of sign stems from the fact that for a Riemannian ambient metric, the unit normal vector $\boldsymbol{n}$ is spacelike and the orthogonal projector is $\gamma^{\alpha}{}_{\beta} = \delta^{\alpha}{}_{\beta} - n^{\alpha} n_{\beta}$ instead of $\gamma^{\alpha}{}_{\beta} = \delta^{\alpha}{}_{\beta} + n^{\alpha} n_{\beta}$ [the latter form has been used explicitly in the calculation leading to Eq. (3.72)]. Moreover, in dimension 2, formula (1) can be simplified by letting appear the principal curvatures κ_1 and κ_2 of Σ (cf. Sect. 3.3.4). Indeed, $\boldsymbol{K}$ can be diagonalized in an orthonormal basis (with respect to γ) so that $K_{ij} = \mathrm{diag}(\kappa_1, \kappa_2)$ and $K^{ij} = \mathrm{diag}(\kappa_1, \kappa_2)$. Consequently, $K = \kappa_1 + \kappa_2$ and $K_{ij} K^{ij} = \kappa_1^2 + \kappa_2^2$ and Eq. (3.75) becomes

$$R = 2\kappa_1 \kappa_2 \quad (\boldsymbol{g} \text{ Euclidean, } \Sigma \text{ dimension 2}). \tag{3.77}$$

Example 3.6 We may check the Theorema Egregium (3.76) for the examples of Sect. 3.3.5. It is trivial for the plane, since each term vanishes separately. For the cylinder of radius a, $R = 0$, $K = -1/a$ [Eq. (3.30)], $K_{ij} K^{ij} = 1/a^2$ [Eq. (3.29)], so that Eq. (3.76) is satisfied. For the sphere of radius a, $R = 2/a^2$ [Eq. (3.32)], $K = -2/a$ [Eq. (3.37)], $K_{ij} K^{ij} = 2/a^2$ [Eq. (3.36)], so that Eq. (3.76) is satisfied as well.

Example 3.7 Let us check the 4-dimensional scalar Gauss relation (3.75) on the case of the hyperbolic 3-space embedded in Minkowski spacetime (Example 3.4). Since Minkowski spacetime is flat, ${}^4R = 0$, ${}^4R_{\mu\nu} = 0$ and the left-hand side of Eq. (3.75) identically vanishes. On the right-hand side, $R = -6/b^2$ [Eq. (3.44)], $K^2 = 9/b^2$ [Eq. (3.48)] and $K_{ij} K^{ij} = \gamma_{ij} \gamma^{ij} / b^2 = 3/b^2$ [Eq. (3.47)]. Thus Eq. (3.75) is satisfied.

3.5.2 Codazzi Relation

Let us at present apply the Ricci identity (2.68) to the normal vector $\boldsymbol{n}$ (or more precisely to any extension of $\boldsymbol{n}$ around Σ, cf. Sect. 3.4.2):

$$\left(\nabla_\alpha \nabla_\beta - \nabla_\beta \nabla_\alpha\right) n^\gamma = {}^4R^\gamma{}_{\mu\alpha\beta} n^\mu . \tag{3.78}$$

If we project this relation onto Σ, we get

$$\gamma^\mu{}_\alpha \gamma^\nu{}_\beta \gamma^\gamma{}_\rho \, {}^4R^\rho{}_{\sigma\mu\nu} n^\sigma = \gamma^\mu{}_\alpha \gamma^\nu{}_\beta \gamma^\gamma{}_\rho \left(\nabla_\mu \nabla_\nu n^\rho - \nabla_\nu \nabla_\mu n^\rho\right) . \tag{3.79}$$

Now, from Eq. (3.64),

$$\begin{aligned}\gamma^{\mu}{}_{\alpha}\gamma^{\nu}{}_{\beta}\gamma^{\gamma}{}_{\rho}\nabla_{\mu}\nabla_{\nu}n^{\rho} &= \gamma^{\mu}{}_{\alpha}\gamma^{\nu}{}_{\beta}\gamma^{\gamma}{}_{\rho}\nabla_{\mu}\left(-K^{\rho}_{\nu} - a^{\rho}n_{\nu}\right)\\ &= -\gamma^{\mu}{}_{\alpha}\gamma^{\nu}{}_{\beta}\gamma^{\gamma}{}_{\rho}\left(\nabla_{\mu}K^{\rho}{}_{\nu} + \nabla_{\mu}a^{\rho}n_{\nu} + a^{\rho}\nabla_{\mu}n_{\nu}\right)\\ &= -D_{\alpha}K^{\gamma}{}_{\beta} + a^{\gamma}K_{\alpha\beta},\end{aligned} \tag{3.80}$$

where we have used Eq. (3.68), as well as $\gamma^{\nu}_{\beta}n_{\nu} = 0$, $\gamma^{\gamma}_{\rho}a^{\rho} = a^{\gamma}$, and $\gamma^{\mu}_{\alpha}\gamma^{\nu}_{\beta}\nabla_{\mu}n_{\nu} = -K_{\alpha\beta}$ to get the last line. After permutation of the indices α and β and subtraction from Eq. (3.80), taking into account the symmetry of $K_{\alpha\beta}$, we see that Eq. (3.79) becomes

$$\boxed{\gamma^{\gamma}_{\rho}n^{\sigma}\gamma^{\mu}_{\alpha}\gamma^{\nu}_{\beta}{}^{4}R^{\rho}_{\sigma\mu\nu} = D_{\beta}K^{\gamma}_{\alpha} - D_{\alpha}K^{\gamma}_{\beta}}. \tag{3.81}$$

This is the ***Codazzi relation***, also called ***Codazzi–Mainardi relation*** in the mathematical literature [11].

Remark 3.9 Thanks to the symmetries of the Riemann tensor (cf. Sect. 2.4.3), changing the index contracted with $\boldsymbol{n}$ in Eq. (3.81) (for instance considering $n_{\rho}\gamma^{\gamma\sigma}\gamma^{\mu}{}_{\alpha}\gamma^{\nu}{}_{\beta}{}^{4}R^{\rho}{}_{\sigma\mu\nu}$ or $\gamma^{\gamma}{}_{\rho}\gamma^{\sigma}{}_{\alpha}n^{\mu}\gamma^{\nu}{}_{\beta}{}^{4}R^{\rho}{}_{\sigma\mu\nu}$) would not give an independent relation: at most it would result in a change of sign of the right-hand side.

Contracting the Codazzi relation on the indices α and γ yields to

$$\gamma^{\mu}{}_{\rho}n^{\sigma}\gamma^{\nu}{}_{\beta}{}^{4}R^{\rho}{}_{\sigma\mu\nu} = D_{\beta}K - D_{\mu}K^{\mu}{}_{\beta},$$

with $\gamma^{\mu}_{\rho}n^{\sigma}\gamma^{\nu}_{\beta}{}^{4}R^{\rho}_{\sigma\mu\nu} = (\delta^{\mu}_{\rho} + n^{\mu}n_{\rho})n^{\sigma}\gamma^{\nu}_{\beta}{}^{4}R^{\rho}_{\sigma\mu\nu} = n^{\sigma}\gamma^{\nu}_{\beta}{}^{4}R_{\sigma\nu} + \gamma^{\nu}_{\beta}{}^{4}R^{\rho}_{\sigma\mu\nu}n_{\rho}n^{\sigma}n^{\mu}$. Now, from the antisymmetry of the Riemann tensor with respect to its first two indices [Eq. (2.73)], the last term vanishes, so that one is left with

$$\boxed{\gamma^{\mu}{}_{\alpha}n^{\nu}{}^{4}R_{\mu\nu} = D_{\alpha}K - D_{\mu}K^{\mu}{}_{\alpha}}. \tag{3.82}$$

We shall call this equation the ***contracted Codazzi relation***.

Example 3.8 The Codazzi relation is trivially satisfied by the three examples of Sect. 3.3.5 because the Riemann tensor vanishes for the Euclidean space $\mathbb{R}^3$ and for each of the considered surfaces, either $\boldsymbol{K} = 0$ (plane) or $\boldsymbol{K}$ is constant on Σ, in the sense that $\boldsymbol{DK} = 0$.

Example 3.9 Regarding the hyperbolic 3-space in Minkowski spacetime (Example 3.4), the Codazzi relation (3.81) is satisfied because ${}^{4}R^{\rho}_{\sigma\mu\nu} = 0$ (flat spacetime) and, from Eqs. (3.47) and (3.14), $\boldsymbol{DK} = -\boldsymbol{D}(b^{-1}\boldsymbol{\gamma}) = -b^{-1}\boldsymbol{D\gamma} = 0$.

References

1. Poisson, E.: A Relativist's Toolkit, The Mathematics of Black-Hole Mechanics. Cambridge University Press, Cambridge (2004). http://www.physics.uoguelph.ca/poisson/toolkit/
2. Carroll, S.M.: Spacetime and Geometry: An Introduction to General Relativity. Addison Wesley (Pearson Education), San Fransisco (2004). http://preposterousuniverse.com/spacetimeandgeometry/
3. Straumann, N.: General Relavity, With Applications to Astrophysics. Springer, Berlin (2004)
4. Misner, C.W., Thorne, K.S., Wheeler, J.A.: Gravitation. Freeman, New York (1973)
5. Wald, R.M.: General relativity. University of Chicago Press, Chicago (1984)
6. Deruelle, N.: General Relativity: A Primer. Lectures at Institut Henri Poincaré, Paris (2006). http://www.luth.obspm.fr/IHP06/
7. Hawking, S.W., Ellis, G.F.R.: The large scale structure of space-time. Cambridge University Press, Cambridge (1973)
8. Carter, B.: Outer curvature and conformal geometry of an imbedding. J. Geom. Phys. **8**, 53 (1992)
9. Carter, B.: Basic brane theory. Class. Quantum Grav. **9**, S19 (1992)
10. Carter, B.: Extended tensorial curvature analysis for embeddings and foliations. Contemp. Math. **203**, 207 (1997)
11. Berger, M.: A Panoramic View of Riemannian Geometry. Springer, Berlin (2003)
12. Berger, M., Gostiaux, B.: Géométrie Différentielle: Variétés, Courbes et Surfaces. Presses Universitaires de France, Paris (1987)

Chapter 4
Geometry of Foliations

Abstract Whereas the previous chapter focused on a single hypersurface Σ embedded in the spacetime $(\mathscr{M}, \boldsymbol{g})$, we consider here a continuous set of hypersurfaces $(\Sigma_t)_{t\in\mathbb{R}}$ that covers the manifold $\mathscr{M}$. We introduce the concepts of *lapse function*, *normal evolution vector* and *Eulerian observer*. We consider the evolution of the 3-metric along the normal to the slices Σ_t and we compute the last part of the 3+1 decomposition of the Riemann tensor, complementary to the Gauss and Codazzi equations obtained in Chap. 3.

4.1 Introduction

As already mentioned, the 3+1 formalism for general relativity is based on a foliation of spacetime by a 1-parameter family of spacelike hypersurfaces. This is possible for a wide class of spacetimes, the so-called *globally hyperbolic spacetimes*, which we introduce in Sect. 4.2. Actually this class covers most of the spacetimes of astrophysical or cosmological interest. Again the title of this chapter is "Geometry...", since as in Chap. 3, all the results are independent of the Einstein equation. All the results are also independent on the choice of coordinates (x^i) in each slice of the foliation.

4.2 Globally Hyperbolic Spacetimes and Foliations

4.2.1 Globally Hyperbolic Spacetimes

A ***Cauchy surface*** is a spacelike hypersurface Σ in $\mathscr{M}$ such that each causal (i.e. timelike or null) curve without end point intersects Σ once and only once [1, 2]. Equivalently, Σ is a Cauchy surface iff its domain of dependence is the whole spacetime $\mathscr{M}$. Not all spacetimes admit a Cauchy surface. For instance spacetimes with

É. Gourgoulhon, *3+1 Formalism in General Relativity*, Lecture Notes in Physics 846,
DOI: 10.1007/978-3-642-24525-1_4,

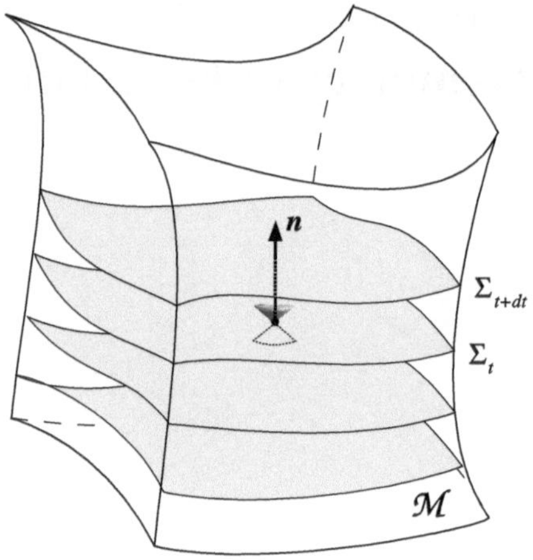

Fig. 4.1 Foliation of the spacetime $\mathcal{M}$ by a family of spacelike hypersurfaces $(\Sigma_t)_{t\in\mathbb{R}}$

closed timelike curves do not. Other examples are provided in Ref. [3]. A spacetime $(\mathcal{M}, \boldsymbol{g})$ that admits a Cauchy surface Σ is said to be *globally hyperbolic*. The name *globally hyperbolic* stems from the fact that the scalar wave equation is well posed in these spacetimes (see e.g. [4]).

The topology of a globally hyperbolic spacetime $\mathcal{M}$ is necessarily $\Sigma \times \mathbb{R}$ (where Σ is the Cauchy surface entering in the definition of global hyperbolicity).

Remark 4.1 The original definition of a globally hyperbolic spacetime, given by Leray in 1953 [5] is actually more technical that the one given above, but the latter has been shown to be equivalent to the original one [1] (see e.g. Ref. [4, 6]).

4.2.2 Definition of a Foliation

Any globally hyperbolic spacetime $(\mathcal{M}, \boldsymbol{g})$ can be foliated by a family of *spacelike* hypersurfaces $(\Sigma_t)_{t\in\mathbb{R}}$. By ***foliation*** or ***slicing***, it is meant that there exists a smooth scalar field $\hat{t}$ on $\mathcal{M}$, which is regular (in the sense that its gradient never vanishes), such that each hypersurface is a level surface of this scalar field:

$$\forall t \in \mathbb{R}, \quad \Sigma_t := \left\{p \in \mathcal{M}, \hat{t}(p) = t\right\}. \tag{4.1}$$

Since $\hat{t}$ is regular, the hypersurfaces Σ_t are non-intersecting:

$$\Sigma_t \cap \Sigma_{t'} = \emptyset \quad \text{for } t \neq t'. \tag{4.2}$$

In the following, we do no longer distinguish between t and $\hat{t}$, i.e. we skip the hat in the name of the scalar field. Each hypersurface Σ_t is called a ***leaf*** or a ***slice*** of the foliation. We assume that all Σ_t's are spacelike and that the foliation covers $\mathcal{M}$ (cf. Fig. 4.1):

$$\mathscr{M} = \bigcup_{t\in\mathbb{R}} \Sigma_t. \tag{4.3}$$

4.3 Foliation Kinematics

4.3.1 Lapse Function

As already noticed in Sect. 3.3.2, the timelike and future-directed unit vector $\boldsymbol{n}$ normal to the slice Σ_t is necessarily collinear to the vector $\vec{\nabla} t$ associated with the gradient 1-form ∇t. Hence we may write

$$\boxed{\boldsymbol{n} := -N\vec{\nabla} t} \tag{4.4}$$

with

$$N := \left(-\vec{\nabla} t \cdot \vec{\nabla} t\right)^{-1/2} = \left(-\langle \nabla t, \vec{\nabla} t\rangle\right)^{-1/2}. \tag{4.5}$$

The minus sign in (4.4) is chosen so that the vector $\boldsymbol{n}$ is future-oriented if the scalar field t is increasing towards the future. Notice that the value of N ensures that $\boldsymbol{n}$ is a unit vector:

$$\boldsymbol{n} \cdot \boldsymbol{n} = -1. \tag{4.6}$$

The scalar field N hence defined is called the ***lapse function***. The name *lapse* has been coined by Wheeler in 1964 [7].

Remark 4.2 In most of the numerical relativity literature, the lapse function is denoted by the letter α instead of N. We adopt here the same notation as ADM [8], MTW [9] and Choquet–Bruhat [4].

Notice that by construction [Eq. (4.5)],

$$N > 0. \tag{4.7}$$

In particular, the lapse function never vanishes for a regular foliation. Equation (4.4) also says that $-N$ is the proportionality factor between the gradient 1-form ∇t and the 1-form $\underline{\boldsymbol{n}}$ associated to the vector $\boldsymbol{n}$ by the metric duality:

$$\boxed{\underline{\boldsymbol{n}} = -N\nabla t}. \tag{4.8}$$

4.3.2 Normal Evolution Vector

Let us define the ***normal evolution vector*** as the timelike vector normal to Σ_t such that

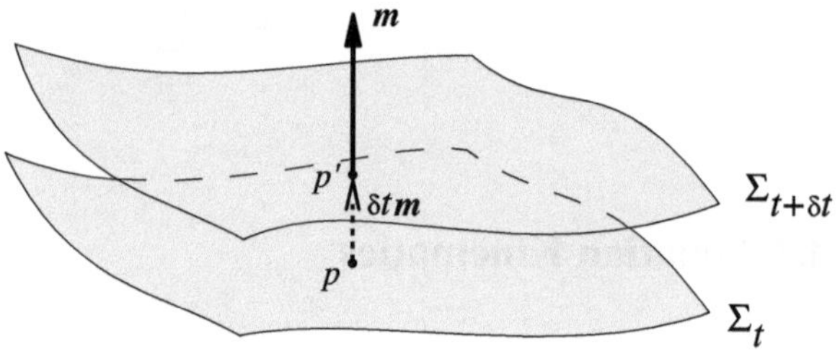

Fig. 4.2 The point p' deduced from $p \in \Sigma_t$ by the displacement $\delta t\boldsymbol{m}$ belongs to $\Sigma_{t+\delta t}$, i.e. the hypersurface Σ_t is transformed to $\Sigma_{t+\delta t}$ by the vector field $\delta t\boldsymbol{m}$ (Lie dragging)

$$\boxed{\boldsymbol{m} := N\boldsymbol{n}}. \tag{4.9}$$

Since $\boldsymbol{n}$ is a unit vector, the scalar square of $\boldsymbol{m}$ is

$$\boldsymbol{m}\cdot\boldsymbol{m} = -N^2. \tag{4.10}$$

Besides, we have

$$\langle \boldsymbol{\nabla} t, \boldsymbol{m}\rangle = N\langle \boldsymbol{\nabla} t, \boldsymbol{n}\rangle = N^2 \underbrace{(-\langle \boldsymbol{\nabla} t, \vec{\boldsymbol{\nabla}} t\rangle)}_{N^{-2}} = 1,$$

where we have used Eqs. (4.4) and (4.5). Hence

$$\boxed{\langle \boldsymbol{\nabla} t, \boldsymbol{m}\rangle = \boldsymbol{\nabla}_{\boldsymbol{m}} t = m^\mu \nabla_\mu t = 1}. \tag{4.11}$$

This relation means that the normal vector $\boldsymbol{m}$ is "adapted" to the scalar field t, contrary to the normal vector $\boldsymbol{n}$. A geometrical consequence of this property is that the hypersurface $\Sigma_{t+\delta t}$ can be obtained from the neighbouring hypersurface Σ_t by the small displacement $\delta t\boldsymbol{m}$ of each point of Σ_t. Indeed consider some point p in Σ_t and displace it by the infinitesimal vector $\delta t\boldsymbol{m}$ to the point $p' = p + \delta t\boldsymbol{m}$ (cf. Fig. 4.2). From the very definition of the gradient 1-form $\boldsymbol{\nabla} t$, the value of the scalar field t at p' is

$$\begin{aligned} t(p') &= t(p+\delta t\boldsymbol{m}) = t(p) + \langle \boldsymbol{\nabla} t, \delta t\boldsymbol{m}\rangle = t(p) + \delta t \underbrace{\langle \boldsymbol{\nabla} t, \boldsymbol{m}\rangle}_{1} \\ &= t(p) + \delta t. \end{aligned}$$

This last equality shows that $p' \in \Sigma_{t+\delta t}$. Hence the vector $\delta t\boldsymbol{m}$ carries the hypersurface Σ_t into the neighbouring one $\Sigma_{t+\delta t}$. One says equivalently that the hypersurfaces (Σ_t) are ***Lie dragged*** by the vector $\boldsymbol{m}$. This justifies the name *normal evolution vector* given to $\boldsymbol{m}$.

An immediate consequence of the Lie dragging of the hypersurfaces Σ_t by the vector $\boldsymbol{m}$ is that the Lie derivative along $\boldsymbol{m}$ of any vector tangent to Σ_t is also a vector tangent to Σ_t:

$$\boxed{\forall \boldsymbol{v} \in \mathcal{T}(\Sigma_t), \quad \mathcal{L}_{\boldsymbol{m}} \boldsymbol{v} \in \mathcal{T}(\Sigma_t)}. \tag{4.12}$$

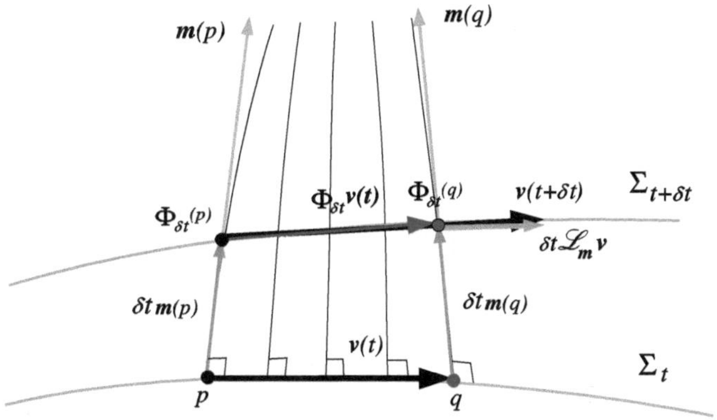

Fig. 4.3 Geometrical construction showing that $\mathscr{L}_m v \in \mathscr{T}(\Sigma_t)$ for any vector v tangent to the hypersurface Σ_t : on Σ_t, a vector can be identified to a infinitesimal displacement between two points, p and q say. These points are transported onto the neighbouring hypersurface $\Sigma_{t+\delta t}$ along the field lines of the vector field m(thin lines on the figure) by the diffeomorphism $\Phi_{\delta t}$ associated with m: the displacement between p and $\Phi_{\delta t}(p)$ is the vector $\delta t m$. The couple of points $(\Phi_{\delta t}(p), \Phi_{\delta t}(q))$ defines the vector $\Phi_{\delta t} v(t)$, which is tangent to $\Sigma_{t+\delta t}$ since both points $\Phi_{\delta t}(p)$ and $\Phi_{\delta t}(q)$ belong to $\Sigma_{t+\delta t}$. The Lie derivative of v along m is then defined by the difference between the value of the vector field v at the point $\Phi_{\delta t}(p)$, i.e. $v(t+\delta t)$, and the vector transported from Σ_t along m's field lines, i.e. $\Phi_{\delta t} v(t)$: $\mathscr{L}_m v(t+\delta t) = \lim_{\delta t \to 0}[v(t+\delta t) - \Phi_{\delta t} v(t)]/\delta t$. Since both vectors $v(t+\delta t)$ and $\Phi_{\delta t} v(t)$ are in $\mathscr{T}(\Sigma_{t+\delta t})$, it follows then that $\mathscr{L}_m v(t+\delta t) \in \mathscr{T}(\Sigma_{t+\delta t})$

This is obvious from the geometric definition of the Lie derivative (cf. Sect. 2.5 and Fig. 4.3).

Example 4.1 **Hyperboloidal slicing of Minkowski spacetime** Let $(\mathscr{M}, g)$ be the Minkowski spacetime. Given some Minkowskian coordinates $(X^\alpha) = (w, x, y, z)$, i.e. coordinates such that $g_{\alpha\beta} = \mathrm{diag}(-1, 1, 1, 1)$, and some constant $b > 0$ having the dimension of a length, we consider the foliation $(\Sigma_t)_{t\in\mathbb{R}}$ defined by the iso-surfaces of the scalar field

$$t = w - \sqrt{b^2 + x^2 + y^2 + z^2}. \tag{4.13}$$

Each Σ_t is a copy of the hyperbolic 3-space discussed in Example 3.4, the latter corresponding to $t = 0$ (cf. Fig. 4.4). Indeed, Σ_t is nothing but the image of Σ_0 by the translation $w \mapsto w + t$ along the w-axis. Note that each Σ_t, while remaining spacelike, becomes asymptotic to the future light cone emerging from the point $(w, x, y, z) = (t, 0, 0, 0)$ as $x \to \pm\infty$. The components of the gradient of t with respect to the coordinates $(X^\alpha) = (w, x, y, z)$ are $\nabla_\alpha t = \partial t/\partial X^\alpha$, i.e.

$$\nabla_\alpha t = \left(1, -\frac{x}{\sqrt{b^2+x^2+y^2+z^2}}, -\frac{y}{\sqrt{b^2+x^2+y^2+z^2}}, -\frac{z}{\sqrt{b^2+x^2+y^2+z^2}}\right).$$

The value of the lapse function is then easily deduced from Eq. (4.5):

$$N = \left[\left(\frac{\partial t}{\partial w}\right)^2 - \left(\frac{\partial t}{\partial x}\right)^2 - \left(\frac{\partial t}{\partial y}\right)^2 - \left(\frac{\partial t}{\partial z}\right)^2\right]^{-1/2},$$

yielding

$$N = \sqrt{1 + \frac{x^2+y^2+z^2}{b^2}}. \tag{4.14}$$

The unit normal to Σ_t is obtained from Eq. (4.8):

$$n_\alpha = \frac{1}{b}\left(-\sqrt{b^2+x^2+y^2+z^2}, x, y, z\right) = \left(-N, \frac{x}{b}, \frac{y}{b}, \frac{z}{b}\right) \tag{4.15}$$

$$n^\alpha = \frac{1}{b}\left(\sqrt{b^2+x^2+y^2+z^2}, x, y, z\right) = \left(N, \frac{x}{b}, \frac{y}{b}, \frac{z}{b}\right). \tag{4.16}$$

For $t=0$, we recover expression (3.45) since in this case $\sqrt{b^2+x^2+y^2+z^2} = w$. The vector $\boldsymbol{n}$ is represented at two points A and B on Σ_0 in Fig. 4.4.

4.3.3 Eulerian Observers

Since $\boldsymbol{n}$ is a unit timelike vector, it can be regarded as the 4-velocity of some observer. We call such observer an ***Eulerian observer***. The worldlines of the Eulerian observers are thus orthogonal to the hypersurfaces Σ_t. Physically, this means that the hypersurface Σ_t is *locally* the set of events that are simultaneous from the point of view of the Eulerian observer, according to Einstein-Poincaré simultaneity convention.

Remark 4.3 The Eulerian observers are sometimes called ***fiducial observers*** (e.g. [10]). In the special case of axisymmetric and stationary spacetimes, they are also called ***locally nonrotating observers*** [11] or ***zero-angular-momentum observers*** (***ZAMO***) [10].

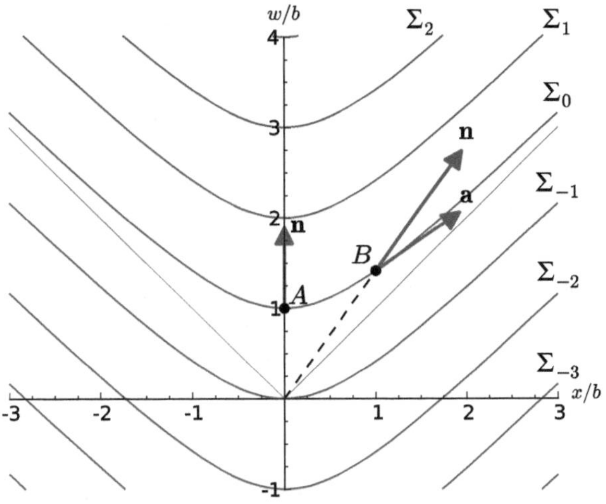

Fig. 4.4 Slicing of Minkowski spacetime by a family $(\Sigma_t)_{t\in\mathbb{R}}$ of hypersurfaces isometric to the hyperbolic space $\mathbb{H}^3$. Σ_0 is the same hypersurface as that depicted in Fig. 3.5, with the dimension along y suppressed

Let us consider two close events p and p' on the worldline of some Eulerian observer. Let t be the "coordinate time" of the event p and $t + \delta t$ ($\delta t > 0$) that of p', in the sense that $p \in \Sigma_t$ and $p' \in \Sigma_{t+\delta t}$. Then $p' = p + \delta t \boldsymbol{m}$, as above. The proper time $\delta\tau$ between the events p and p', as measured the Eulerian observer, is given by the metric length of the vector linking p and p':

$$\delta\tau = \sqrt{-\boldsymbol{g}(\delta t \boldsymbol{m}, \delta t \boldsymbol{m})} = \sqrt{-\boldsymbol{g}(\boldsymbol{m}, \boldsymbol{m})}\, \delta t.$$

Since $\boldsymbol{g}(\boldsymbol{m}, \boldsymbol{m}) = -N^2$ [Eq. (4.10)] and $N > 0$ [Eq. (4.7)], we get

$$\boxed{\delta\tau = N\delta t}. \tag{4.17}$$

This equality justifies the name *lapse function* given to N: N relates the "coordinate time" t labelling the leaves of the foliation to the physical time τ measured by the Eulerian observer.

The 4-acceleration of the Eulerian observer is the covariant derivative of the 4-velocity $\boldsymbol{n}$ along itself:

$$\boldsymbol{a} = \boldsymbol{\nabla}_{\boldsymbol{n}} \boldsymbol{n}. \tag{4.18}$$

As already noticed in Sect. 3.4.2, the vector $\boldsymbol{a}$ is orthogonal to $\boldsymbol{n}$ and hence tangent to Σ_t. Moreover, it can be expressed in terms of the spatial gradient of the lapse function. Indeed, by means Eq. (4.8), we have

$$a_\alpha = n^\mu \nabla_\mu n_\alpha = -n^\mu \nabla_\mu (N \nabla_\alpha t) = -n^\mu \nabla_\mu N \nabla_\alpha t - N n^\mu \underbrace{\nabla_\mu \nabla_\alpha t}_{\nabla_\alpha \nabla_\mu t}$$

$$= \frac{1}{N} n_\alpha n^\mu \nabla_\mu N + N n^\mu \nabla_\alpha \left(-\frac{1}{N} n_\mu \right) = \frac{1}{N} n_\alpha n^\mu \nabla_\mu N + \frac{1}{N} \nabla_\alpha N \underbrace{n^\mu n_\mu}_{-1} - \underbrace{n^\mu \nabla_\alpha n_\mu}_{0}$$

$$= \frac{1}{N} \left(\nabla_\alpha N + n_\alpha n^\mu \nabla_\mu N \right) = \frac{1}{N} \gamma^\mu_{\ \alpha} \nabla_\mu N = \frac{1}{N} D_\alpha N = D_\alpha \ln N,$$

where we have used the torsion-free character of the connection $\boldsymbol{\nabla}$ to write $\nabla_\mu \nabla_\alpha t = \nabla_\alpha \nabla_\mu t$, as well as the expression (3.53) of the orthogonal projector onto Σ_t, $\vec{\boldsymbol{\gamma}}$, and the relation (3.68) between $\boldsymbol{\nabla}$ and $\boldsymbol{D}$ derivatives. Thus we have

$$\boxed{\underline{\boldsymbol{a}} = \boldsymbol{D} \ln N} \quad \text{and} \quad \boxed{\boldsymbol{a} = \vec{\boldsymbol{D}} \ln N}. \tag{4.19}$$

Thus, the 4-acceleration of the Eulerian observer appears to be nothing but the gradient within $(\Sigma_t, \boldsymbol{\gamma})$ of the logarithm of the lapse function. Notice that since a spatial gradient is always tangent to Σ_t, we recover immediately from formula (4.19) that $\boldsymbol{n} \cdot \boldsymbol{a} = 0$.

Remark 4.4 Because they are hypersurface-orthogonal, the congruence formed by all the Eulerian observers' worldlines has a vanishing vorticity, hence the name "non-rotating" observer given sometimes to the Eulerian observer (cf. Remark 4.3).

Example 4.2 Let us pursue with Example 4.1. With respect to the Minkowskian coordinates $(X^\alpha) = (w, x, y, z)$, the components of the covariant derivative of $\boldsymbol{n}$ are $\nabla_\beta n_\alpha = \partial n^\alpha / \partial X^\beta$. Using expression (4.15) for n_α, we get (α = row index, β = column index)

$$\nabla_\beta n_\alpha = \frac{1}{b} \begin{pmatrix} 0 & -\frac{x}{bN} & -\frac{y}{bN} & -\frac{z}{bN} \\ 0 & 1 & 0 & 0 \\ 0 & 0 & 1 & 0 \\ 0 & 0 & 0 & 1 \end{pmatrix}. \tag{4.20}$$

Note that we do not recover Eq. (3.46) for $t = 0$, since in Example 3.4, the normal $\boldsymbol{n}$ was extended away from Σ_0 in such a way that $\boldsymbol{n} \cdot \boldsymbol{n} \neq -1$. Here, by construction, the field $\boldsymbol{n}$ obeys $\boldsymbol{n} \cdot \boldsymbol{n} = -1$ everywhere in $\mathscr{M}$. Hence the two fields differ outside Σ_0. From Eqs. (4.20) and (4.16) we deduce the 4-acceleration of the Eulerian observer via Eq. (4.18):

$$a_\alpha = \frac{1}{b^2} \left(-\frac{x^2 + y^2 + z^2}{bN}, x, y, z \right). \tag{4.21}$$

We verify, via (4.16), that $a_\mu n^\mu = 0$, as it should be. The vector $\boldsymbol{a}$ is represented at point B in Fig. 4.4; it vanishes at point A.

From the expression (4.14) of N, we have

$$\nabla_\alpha N = \left(\frac{\partial N}{\partial w}, \frac{\partial N}{\partial x}, \frac{\partial N}{\partial Y}, \frac{\partial N}{\partial z}\right) = \frac{1}{b^2 N}(0, x, y, z). \tag{4.22}$$

Now

$$D_\alpha N = \gamma^\mu_{\ \alpha} \nabla_\mu N = (\delta^\mu_{\ \alpha} + n^\mu n_\alpha)\nabla_\mu N = \nabla_\alpha N + n^\mu \nabla_\mu N n_\alpha.$$

with $n^\mu \nabla_\mu N = (x^2+y^2+z^2)/(b^2 N)$, thanks to Eqs. (4.16) and (4.22). Hence, using expression (4.15) for n_α, we get

$$D_\alpha \ln N = \frac{1}{N} D_\alpha N = \frac{1}{b^2}\left(-\frac{x^2+y^2+z^2}{bN}, x, y, z\right).$$

Comparing with Eq. (4.21) we conclude that formula (4.19) holds.

4.3.4 Gradients of n and m

Substituting Eq. (4.19) for $\underline{\boldsymbol{a}}$ into Eq. (3.63) leads to the following relation between the extrinsic curvature tensor, the gradient of $\underline{\boldsymbol{n}}$ and the spatial gradient of the lapse function:

$$\boxed{\nabla \underline{\boldsymbol{n}} = -\boldsymbol{K} - \boldsymbol{D} \ln N \otimes \underline{\boldsymbol{n}}}, \tag{4.23}$$

or, in components[1]:

$$\boxed{\nabla_\beta n_\alpha = -K_{\alpha\beta} - D_\alpha \ln N \, n_\beta}. \tag{4.24}$$

The covariant derivative of the normal evolution vector is deduced from $\nabla \underline{\boldsymbol{m}} = \nabla(N\underline{\boldsymbol{n}}) = N\nabla\underline{\boldsymbol{n}} + \underline{\boldsymbol{n}} \otimes \nabla N$. We get

$$\boxed{\nabla \boldsymbol{m} = -N\vec{\boldsymbol{K}} - \vec{\boldsymbol{D}} N \otimes \underline{\boldsymbol{n}} + \boldsymbol{n} \otimes \nabla N}, \tag{4.25}$$

or, in components:

$$\boxed{\nabla_\beta m^\alpha = -NK^\alpha_{\ \beta} - D^\alpha N n_\beta + n^\alpha \nabla_\beta N}. \tag{4.26}$$

[1] Recall that $(\nabla \underline{\boldsymbol{n}})_{\alpha\beta} = \nabla_\beta n_\alpha$ (cf. Remark 2.11).

Example 4.3 Let us use expression (4.24) in the form $K_{\alpha\beta} = -\nabla_\beta n_\alpha - a_\alpha\, n_\beta$, along with (4.20), (4.21) and (4.15) to evaluate the extrinsic curvature tensor for the hyperboloidal slicing considered in Examples 4.1 and 4.2. We find

$$K_{\alpha\beta} = \frac{1}{b^3}\begin{pmatrix} -x^2-y^2-z^2 & xbN & ybN & zbN \\ xbN & -b^2-x^2 & -xy & -xz \\ ybN & -xy & -b^2-y^2 & -yz \\ zbN & -xz & -yz & -b^2-z^2 \end{pmatrix}. \tag{4.27}$$

On the other side, from $\gamma_{\alpha\beta} = g_{\alpha\beta} + n_\alpha n_\beta$ [Eq. (3.58)] and expression (4.15) for n_α, we have

$$\gamma_{\alpha\beta} = \frac{1}{b^2}\begin{pmatrix} x^2+y^2+z^2 & -xbN & -ybN & -zbN \\ -xbN & b^2+x^2 & xy & xz \\ -ybN & xy & b^2+y^2 & yz \\ -zbN & xz & yz & b^2+z^2 \end{pmatrix}. \tag{4.28}$$

Comparing with (4.27), we conclude that, for the considered foliation,

$$K_{\alpha\beta} = -\frac{1}{b}\gamma_{\alpha\beta}. \tag{4.29}$$

We thus recover Eq. (3.47) established for Σ_0. This is not surprising since each Σ_t is the image of Σ_0 by the translation $w \mapsto w + t$. The latter being an isometry of Minkowski spacetime, the property (3.47) linking the first and second fundamental forms of Σ_0 must hold for any hypersurface Σ_t.

4.3.5 Evolution of the 3-Metric

The evolution of Σ_t's metric $\boldsymbol{\gamma}$ is naturally given by the Lie derivative of $\boldsymbol{\gamma}$ along the normal evolution vector $\boldsymbol{m}$ (see Sec. 2.5). By means of Eqs. (2.92) and (4.26), we get

$$\begin{aligned}
\mathcal{L}_{\boldsymbol{m}}\gamma_{\alpha\beta} &= m^\mu\nabla_\mu\gamma_{\alpha\beta} + \gamma_{\mu\beta}\nabla_\alpha m^\mu + \gamma_{\alpha\mu}\nabla_\beta m^\mu \\
&= Nn^\mu\nabla_\mu(n_\alpha n_\beta) - \gamma_{\mu\beta}\left(NK^\mu_{\ \ \alpha} + D^\mu N n_\alpha - n^\mu\nabla_\alpha N\right) \\
&\quad - \gamma_{\alpha\mu}\left(NK^\mu_{\ \ \beta} + D^\mu N n_\beta - n^\mu\nabla_\beta N\right) \\
&= N(\underbrace{n^\mu\nabla_\mu n_\alpha}_{a_\alpha = N^{-1}D_\alpha N}\, n_\beta + n_\alpha\, \underbrace{n^\mu\nabla_\mu n_\beta}_{a_\beta = N^{-1}D_\beta N}) - NK_{\beta\alpha} - D_\beta N n_\alpha - NK_{\alpha\beta} - D_\alpha N n_\beta \\
&= -2NK_{\alpha\beta}.
\end{aligned}$$

Hence the simple result:

$$\boxed{\mathcal{L}_{\boldsymbol{m}} \boldsymbol{\gamma} = -2N\boldsymbol{K}}. \tag{4.30}$$

One can deduce easily from this relation the value of the Lie derivative of the 3-metric along the unit normal $\boldsymbol{n}$. Indeed, since $\boldsymbol{m} = N\boldsymbol{n}$,

$$\begin{aligned}
\mathcal{L}_{\boldsymbol{m}} \gamma_{\alpha\beta} &= \mathcal{L}_{N\boldsymbol{n}} \gamma_{\alpha\beta} \\
&= N n^{\mu} \nabla_{\mu} \gamma_{\alpha\beta} + \gamma_{\mu\beta} \nabla_{\alpha} (N n^{\mu}) + \gamma_{\alpha\mu} \nabla_{\beta} (N n^{\mu}) \\
&= N n^{\mu} \nabla_{\mu} \gamma_{\alpha\beta} + \underbrace{\gamma_{\mu\beta} n^{\mu}}_{0} \nabla_{\alpha} N + N \gamma_{\mu\beta} \nabla_{\alpha} n^{\mu} + \underbrace{\gamma_{\alpha\mu} n^{\mu}}_{0} \nabla_{\beta} N + N \gamma_{\alpha\mu} \nabla_{\beta} n^{\mu} \\
&= N \mathcal{L}_{\boldsymbol{n}} \gamma_{\alpha\beta}.
\end{aligned}$$

Hence

$$\mathcal{L}_{\boldsymbol{n}} \boldsymbol{\gamma} = \frac{1}{N} \mathcal{L}_{\boldsymbol{m}} \boldsymbol{\gamma}. \tag{4.31}$$

Consequently, Eq. (4.30) leads to

$$\boxed{\boldsymbol{K} = -\frac{1}{2} \mathcal{L}_{\boldsymbol{n}} \boldsymbol{\gamma}}. \tag{4.32}$$

This equation sheds some new light on the extrinsic curvature tensor $\boldsymbol{K}$. In addition to being the projection on Σ_t of the gradient of the unit normal to Σ_t [cf. Eq. (3.65)],

$$\boldsymbol{K} = -\vec{\boldsymbol{\gamma}}^{*} \nabla \underline{\boldsymbol{n}}, \tag{4.33}$$

as well as the measure of the difference between $\boldsymbol{D}$-derivatives and $\boldsymbol{\nabla}$-derivatives for vectors tangent to Σ_t [cf. Eq. (3.70)],

$$\forall (\boldsymbol{u}, \boldsymbol{v}) \in \mathcal{T}(\Sigma)^2, \ \boldsymbol{K}(\boldsymbol{u}, \boldsymbol{v}) \boldsymbol{n} = \boldsymbol{D}_{\boldsymbol{u}} \boldsymbol{v} - \boldsymbol{\nabla}_{\boldsymbol{u}} \boldsymbol{v}, \tag{4.34}$$

$\boldsymbol{K}$ is also minus one half the Lie derivative of Σ_t's metric along the unit timelike normal.

Remark 4.5 In numerous numerical relativity articles, Eq. (4.32) is used to *define* the extrinsic curvature tensor of the hypersurface Σ_t. It is worth to keep in mind that this equation has a meaning only because Σ_t is member of a foliation. Indeed the right-hand side is the derivative of the induced metric in a direction which is not parallel to the hypersurface and therefore this quantity could not be defined for a single hypersurface, as considered in Chap. 3. On the contrary, the definition adopted here, i.e. Eq. (3.20), is valid even for a single hypersurface.

4.3.6 Evolution of the Orthogonal Projector

Let us now evaluate the Lie derivative of the orthogonal projector onto Σ_t along the normal evolution vector. Using Eqs. (2.92) and (4.26), we have

$$\begin{aligned}
\mathcal{L}_{\boldsymbol{m}} \gamma^\alpha{}_\beta &= m^\mu \nabla_\mu \gamma^\alpha{}_\beta - \gamma^\mu{}_\beta \nabla_\mu m^\alpha + \gamma^\alpha{}_\mu \nabla_\beta m^\mu \\
&= N n^\mu \nabla_\mu (n^\alpha n_\beta) + \gamma^\mu{}_\beta \left(N K^\alpha{}_\mu + D^\alpha N n_\mu - n^\alpha \nabla_\mu N \right) \\
&\quad - \gamma^\alpha{}_\mu \left(N K^\mu{}_\beta + D^\mu N n_\beta - n^\mu \nabla_\beta N \right) \\
&= N \big(\underbrace{n^\mu \nabla_\mu n^\alpha}_{N^{-1} D^\alpha N} \, n_\beta + n^\alpha \underbrace{n^\mu \nabla_\mu n_\beta}_{N^{-1} D_\beta N} \big) + N K^\alpha{}_\beta - n^\alpha D_\beta N - N K^\alpha{}_\beta - D^\alpha N n_\beta \\
&= 0,
\end{aligned}$$

i.e.

$$\boxed{\mathcal{L}_{\boldsymbol{m}} \vec{\boldsymbol{\gamma}} = 0}. \tag{4.35}$$

An important consequence of this is that the Lie derivative along $\boldsymbol{m}$ of any tensor field $\boldsymbol{T}$ tangent to Σ_t is a tensor field tangent to Σ_t:

$$\boxed{\boldsymbol{T} \text{ tangent to } \Sigma_t \Longrightarrow \mathcal{L}_{\boldsymbol{m}} \boldsymbol{T} \text{ tangent to } \Sigma_t}. \tag{4.36}$$

Indeed a distinctive feature of a tensor field tangent to Σ_t is

$$\vec{\boldsymbol{\gamma}}^* \boldsymbol{T} = \boldsymbol{T}. \tag{4.37}$$

Assume for instance that $\boldsymbol{T}$ is a tensor field of type (1,1). Then the above equation writes [cf. Eq. (3.60)]

$$\gamma^\alpha{}_\mu \gamma^\nu{}_\beta T^\mu{}_\nu = T^\alpha{}_\beta.$$

Taking the Lie derivative along $\boldsymbol{m}$ of this relation, employing the Leibniz rule and making use of Eq. (4.35), leads to

$$\begin{aligned}
&\mathcal{L}_{\boldsymbol{m}} \left(\gamma^\alpha{}_\mu \gamma^\nu{}_\beta T^\mu{}_\nu \right) = \mathcal{L}_{\boldsymbol{m}} T^\alpha{}_\beta \\
&\underbrace{\mathcal{L}_{\boldsymbol{m}} \gamma^\alpha{}_\mu}_{0} \, \gamma^\nu{}_\beta T^\mu{}_\nu + \gamma^\alpha{}_\mu \underbrace{\mathcal{L}_{\boldsymbol{m}} \gamma^\nu{}_\beta}_{0} \, T^\mu{}_\nu + \gamma^\alpha{}_\mu \gamma^\nu{}_\beta \mathcal{L}_{\boldsymbol{m}} T^\mu{}_\nu = \mathcal{L}_{\boldsymbol{m}} T^\alpha{}_\beta \\
&\vec{\boldsymbol{\gamma}}^* \mathcal{L}_{\boldsymbol{m}} \boldsymbol{T} = \mathcal{L}_{\boldsymbol{m}} \boldsymbol{T}.
\end{aligned}$$

This shows that $\mathcal{L}_{\boldsymbol{m}} \boldsymbol{T}$ is tangent to Σ_t. The proof is readily extended to any type of tensor field tangent to Σ_t. Notice that the property (4.36) generalizes that obtained for vectors in Sec. 4.3.2 [cf. Eq. (4.12)].

Remark 4.6 An illustration of property (4.36) is provided by Eq. (4.30), which says that $\mathcal{L}_{\boldsymbol{m}}\boldsymbol{\gamma}$ is $-2N\boldsymbol{K}$: $\boldsymbol{K}$ being tangent to Σ_t, we have immediately that $\mathcal{L}_{\boldsymbol{m}}\boldsymbol{\gamma}$ is tangent to Σ_t.

Remark 4.7 Contrary to $\mathcal{L}_{\boldsymbol{n}}\boldsymbol{\gamma}$ and $\mathcal{L}_{\boldsymbol{m}}\boldsymbol{\gamma}$, which are related by Eq. (4.31), $\mathcal{L}_{\boldsymbol{n}}\vec{\boldsymbol{\gamma}}$ and $\mathcal{L}_{\boldsymbol{m}}\vec{\boldsymbol{\gamma}}$ are not proportional. Indeed a calculation similar to that which lead to Eq. (4.31) gives

$$\mathcal{L}_{\boldsymbol{n}}\vec{\boldsymbol{\gamma}} = \frac{1}{N}\mathcal{L}_{\boldsymbol{m}}\vec{\boldsymbol{\gamma}} + \boldsymbol{n} \otimes \boldsymbol{D}\ln N.$$

Therefore the property $\mathcal{L}_{\boldsymbol{m}}\vec{\boldsymbol{\gamma}} = 0$ implies

$$\mathcal{L}_{\boldsymbol{n}}\vec{\boldsymbol{\gamma}} = \boldsymbol{n} \otimes \boldsymbol{D}\ln N \neq 0. \tag{4.38}$$

Hence the privileged role played by $\boldsymbol{m}$ regarding the evolution of the hypersurfaces Σ_t is not shared by $\boldsymbol{n}$; this merely reflects that the hypersurfaces are Lie dragged by $\boldsymbol{m}$, not by $\boldsymbol{n}$.

4.4 Last Part of the 3+1 Decomposition of the Riemann Tensor

4.4.1 Last Non Trivial Projection of the Spacetime Riemann Tensor

In Chap. 3, we have formed the fully projected part of the spacetime Riemann tensor, i.e. $\vec{\boldsymbol{\gamma}}^{*\,4}\mathbf{Riem}$, yielding the Gauss equation [Eq. (3.73)], as well as the part projected three times onto Σ_t and once along the normal $\boldsymbol{n}$, yielding the Codazzi equation [Eq. (3.81)]. These two decompositions involve only fields tangents to Σ_t and their derivatives in directions parallel to Σ_t, namely $\boldsymbol{\gamma}$, $\boldsymbol{K}$, $\mathbf{Riem}$ and $\boldsymbol{DK}$. This is why they could be meaningful for a single hypersurface. In the present section, we form the projection of the spacetime Riemann tensor twice onto Σ_t and twice along $\boldsymbol{n}$. As we shall see, this involves a derivative of $\boldsymbol{K}$ in the direction normal to the hypersurface, which makes sense only for a foliation.

As for the Codazzi equation, the starting point of the calculation is the Ricci identity applied to the vector $\boldsymbol{n}$, i.e. Eq. (3.78). But instead of projecting it totally onto Σ_t, let us project it only twice onto Σ_t and once along $\boldsymbol{n}$:

$$\gamma_{\alpha\mu} n^{\sigma} \gamma^{\nu}{}_{\beta}\, {}^4R^{\mu}{}_{\rho\nu\sigma} n^{\rho} = \gamma_{\alpha\mu} n^{\sigma} \gamma^{\nu}{}_{\beta} (\nabla_{\nu}\nabla_{\sigma} n^{\mu} - \nabla_{\sigma}\nabla_{\nu} n^{\mu}).$$

Substituting Eq. (4.24) for $\boldsymbol{\nabla n}$, we get successively

$$
\begin{aligned}
\gamma_{\alpha\mu} n^\rho \gamma^\nu{}_\beta n^\sigma {}^4R^\mu{}_{\rho\nu\sigma} &= \gamma_{\alpha\mu} n^\sigma \gamma^\nu{}_\beta \left[-\nabla_\nu (K^\mu{}_\sigma + D^\mu \ln N \, n_\sigma) + \nabla_\sigma (K^\mu{}_\nu + D^\mu \ln N \, n_\nu) \right] \\
&= \gamma_{\alpha\mu} n^\sigma \gamma^\nu{}_\beta \big[-\nabla_\nu K^\mu{}_\sigma - \nabla_\nu n_\sigma D^\mu \ln N - n_\sigma \nabla_\nu D^\mu \ln N \\
&\quad + \nabla_\sigma K^\mu{}_\nu + \nabla_\sigma n_\nu D^\mu \ln N + n_\nu \nabla_\sigma D^\mu \ln N \big] \\
&= \gamma_{\alpha\mu} \gamma^\nu{}_\beta \left[K^\mu{}_\sigma \nabla_\nu n^\sigma + \nabla_\nu D^\mu \ln N + n^\sigma \nabla_\sigma K^\mu{}_\nu + D_\nu \ln N D^\mu \ln N \right] \\
&= -K_{\alpha\sigma} K^\sigma{}_\beta + D_\beta D_\alpha \ln N + \gamma^\mu{}_\alpha \gamma^\nu{}_\beta n^\sigma \nabla_\sigma K_{\mu\nu} + D_\alpha \ln N D_\beta \ln N \\
&= -K_{\alpha\sigma} K^\sigma{}_\beta + \frac{1}{N} D_\beta D_\alpha N + \gamma^\mu{}_\alpha \gamma^\nu{}_\beta n^\sigma \nabla_\sigma K_{\mu\nu}.
\end{aligned}
\tag{4.39}
$$

Note that we have used $K^\mu{}_\sigma n^\sigma = 0$, $n^\sigma \nabla_\nu n_\sigma = 0$, $n_\sigma n^\sigma = -1$, $n^\sigma \nabla_\sigma n_\nu = D_\nu \ln N$ and $\gamma^\nu{}_\beta n_\nu = 0$ to get the third equality. Let us now show that the term $\gamma^\mu{}_\alpha \gamma^\nu{}_\beta n^\sigma \nabla_\sigma K_{\mu\nu}$ is related to $\mathcal{L}_{\boldsymbol{m}} \boldsymbol{K}$. Indeed, from the expression (2.92) of the Lie derivative:

$$
\mathcal{L}_{\boldsymbol{m}} K_{\alpha\beta} = m^\mu \nabla_\mu K_{\alpha\beta} + K_{\mu\beta} \nabla_\alpha m^\mu + K_{\alpha\mu} \nabla_\beta m^\mu .
$$

Substituting Eq. (4.26) for $\nabla_\alpha m^\mu$ and $\nabla_\beta m^\mu$ leads to

$$
\mathcal{L}_{\boldsymbol{m}} K_{\alpha\beta} = N n^\mu \nabla_\mu K_{\alpha\beta} - 2N K_{\alpha\mu} K^\mu{}_\beta - K_{\alpha\mu} D^\mu N \, n_\beta - K_{\beta\mu} D^\mu N \, n_\alpha .
$$

Let us project this equation onto $\boldsymbol{\Sigma}_t$, i.e. let us apply the operator $\vec{\boldsymbol{\gamma}}^*$ to both sides. Using the property $\vec{\boldsymbol{\gamma}}^* \mathcal{L}_{\boldsymbol{m}} \boldsymbol{K} = \mathcal{L}_{\boldsymbol{m}} \boldsymbol{K}$, which stems from the fact that $\mathcal{L}_{\boldsymbol{m}} \boldsymbol{K}$ is tangent to Σ_t since $\boldsymbol{K}$ is [property (4.36)], we get

$$
\mathcal{L}_{\boldsymbol{m}} K_{\alpha\beta} = N \gamma^\mu{}_\alpha \gamma^\nu{}_\beta n^\sigma \nabla_\sigma K_{\mu\nu} - 2N K_{\alpha\mu} K^\mu{}_\beta . \tag{4.40}
$$

Extracting $\gamma^\mu{}_\alpha \gamma^\nu{}_\beta n^\sigma \nabla_\sigma K_{\mu\nu}$ from this relation and plugging it into Eq. (4.39) results in

$$
\boxed{\gamma_{\alpha\mu} n^\rho \gamma^\nu{}_\beta n^\sigma {}^4R^\mu{}_{\rho\nu\sigma} = \frac{1}{N} \mathcal{L}_{\boldsymbol{m}} K_{\alpha\beta} + \frac{1}{N} D_\alpha D_\beta N + K_{\alpha\mu} K^\mu{}_\beta}. \tag{4.41}
$$

Note that we have written $D_\beta D_\alpha N = D_\alpha D_\beta N$ (**D** has no torsion). Equation (4.41) is the relation we sought. It is sometimes called the ***Ricci equation*** [not to be confused with the *Ricci identity* (2.68)]. Together with the Gauss equation (3.73) and the Codazzi equation (3.81), it completes the 3+1 decomposition of the spacetime Riemann tensor. Indeed the part projected three times along $\boldsymbol{n}$ vanish identically, since ${}^4\mathbf{Riem}(\underline{\boldsymbol{n}}, \boldsymbol{n}, \boldsymbol{n}, .) = 0$ and ${}^4\mathbf{Riem}(., \boldsymbol{n}, \boldsymbol{n}, \boldsymbol{n}) = 0$ thanks to the partial antisymmetry of the Riemann tensor. Accordingly one can project ${}^4\mathbf{Riem}$ at most twice along $\boldsymbol{n}$ to get some non-vanishing result.

It is worth to note that the left-hand side of the Ricci equation (4.41) is a term which appears in the contracted Gauss equation (3.74). Therefore, by combining the two equations, we get a formula which does no longer contain the spacetime Riemann tensor, but only the spacetime Ricci tensor:

$$
\boxed{\gamma^\mu{}_\alpha \gamma^\nu{}_\beta {}^4R_{\mu\nu} = -\frac{1}{N} \mathcal{L}_{\boldsymbol{m}} K_{\alpha\beta} - \frac{1}{N} D_\alpha D_\beta N + R_{\alpha\beta} + K K_{\alpha\beta} - 2 K_{\alpha\mu} K^\mu{}_\beta}, \tag{4.42}
$$

or in index-free notation:

$$\boxed{\vec{\boldsymbol{\gamma}}^{*4}\boldsymbol{R} = -\frac{1}{N}\mathscr{L}_{\boldsymbol{m}}\boldsymbol{K} - \frac{1}{N}\boldsymbol{D}\boldsymbol{D}N + \boldsymbol{R} + K\boldsymbol{K} - 2\boldsymbol{K}\cdot\vec{\boldsymbol{K}}}. \tag{4.43}$$

4.4.2 3+1 Expression of the Spacetime Scalar Curvature

Let us take the trace of Eq. (4.43) with respect to the metric $\boldsymbol{\gamma}$. This amounts to contracting Eq. (4.42) with $\gamma^{\alpha\beta}$. In the left-hand side, we have $\gamma^{\alpha\beta}\gamma^{\mu}{}_{\alpha}\gamma^{\nu}{}_{\beta} = \gamma^{\mu\nu}$ and in the right-hand we can limit the range of variation of the indices to $\{1, 2, 3\}$ since all the involved tensors are spatial ones [including $\mathscr{L}_{\boldsymbol{m}}\boldsymbol{K}$, thanks to the property (4.36)]. Hence

$$\gamma^{\mu\nu\,4}R_{\mu\nu} = -\frac{1}{N}\gamma^{ij}\mathscr{L}_{\boldsymbol{m}}K_{ij} - \frac{1}{N}D_iD^iN + R + K^2 - 2K_{ij}K^{ij}. \tag{4.44}$$

Now $\gamma^{\mu\nu\,4}R_{\mu\nu} = (g^{\mu\nu} + n^{\mu}n^{\nu})\,{}^4R_{\mu\nu} = {}^4R + {}^4R_{\mu\nu}n^{\mu}n^{\nu}$ and

$$-\gamma^{ij}\mathscr{L}_{\boldsymbol{m}}K_{ij} = -\mathscr{L}_{\boldsymbol{m}}(\underbrace{\gamma^{ij}K_{ij}}_{K}) + K_{ij}\mathscr{L}_{\boldsymbol{m}}\gamma^{ij}, \tag{4.45}$$

with $\mathscr{L}_{\boldsymbol{m}}\gamma^{ij}$ evaluated from the very definition of the inverse 3-metric:

$$\begin{aligned}
&\gamma_{ik}\gamma^{kj} = \delta^j{}_i \\
\Rightarrow &\mathscr{L}_{\boldsymbol{m}}\gamma_{ik}\gamma^{kj} + \gamma_{ik}\mathscr{L}_{\boldsymbol{m}}\gamma^{kj} = 0 \\
\Rightarrow &\gamma^{il}\gamma^{kj}\mathscr{L}_{\boldsymbol{m}}\gamma_{lk} + \underbrace{\gamma^{il}\gamma_{lk}}_{\delta^i{}_k}\mathscr{L}_{\boldsymbol{m}}\gamma^{lj} = 0 \\
\Rightarrow &\mathscr{L}_{\boldsymbol{m}}\gamma^{ij} = -\gamma^{ik}\gamma^{jl}\mathscr{L}_{\boldsymbol{m}}\gamma_{kl} \\
\Rightarrow &\mathscr{L}_{\boldsymbol{m}}\gamma^{ij} = 2N\gamma^{ik}\gamma^{kl}K_{kl} \\
\Rightarrow &\boxed{\mathscr{L}_{\boldsymbol{m}}\gamma^{ij} = 2NK^{ij}},
\end{aligned} \tag{4.46}$$

where we have used Eq. (4.30). Plugging Eq. (4.46) into Eq. (4.45) gives

$$-\gamma^{ij}\mathscr{L}_{\boldsymbol{m}}K_{ij} = -\mathscr{L}_{\boldsymbol{m}}K + 2NK_{ij}K^{ij}. \tag{4.47}$$

Consequently Eq. (4.44) becomes

$$\boxed{{}^4R + {}^4R_{\mu\nu}n^{\mu}n^{\nu} = R + K^2 - \frac{1}{N}\mathscr{L}_{\boldsymbol{m}}K - \frac{1}{N}D_iD^iN}. \tag{4.48}$$

It is worth to combine with equation with the scalar Gauss relation (3.75) to get rid of the Ricci tensor term ${}^4R_{\mu\nu}n^{\mu}n^{\nu}$ and obtain an equation which involves only the spacetime scalar curvature ${}^4R$:

$$\boxed{{}^4R = R + K^2 + K_{ij}K^{ij} - \frac{2}{N}\mathcal{L}_m K - \frac{2}{N}D_iD^iN}. \tag{4.49}$$

Example 4.4 Let us check Eq. (4.49) on the hyperboloidal slicing of Minkowski spacetime considered in Examples 4.1, 4.2 and 4.3. We have ${}^4R = 0$ for $(\mathcal{M}, \boldsymbol{g})$ is Minkowski spacetime. On the right-hand side, $R = -6/b^2$ [Eq. (3.44)], $K^2 = 9/b^2$ [Eq. (3.48)], $K_{ij}K^{ij} = \gamma_{ij}\gamma^{ij}/b^2 = 3/b^2$ [Eq. (4.29)] and $\mathcal{L}_m K = \mathcal{L}_m(-3/b) = 0$ [Eq. (3.48)], so that

$$R + K^2 + K_{ij}K^{ij} - \frac{2}{N}\mathcal{L}_m K = \frac{6}{b^2}. \tag{4.50}$$

There remains to evaluate $2N^{-1}D_iD^iN$. To this aim, let us use the coordinates $(x^i) = (\rho, \theta, \varphi)$ introduced on Σ_0 by Eq. (3.40)–(3.41). In these coordinates, the lapse function takes the simple form

$$N = \cosh\rho. \tag{4.51}$$

Indeed, the last equation of (3.41) and (3.38) imply

$$\cosh\rho = \frac{w}{b} = \sqrt{1 + \frac{x^2 + y^2 + z^2}{b^2}},$$

which coincides with expression (4.14) for N. The Laplacian of N can be computed via a standard formula which follows readily from (2.65):

$$D_iD^iN = \frac{1}{\sqrt{\gamma}}\frac{\partial}{\partial x^i}\left(\sqrt{\gamma}\gamma^{ij}\frac{\partial N}{\partial x^j}\right), \tag{4.52}$$

where γ is the determinant of the metric components (γ_{ij}). Given (3.43), we have $\sqrt{\gamma} = b^3\sinh^2\rho\sin\theta$. Since γ^{ij} is diagonal [cf. Eq. (3.43)] and N depends only on ρ, we have

$$D_iD^iN = \frac{1}{b^3\sinh^2\rho\sin\theta}\frac{\partial}{\partial\rho}\left(b^3\sinh^2\rho\sin\theta\frac{1}{b^2}\underbrace{\frac{\partial N}{\partial\rho}}_{\sinh\rho}\right)$$
$$= \frac{1}{b^2\sinh^2\rho}\frac{\partial}{\partial\rho}\left(\sinh^3\rho\right) = \frac{3}{b^2}\cosh\rho.$$

Hence

$$-\frac{2}{N}D_iD^iN = -\frac{6}{b^2}.$$

In view of Eq. (4.50) and ${}^4R = 0$ we conclude that Eq. (4.49) is satisfied.

References

1. Geroch, R.: Domain of Dependence. J. Math. Phys. **11**, 437 (1970)
2. Hawking, S.W., Ellis, G.F.R.: The large scale structure of space-time. Cambridge University Press, Cambridge (1973)
3. Friedman, J.L.: The Cauchy Problem on Spacetimes That Are Not Globally Hyperbolic. in Ref. [12], p. 331
4. Choquet-Bruhat, Y.: General Relativity and Einstein's Equations. Oxford Univ. Press, New York (2009)
5. Leray, J.: Hyperbolic differential equations, lecture notes.. Institute of Advanced Studies, Princeton (1953)
6. Choquet-Bruhat, Y., York, J.W.: The cauchy problem. In: A., H.e.l.d. (eds) General Relativity and Gravitation, one hundred Years after the Birth of Albert Einstein Vol 1., pp. 99. Plenum Press, New York (1980)
7. Wheeler, J.A.: Geometrodynamics and the issue of the final state. In: DeWitt, C., DeWitt, B.S. (eds) Relativity, Groups and Topology, pp. 316. Gordon and Breach, New York (1964)
8. Arnowitt, R., Deser, S., Misner, C.W.: The Dynamics of General Relativity. In: Witten, L. (ed) Gravitation: An Introduction to Current Research, p. 227. Wiley, New York (1962). http://arxiv.org/abs/gr-qc/0405109
9. Misner, C.W., Thorne, K.S., Wheeler, J.A.: Gravitation. Freeman, New York (1973)
10. Thorne, K.S., Macdonald, D.: Electrodynamics in curved spacetime: 3+1 formulation. Mon. Not. R. Astron. Soc. **198**, 339 (1982)
11. Bardeen, J.M.: A variational principle for rotating stars in general relativity. Astrophys. J. **162**, 71 (1970)
12. Chruściel, P.T., Friedrich, H. (eds): The Einstein equations and the large scale behavior of gravitational fields—50 years of the Cauchy problem in general relativity. Birkhäuser Verlag, Basel (2004)

References

[illegible]

Chapter 5
3+1 Decomposition of Einstein Equation

Abstract The fundamental equation for general relativity, the *Einstein equation*, is decomposed orthogonally with respect to a 3+1 foliation of spacetime. Then we introduce spatial coordinates on the hypersurfaces forming the foliation, thereby introducing the famous *shift vector*. This enables one to turn the 3+1 Einstein equation into a system of partial-differential equations. This system can be formulated as a *Cauchy problem with constraints* and we discuss briefly the known existence and uniqueness results regarding it. Finally we discuss the ADM Hamiltonian approach to general relativity, which is based on the 3+1 decomposition.

5.1 Einstein Equation in 3+1 Form

5.1.1 The Einstein Equation

After the first two chapters devoted to the geometry of hypersurfaces and foliations, we are now back to physics: we consider a spacetime $(\mathscr{M}, \boldsymbol{g})$ such that $\boldsymbol{g}$ obeys the ***Einstein equation***:

$$\boxed{{}^4\boldsymbol{R} - \frac{1}{2}{}^4R\boldsymbol{g} = 8\pi \boldsymbol{T}}, \tag{5.1}$$

where ${}^4\boldsymbol{R}$ is the Ricci tensor associated with $\boldsymbol{g}$ [cf. Eq. (2.75)], ${}^4R$ the corresponding Ricci scalar, and $\boldsymbol{T}$ the matter stress-energy tensor. We are using units in which Newton's gravitational constant G is set to unity. Otherwise, the coefficient in front of $\boldsymbol{T}$ in Eq. (5.1) should read $8\pi G$, and even $8\pi G/c^4$ if we are relaxing $c = 1$.

We shall also use the equivalent form

$${}^4\boldsymbol{R} = 8\pi\left(\boldsymbol{T} - \frac{1}{2}T\boldsymbol{g}\right), \tag{5.2}$$

É. Gourgoulhon, *3+1 Formalism in General Relativity*, Lecture Notes in Physics 846,
DOI: 10.1007/978-3-642-24525-1_5, © Springer-Verlag Berlin Heidelberg 2012

where $T := g^{\mu\nu} T_{\mu\nu}$ stands for the trace (with respect to $\boldsymbol{g}$) of $\boldsymbol{T}$.

Remark 5.1 We are considering the Einstein equation without any cosmological constant. Taking into account a non-vanishing cosmological Λ, the Einstein equation (5.1) should be written

$$ {}^4\boldsymbol{R} - \frac{1}{2}{}^4R\boldsymbol{g} + \Lambda\boldsymbol{g} = 8\pi\boldsymbol{T}. \tag{5.3}$$

We limit ourselves to $\Lambda = 0$, but all results could be easily generalized to accommodate for $\Lambda \neq 0$.

Let us assume that the spacetime $(\mathscr{M}, \boldsymbol{g})$ is globally hyperbolic (cf. Sect. 4.2.1) and let $(\Sigma_t)_{t\in\mathbb{R}}$ be a foliation of $\mathscr{M}$ by a family of spacelike hypersurfaces. The 3+1 formalism for general relativity amounts to projecting the Einstein equation (5.1) onto Σ_t and perpendicularly to Σ_t. To this aim let us first consider the 3+1 decomposition of the stress-energy tensor.

5.1.2 3+1 Decomposition of the Stress-Energy Tensor

From the very definition of a stress-energy tensor, the ***matter energy density*** as measured by the Eulerian observer introduced in Sect. 4.3.3 is

$$ \boxed{E := \boldsymbol{T}(\boldsymbol{n}, \boldsymbol{n})}\,. \tag{5.4}$$

This follows from the fact that the 4-velocity of the Eulerian observer is the unit normal vector $\boldsymbol{n}$ of the hypersurfaces Σ_t.

Similarly, also from the very definition of a stress-energy tensor, the ***matter momentum density*** as measured by the Eulerian observer is the linear form

$$ \boxed{\boldsymbol{p} := -\boldsymbol{T}(\vec{\boldsymbol{\gamma}}(.), \boldsymbol{n})}\,, \tag{5.5}$$

i.e. the linear form defined by

$$ \forall \boldsymbol{v} \in \mathscr{T}_p(\mathscr{M}), \quad \langle \boldsymbol{p}, \boldsymbol{v}\rangle = -\boldsymbol{T}(\vec{\boldsymbol{\gamma}}(\boldsymbol{v}), \boldsymbol{n}). \tag{5.6}$$

In components:

$$ p_\alpha = -T_{\mu\nu}\gamma^\mu{}_\alpha n^\nu. \tag{5.7}$$

Notice that, thanks to the projector $\vec{\boldsymbol{\gamma}}$, $\boldsymbol{p}$ is a linear form tangent to Σ_t.

Remark 5.2 The momentum density $\boldsymbol{p}$ is often denoted $\boldsymbol{j}$. Here we reserve the latter for electric current density.

If we permute the role of $\vec{\boldsymbol{\gamma}}$ and $\boldsymbol{n}$ in (5.5), we obtain the ***energy flux 1-form*** as measured by the Eulerian observer:

$$\boxed{\boldsymbol{\varphi} := -\boldsymbol{T}(\boldsymbol{n}, \vec{\boldsymbol{\gamma}}(.))}\,. \tag{5.8}$$

Given an elementary (2-dimensional) surface of area $\mathrm{d}\boldsymbol{S}$ in Σ_t, $\boldsymbol{\varphi}$ gives the energy $\mathrm{d}e$ that crosses the surface during the Eulerian observer's proper time $\mathrm{d}\tau$ according to

$$\frac{\mathrm{d}e}{\mathrm{d}\tau} = \langle \boldsymbol{\varphi}, \mathrm{d}\boldsymbol{S} \rangle. \tag{5.9}$$

A fundamental property of the stress-energy tensor $\boldsymbol{T}$ is to be *symmetric*. As a consequence:

$$\boxed{\boldsymbol{\varphi} = \boldsymbol{p}}\,. \tag{5.10}$$

Remark 5.3 In units where c is not set to unity, this formula should read $\boldsymbol{\varphi} = c^2\boldsymbol{p}$. It can be interpreted as an expression of the equivalence between mass and energy in relativity.

Finally, still from the very definition of a stress-energy tensor, the ***matter stress tensor*** as measured by the Eulerian observer is the bilinear form

$$\boxed{\boldsymbol{S} := \vec{\boldsymbol{\gamma}}^{*}\boldsymbol{T}}\,, \tag{5.11}$$

or, in components,

$$S_{\alpha\beta} = T_{\mu\nu}\gamma^{\mu}{}_{\alpha}\gamma^{\nu}{}_{\beta}. \tag{5.12}$$

As for $\boldsymbol{p}$, $\boldsymbol{S}$ is a tensor field tangent to Σ_t. Let us recall the physical interpretation of the stress tensor $\boldsymbol{S}$: given two spacelike unit vectors $\boldsymbol{e}$ and $\boldsymbol{e}'$ (possibly equal) in the rest frame of the Eulerian observer (i.e. two unit vectors orthogonal to $\boldsymbol{n}$), $S(\boldsymbol{e}, \boldsymbol{e}')$ is the force in the direction $\boldsymbol{e}$ acting on the unit surface whose normal is $\boldsymbol{e}'$. Let us denote by S the trace of $\boldsymbol{S}$ with respect to the metric $\boldsymbol{\gamma}$ (or equivalently with respect to the metric $\boldsymbol{g}$, since $\boldsymbol{S}$ is tangent to Σ_t):

$$\boxed{S := \gamma^{ij}S_{ij} = g^{\mu\nu}S_{\mu\nu}}\,. \tag{5.13}$$

The knowledge of $(E, \boldsymbol{p}, \boldsymbol{S})$ is sufficient to reconstruct $\boldsymbol{T}$ since

$$\boxed{\boldsymbol{T} = \boldsymbol{S} + \underline{\boldsymbol{n}} \otimes \boldsymbol{p} + \boldsymbol{p} \otimes \underline{\boldsymbol{n}} + E\underline{\boldsymbol{n}} \otimes \underline{\boldsymbol{n}}}\,. \tag{5.14}$$

This formula is easily established by substituting Eq. (3.53) for $\gamma^{\alpha}{}_{\beta}$ into Eq. (5.12) and expanding the result. It constitutes the 3+1 decomposition of $\boldsymbol{T}$.

Remark 5.4 Expression (5.14) uses the fact that $\varphi = p$ [Eq. (5.10)], otherwise it should read $\boldsymbol{T} = \boldsymbol{S} + \underline{\boldsymbol{n}} \otimes \boldsymbol{\varphi} + \boldsymbol{p} \otimes \underline{\boldsymbol{n}} + E\underline{\boldsymbol{n}} \otimes \underline{\boldsymbol{n}}$.

Taking the trace of Eq. (5.14) with respect to the metric $\boldsymbol{g}$ yields

$$T = S + 2\underbrace{\langle \boldsymbol{p}, \boldsymbol{n}\rangle}_{0} + E\underbrace{\langle \underline{\boldsymbol{n}}, \boldsymbol{n}\rangle}_{-1},$$

hence

$$T = S - E. \tag{5.15}$$

5.1.3 Projection of the Einstein Equation

With the above 3+1 decomposition of the stress-energy tensor and the 3+1 decompositions of the spacetime Ricci tensor obtained in Chaps. 3 and 4, we are fully equipped to perform the projection of the Einstein equation (5.1) onto the hypersurface Σ_t and along its normal. There are only three possibilities:

5.1.3.1 Full Projection onto Σ_t

This amounts to applying the operator $\vec{\boldsymbol{\gamma}}^*$ to the Einstein equation. It is convenient to take the version (5.2) of the latter; we get

$$\vec{\boldsymbol{\gamma}}^{*\,4}\boldsymbol{R} = 8\pi\left(\vec{\boldsymbol{\gamma}}^{*}\boldsymbol{T} - \frac{1}{2}T\vec{\boldsymbol{\gamma}}^{*}\boldsymbol{g}\right). \tag{5.16}$$

$\vec{\boldsymbol{\gamma}}^{*\,4}\boldsymbol{R}$ is given by Eq. (4.43) (combination of the contracted Gauss equation with the Ricci equation), $\vec{\boldsymbol{\gamma}}^{*}\boldsymbol{T}$ is by definition $\boldsymbol{S}$, $T = S - E$ [Eq. (5.15)], and $\vec{\boldsymbol{\gamma}}^{*}\boldsymbol{g}$ is simply $\boldsymbol{\gamma}$. Therefore

$$-\frac{1}{N}\mathcal{L}_{\boldsymbol{m}}\boldsymbol{K} - \frac{1}{N}\boldsymbol{D}\boldsymbol{D}N + \boldsymbol{R} + K\boldsymbol{K} - 2\boldsymbol{K}\cdot\vec{\boldsymbol{K}} = 8\pi\left[\boldsymbol{S} - \frac{1}{2}(S-E)\boldsymbol{\gamma}\right],$$

or equivalently

$$\boxed{\mathcal{L}_{\boldsymbol{m}}\boldsymbol{K} = -\boldsymbol{D}\boldsymbol{D}N + N\left\{\boldsymbol{R} + K\boldsymbol{K} - 2\boldsymbol{K}\cdot\vec{\boldsymbol{K}} + 4\pi\left[(S-E)\boldsymbol{\gamma} - 2\boldsymbol{S}\right]\right\}}. \tag{5.17}$$

In components:

$$\mathcal{L}_{\boldsymbol{m}}K_{\alpha\beta} = -D_\alpha D_\beta N + N\left\{R_{\alpha\beta} + KK_{\alpha\beta} - 2K_{\alpha\mu}K^{\mu}{}_{\beta} + 4\pi\left[(S-E)\gamma_{\alpha\beta} - 2S_{\alpha\beta}\right]\right\}. \tag{5.18}$$

Notice that each term in the above equation is a tensor field tangent to Σ_t. For $\mathcal{L}_{\boldsymbol{m}}\boldsymbol{K}$, this results from the fundamental property (4.36) of $\mathcal{L}_{\boldsymbol{m}}$. Consequently, we may restrict to spatial indices without any loss of generality and write Eq. (5.18) as

$$\boxed{\mathcal{L}_{\boldsymbol{m}} K_{ij} = -D_i D_j N + N\left\{R_{ij} + K K_{ij} - 2K_{ik}K^k{}_j + 4\pi\left[(S-E)\gamma_{ij} - 2S_{ij}\right]\right\}}. \tag{5.19}$$

5.1.3.2 Full Projection Perpendicular to Σ_t

This amounts to applying the Einstein equation (5.1), which is an identity between bilinear forms, to the couple $(\boldsymbol{n}, \boldsymbol{n})$; we get, since $\boldsymbol{g}(\boldsymbol{n}, \boldsymbol{n}) = -1$,

$$ {}^4\boldsymbol{R}(\boldsymbol{n}, \boldsymbol{n}) + \frac{1}{2}{}^4R = 8\pi \boldsymbol{T}(\boldsymbol{n}, \boldsymbol{n}). $$

Using the scalar Gauss equation (3.75), and noticing that $\boldsymbol{T}(\boldsymbol{n}, \boldsymbol{n}) = E$ [Eq. (5.4)] yields

$$\boxed{R + K^2 - K_{ij}K^{ij} = 16\pi E}. \tag{5.20}$$

This equation is called the ***Hamiltonian constraint***. The word '*constraint*' will be justified in Sect. 5.4.3 and the qualifier '*Hamiltonian*' in Sect. 5.5.2.

5.1.3.3 Mixed Projection

Finally, let us project the Einstein equation (5.1) once onto Σ_t and once along the normal $\boldsymbol{n}$:

$$ {}^4\boldsymbol{R}(\boldsymbol{n}, \vec{\boldsymbol{\gamma}}(.)) - \frac{1}{2}{}^4R\,\underbrace{\boldsymbol{g}(\boldsymbol{n}, \vec{\boldsymbol{\gamma}}(.))}_{0} = 8\pi \boldsymbol{T}(\boldsymbol{n}, \vec{\boldsymbol{\gamma}}(.)). $$

By means of the contracted Codazzi equation (3.82) and $\boldsymbol{T}(\boldsymbol{n}, \vec{\boldsymbol{\gamma}}(.)) = -\boldsymbol{p}$ [Eq. (5.5)], we get

$$\boxed{\boldsymbol{D} \cdot \vec{\boldsymbol{K}} - \boldsymbol{D}K = 8\pi \boldsymbol{p}}, \tag{5.21}$$

or, in components,

$$\boxed{D_j K^j{}_i - D_i K = 8\pi p_i}. \tag{5.22}$$

This equation is called the ***momentum constraint***. Again, the word '*constraint*' will be justified in Sect. 5.4.

5.1.3.4 Summary

The Einstein equation is equivalent to the system of three equations: (5.17), (5.20) and (5.21). Equation (5.17) is a rank 2 tensorial (bilinear form) equation within Σ_t, involving only symmetric tensors: it has therefore 6 independent components. Equation (5.20) is a scalar equation and Eq. (5.21) is a rank 1 tensorial (linear forms) within Σ_t: it has therefore 3 independent components. The total number of independent components is thus $6 + 1 + 3 = 10$, i.e. the same as the original Einstein equation (5.1).

5.2 Coordinates Adapted to the Foliation

5.2.1 Definition

The system (5.17)+(5.20)+(5.21) is a system of tensorial equations. In order to transform it into a system of partial differential equations (PDE), one must introduce coordinates on the spacetime manifold $\mathscr{M}$, which we have not done yet. Coordinates adapted to the foliation $(\Sigma_t)_{t\in\mathbb{R}}$ are set in the following way. On each hypersurface Σ_t one introduces some coordinate system $(x^i) = (x^1, x^2, x^3)$. If this coordinate system varies smoothly between neighbouring hypersurfaces, then $(x^\alpha) = (t, x^1, x^2, x^3)$ constitutes a well-behaved coordinate system on $\mathscr{M}$. We shall call $(x^i) = (x^1, x^2, x^3)$ the ***spatial coordinates***.

Let us denote by $(\boldsymbol{\partial}_\alpha) = (\boldsymbol{\partial}_t, \boldsymbol{\partial}_i)$ the natural basis of $\mathscr{T}_p(\mathscr{M})$ associated with the coordinates (x^α), i.e. the set of vectors

$$\boldsymbol{\partial}_t := \frac{\partial}{\partial t} \tag{5.23}$$

$$\boldsymbol{\partial}_i := \frac{\partial}{\partial x^i}, \quad i \in \{1, 2, 3\}. \tag{5.24}$$

Notice that the vector $\boldsymbol{\partial}_t$ is tangent to the lines of constant spatial coordinates, i.e. the curves of $\mathscr{M}$ defined by $(x^1 = K^1, x^2 = K^2, x^3{=}K^3)$, where K^1, K^2 and K^3 are three constants (cf. Fig. 5.1). We shall call $\boldsymbol{\partial}_t$ the ***time vector***.

Remark 5.5 $\boldsymbol{\partial}_t$ is not necessarily a timelike vector. This will be discussed further below [Eqs. (5.32)–(5.34)].

For any $i \in \{1, 2, 3\}$, the vector $\boldsymbol{\partial}_i$ is tangent to the lines $t = K^0$, $x^j = K^j$ $(j \neq i)$, where K^0 and K^j $(j \neq i)$ are three constants. Having t constant, these lines

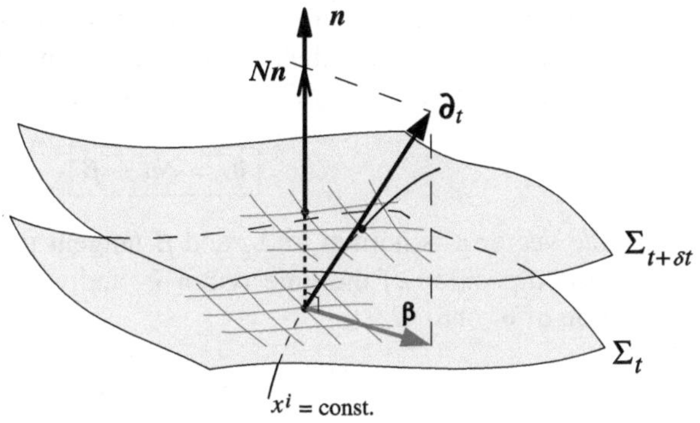

Fig. 5.1 Coordinates (x^i) on the hypersurfaces Σ_t : each line $x^i = \text{const}$ cuts across the foliation $(\Sigma_t)_{t\in\mathbb{R}}$ and defines the time vector $\boldsymbol{\partial}_t$ and the shift vector $\boldsymbol{\beta}$ of the spacetime coordinate system $(x^\alpha) = (t, x^i)$

belong to the hypersurfaces Σ_t. This implies that $\boldsymbol{\partial}_i$ is tangent to Σ_t:

$$\boldsymbol{\partial}_i \in \mathscr{T}_p(\Sigma_t), \quad i \in \{1, 2, 3\}. \tag{5.25}$$

5.2.2 Shift Vector

The dual basis associated with $(\boldsymbol{\partial}_\alpha)$ is the gradient 1-form basis $(\mathbf{d}x^\alpha)$, which is a basis of the space of linear forms $\mathscr{T}_p^*(\mathscr{M})$ (cf. Sect. 2.2.3): In particular, the 1-form $\mathbf{d}t = \boldsymbol{\nabla} t$ is dual to the vector $\boldsymbol{\partial}_t$ [cf. Eqs. (2.20) and (2.14) with $\alpha = \beta = 0$]:

$$\langle \boldsymbol{\nabla} t, \boldsymbol{\partial}_t \rangle = 1. \tag{5.26}$$

Hence the time vector $\boldsymbol{\partial}_t$ obeys to the same property as the normal evolution vector $\boldsymbol{m}$, since $\langle \boldsymbol{\nabla} t, \boldsymbol{m} \rangle = 1$ [Eq. (4.11)]. In particular, $\boldsymbol{\partial}_t$ Lie drags the hypersurfaces Σ_t, as $\boldsymbol{m}$ does (cf. Sect. 4.3.2). In general the two vectors $\boldsymbol{\partial}_t$ and $\boldsymbol{m}$ differ. They coincide only if the coordinates (x^i) are such that the lines $x^i = \text{const}$ are orthogonal to the hypersurfaces Σ_t (cf. Fig. 5.1). The difference between $\boldsymbol{\partial}_t$ and $\boldsymbol{m}$ is called the ***shift vector*** and is denoted $\boldsymbol{\beta}$:

$$\boxed{\boldsymbol{\partial}_t =: \boldsymbol{m} + \boldsymbol{\beta}}\,. \tag{5.27}$$

As for the lapse, the name *shift vector* has been coined by Wheeler [1]. By combining Eqs. (5.26) and (4.11), we get

$$\langle \boldsymbol{\nabla} t, \boldsymbol{\beta} \rangle = \langle \boldsymbol{\nabla} t, \boldsymbol{\partial}_t \rangle - \langle \boldsymbol{\nabla} t, \boldsymbol{m} \rangle = 1 - 1 = 0.$$

This shows that the vector $\boldsymbol{\beta}$ is tangent to the hypersurfaces Σ_t. Since $\boldsymbol{\nabla} t = -N^{-1}\underline{\boldsymbol{n}}$ [Eq. (4.8)], the relation $\langle \boldsymbol{\nabla} t, \boldsymbol{\beta} \rangle = 0$ can be written

$$\boxed{\boldsymbol{n} \cdot \boldsymbol{\beta} = 0}\,. \tag{5.28}$$

The lapse function and the shift vector have been introduced for the first time explicitly, although without their present names, by Choquet–Bruhat in 1956 [2].

It is useful to rewrite Eq. (5.27) by means of the relation $\boldsymbol{m} = N\boldsymbol{n}$ [Eq. (4.9)]:

$$\boxed{\boldsymbol{\partial}_t = N\boldsymbol{n} + \boldsymbol{\beta}}. \tag{5.29}$$

Since the vector $\boldsymbol{n}$ is normal to Σ_t and $\boldsymbol{\beta}$ tangent to Σ_t, Eq. (5.29) can be seen as a 3+1 decomposition of the time vector $\boldsymbol{\partial}_t$ and we may write $\boldsymbol{\beta}$ as the orthogonal projection of $\boldsymbol{\partial}_t$ onto Σ_t:

$$\boldsymbol{\beta} = \vec{\boldsymbol{\gamma}}(\boldsymbol{\partial}_t). \tag{5.30}$$

The scalar square of $\boldsymbol{\partial}_t$ is deduced immediately from Eq. (5.29), taking into account $\boldsymbol{n} \cdot \boldsymbol{n} = -1$ and Eq. (5.28):

$$\boldsymbol{\partial}_t \cdot \boldsymbol{\partial}_t = -N^2 + \boldsymbol{\beta} \cdot \boldsymbol{\beta}. \tag{5.31}$$

Hence we have the following:

$$\boldsymbol{\partial}_t \text{ is timelike} \iff \boldsymbol{\beta} \cdot \boldsymbol{\beta} < N^2, \tag{5.32}$$

$$\boldsymbol{\partial}_t \text{ is null} \iff \boldsymbol{\beta} \cdot \boldsymbol{\beta} = N^2, \tag{5.33}$$

$$\boldsymbol{\partial}_t \text{ is spacelike} \iff \boldsymbol{\beta} \cdot \boldsymbol{\beta} > N^2. \tag{5.34}$$

Remark 5.6 A shift vector that fulfills the condition (5.34) is sometimes called a ***superluminal shift***. Notice that, since a priori the time vector $\boldsymbol{\partial}_t$ is a pure coordinate quantity and is not associated with the 4-velocity of some observer (contrary to $\boldsymbol{m}$, which is proportional to the 4-velocity of the Eulerian observer), there is nothing unphysical in having $\boldsymbol{\partial}_t$ spacelike.

Since $\boldsymbol{\beta}$ is tangent to Σ_t, let us introduce the components of $\boldsymbol{\beta}$ and the metric dual form $\underline{\boldsymbol{\beta}}$ with respect to the spatial coordinates (x^i) according to

$$\boldsymbol{\beta} =: \beta^i \boldsymbol{\partial}_i \quad \text{and} \quad \underline{\boldsymbol{\beta}} =: \beta_i \mathbf{d}x^i. \tag{5.35}$$

Equation (5.29) then shows that the components of the unit normal vector $\boldsymbol{n}$ with respect to the natural basis $(\boldsymbol{\partial}_\alpha)$ are expressible in terms of N and (β^i) as

$$\boxed{n^\alpha = \left(\frac{1}{N}, -\frac{\beta^1}{N}, -\frac{\beta^2}{N}, -\frac{\beta^3}{N}\right)}. \tag{5.36}$$

Notice that the covariant components (i.e. the components of $\underline{\boldsymbol{n}}$ with respect to the basis $(\mathbf{d}x^\alpha)$ of $\mathscr{T}_p^*(\mathscr{M})$) are immediately deduced from the relation $\underline{\boldsymbol{n}} = -N\boldsymbol{\nabla} t$ [Eq. (4.8)]:

$$\boxed{n_\alpha = (-N, 0, 0, 0)}. \tag{5.37}$$

Example 5.1 Let us consider the hyperboloidal slicing of Minkowski spacetime presented in Examples 4.1–4.4. We naturally choose the $\mathbb{H}^3$ coordinates $(x^i) = (\rho, \theta, \varphi)$ introduced in Example 3.4 as the spatial coordinates on each Σ_t. In this Chapter, we shall use a hat to distinguish the adapted coordinates $(x^\alpha) = (t, \rho, \theta, \varphi)$ from the Minkowskian ones, $(X^{\hat{\alpha}}) = (w, x, y, z)$. The two coordinate systems cover the entire manifold $\mathscr{M}$ and are related by

$$\begin{cases} x = b \sinh \rho \sin \theta \cos \varphi \\ y = b \sinh \rho \sin \theta \sin \varphi \\ z = b \sinh \rho \cos \theta \\ w = t + b \cosh \rho . \end{cases} \tag{5.38}$$

These relations are adapted from the system (3.41) by modifying the last equation to take into account the generalization (4.13) of Eq. (3.38) to $t \neq 0$. The time vector $\boldsymbol{\partial}_t$ of the adapted coordinates (x^α) is obtained from the formula

$$\boldsymbol{\partial}_t = \left(\frac{\partial}{\partial t}\right)_{x^i} = \underbrace{\left(\frac{\partial w}{\partial t}\right)_{x^i}}_{1} \frac{\partial}{\partial w} + \underbrace{\left(\frac{\partial x}{\partial t}\right)_{x^i}}_{0} \frac{\partial}{\partial x} + \underbrace{\left(\frac{\partial y}{\partial t}\right)_{x^i}}_{0} \frac{\partial}{\partial y} + \underbrace{\left(\frac{\partial z}{\partial t}\right)_{x^i}}_{0} \frac{\partial}{\partial z}.$$

Since $\partial/\partial w = \boldsymbol{\partial}_w$—the first vector of the natural basis associated with the Minkowskian coordinates, we conclude that

$$\boldsymbol{\partial}_t = \boldsymbol{\partial}_w. \tag{5.39}$$

Accordingly, the components of $\boldsymbol{\partial}_t$ with respect to the Minkowskian coordinates are simply

$$(\boldsymbol{\partial}_t)^{\hat{\alpha}} = (1, 0, 0, 0). \tag{5.40}$$

The vector $\boldsymbol{\partial}_t$ is depicted in Fig. 5.2. The fact that this vector is parallel to the w-axis is not surprising since from (5.38), the line $(\rho, \theta, \varphi) = \text{const}$ is a line $x = \text{const}$.

The components of the shift vector with respect to the Minkowskian coordinates are computed from Eq. (5.29): $\beta^{\hat{\alpha}} = (\boldsymbol{\partial}_t)^{\hat{\alpha}} - N n^{\hat{\alpha}}$. Given the values (4.14) for N and (4.16) for $n^{\hat{\alpha}}$, we obtain

$$\beta^{\hat{\alpha}} = -\frac{1}{b}\left(\frac{x^2 + y^2 + z^2}{b},\ xN,\ yN,\ zN\right). \tag{5.41}$$

The vector $\boldsymbol{\beta}$ is represented in Fig. 5.2.

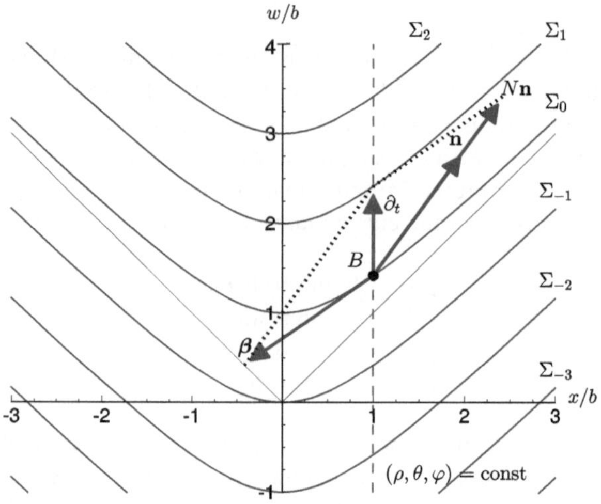

Fig. 5.2 Time vector $\boldsymbol{\partial}_t$ and shift vector $\boldsymbol{\beta}$ corresponding to the choice of the hyperboloidal coordinates $(x^i) = (\rho, \theta, \varphi)$ on each hypersurface Σ_t of the slicing considered in Fig. 4.4

5.2.3 3+1 Writing of the Metric Components

Let us introduce the components γ_{ij} of the 3-metric $\boldsymbol{\gamma}$ with respect to the coordinates (x^i)

$$\boldsymbol{\gamma} =: \gamma_{ij}\mathbf{d}x^i \otimes \mathbf{d}x^j. \tag{5.42}$$

From the definition of $\underline{\boldsymbol{\beta}}$, we have

$$\beta_i = \gamma_{ij}\beta^j. \tag{5.43}$$

The components $g_{\alpha\beta}$ of the metric $\boldsymbol{g}$ with respect to the coordinates (x^α) are defined by

$$\boldsymbol{g} =: g_{\alpha\beta}\mathbf{d}x^\alpha \otimes \mathbf{d}x^\beta. \tag{5.44}$$

Each component can be computed as

$$g_{\alpha\beta} = \boldsymbol{g}(\boldsymbol{\partial}_\alpha, \boldsymbol{\partial}_\beta). \tag{5.45}$$

Accordingly, thanks to Eq. (5.31),

$$g_{00} = \boldsymbol{g}(\boldsymbol{\partial}_t, \boldsymbol{\partial}_t) = \boldsymbol{\partial}_t \cdot \boldsymbol{\partial}_t = -N^2 + \boldsymbol{\beta} \cdot \boldsymbol{\beta} = -N^2 + \beta_i\beta^i \tag{5.46}$$

and, thanks to Eq. (5.27)

$$g_{0i} = \boldsymbol{g}(\boldsymbol{\partial}_t, \boldsymbol{\partial}_i) = (\boldsymbol{m} + \boldsymbol{\beta}) \cdot \boldsymbol{\partial}_i.$$

Now, as noticed above [cf. Eq. (5.25)], the vector $\boldsymbol{\partial}_i$ is tangent to Σ_t, so that $\boldsymbol{m}\cdot\boldsymbol{\partial}_i = 0$. Hence

$$g_{0i} = \boldsymbol{\beta} \cdot \boldsymbol{\partial}_i = \langle \underline{\boldsymbol{\beta}}, \boldsymbol{\partial}_i \rangle = \langle \beta_j \mathbf{d}x^j, \boldsymbol{\partial}_i \rangle = \beta_j \underbrace{\langle \mathbf{d}x^j, \boldsymbol{\partial}_i \rangle}_{\delta^j{}_i} = \beta_i. \tag{5.47}$$

Besides, since $\boldsymbol{\partial}_i$ and $\boldsymbol{\partial}_j$ are tangent to Σ_t,

$$g_{ij} = \boldsymbol{g}(\boldsymbol{\partial}_i, \boldsymbol{\partial}_j) = \boldsymbol{\gamma}(\boldsymbol{\partial}_i, \boldsymbol{\partial}_j) = \gamma_{ij}. \tag{5.48}$$

Collecting Eqs. (5.46), (5.47) and (5.48), we get the following expression of the metric components in terms of 3+1 quantities:

$$\boxed{g_{\alpha\beta} = \begin{pmatrix} g_{00} & g_{0j} \\ g_{i0} & g_{ij} \end{pmatrix} = \begin{pmatrix} -N^2 + \beta_k\beta^k & \beta_j \\ \beta_i & \gamma_{ij} \end{pmatrix}}, \tag{5.49}$$

or, in terms of line elements [using Eq. (5.43)],

$$\boxed{g_{\mu\nu}\mathrm{d}x^\mu \mathrm{d}x^\nu = -N^2\mathrm{d}t^2 + \gamma_{ij}(\mathrm{d}x^i + \beta^i\mathrm{d}t)(\mathrm{d}x^j + \beta^j\mathrm{d}t)}. \tag{5.50}$$

The components of the inverse metric are given by the matrix inverse of (5.49):

$$\boxed{g^{\alpha\beta} = \begin{pmatrix} g^{00} & g^{0j} \\ g^{i0} & g^{ij} \end{pmatrix} = \begin{pmatrix} -\frac{1}{N^2} & \frac{\beta^j}{N^2} \\ \frac{\beta^i}{N^2} & \gamma^{ij} - \frac{\beta^i\beta^j}{N^2} \end{pmatrix}}. \tag{5.51}$$

Indeed, it is easily checked that the matrix product $g^{\alpha\mu}g_{\mu\beta}$ is equal to the identity matrix $\delta^\alpha{}_\beta$.

Remark 5.7 Notice that $g_{ij} = \gamma_{ij}$ but that in general $g^{ij} \neq \gamma^{ij}$.

One can deduce from the above formulæ a simple relation between the determinants of $\boldsymbol{g}$ and $\boldsymbol{\gamma}$. Let us first define the latter ones by

$$\boxed{g := \det(g_{\alpha\beta})}, \tag{5.52}$$

$$\boxed{\gamma := \det(\gamma_{ij})}. \tag{5.53}$$

Notice that g and γ depend upon the choice of the coordinates (x^α). They are not scalar quantities, but scalar *densities*. Using Cramer's rule for expressing the inverse $(g^{\alpha\beta})$ of the matrix $(g_{\alpha\beta})$, we have

$$g^{00} = \frac{C_{00}}{\det(g_{\alpha\beta})} = \frac{C_{00}}{g}, \tag{5.54}$$

where C_{00} is the element (0,0) of the cofactor matrix associated with $(g_{\alpha\beta})$. It is given by $C_{00} = (-1)^0 M_{00} = M_{00}$, where M_{00} is the minor (0,0) of the matrix $(g_{\alpha\beta})$, i.e. the determinant of the 3×3 matrix deduced from $(g_{\alpha\beta})$ by suppressing the first line and the first column. From Eq. (5.49), we read

$$M_{00} = \det(\gamma_{ij}) = \gamma.$$

Hence Eq. (5.54) becomes

$$g^{00} = \frac{\gamma}{g}.$$

Expressing g^{00} from Eq. (5.51) yields then $g = -N^2\gamma$, or equivalently,

$$\boxed{\sqrt{-g} = N\sqrt{\gamma}}. \tag{5.55}$$

Example 5.2 Let us evaluate the metric components within the adapted coordinates $(x^\alpha) = (t, \rho, \theta, \varphi)$, in the case of the hyperboloidal slicing of Minkowski spacetime considered in Example 5.1. We shall compute the components $g_{\alpha\beta}$ from the equality of the line element expressed in both coordinates systems:

$$g_{\mu\nu} \mathrm{d}x^\mu \mathrm{d}x^\nu = g_{\hat{\mu}\hat{\nu}} \mathrm{d}X^{\hat{\mu}} \mathrm{d}X^{\hat{\nu}} = -\mathrm{d}w^2 + \mathrm{d}x^2 + \mathrm{d}y^2 + \mathrm{d}z^2. \tag{5.56}$$

From the last line of (5.38), we have

$$\mathrm{d}w^2 = (\mathrm{d}t + b \sinh\rho\, \mathrm{d}\rho)^2 = \mathrm{d}t^2 + 2b \sinh\rho\, \mathrm{d}t\, \mathrm{d}\rho + b^2 \sinh^2\rho\, \mathrm{d}\rho^2.$$

On the other side, $\mathrm{d}x^2 + \mathrm{d}y^2 + \mathrm{d}z^2$ is given by Eq. (3.42). Accordingly, Eq. (5.56) becomes

$$g_{\mu\nu} \mathrm{d}x^\mu \mathrm{d}x^\nu = -\mathrm{d}t^2 - 2b \sinh\rho\, \mathrm{d}t\, \mathrm{d}\rho + b^2 \left[\mathrm{d}\rho^2 + \sinh^2\rho (\mathrm{d}\theta^2 + \sin^2\theta \mathrm{d}\varphi^2) \right],$$

which can be put in the form

$$\begin{aligned} g_{\mu\nu} \mathrm{d}x^\mu \mathrm{d}x^\nu = & - \cosh^2\rho\, \mathrm{d}t^2 + b^2 \left[\left(\mathrm{d}\rho - \frac{\sinh\rho}{b} \mathrm{d}t \right)^2 \right. \\ & \left. + \sinh^2\rho (\mathrm{d}\theta^2 + \sin^2\theta \mathrm{d}\varphi^2) \right]. \end{aligned} \tag{5.57}$$

By comparing with (5.50), we get the lapse function, the shift vector and the 3-metric:

$$N = \cosh\rho \tag{5.58}$$

$$\beta^i = \left(-\frac{\sinh\rho}{b},\ 0,\ 0\right) \tag{5.59}$$

$$\gamma_{ij} = \mathrm{diag}\left(b^2,\ b^2\sinh^2\rho,\ b^2\sinh^2\rho\sin^2\theta\right). \tag{5.60}$$

In particular, we recover the value of N obtained in Chap. 4 [Eq. (4.51)] and the value of γ_{ij} obtained in Chap. 3 [Eq. (3.43)]. Note that $\boldsymbol{\beta}\cdot\boldsymbol{\beta} = \gamma_{ij}\beta^i\beta^j = \sinh^2\rho$, so that the shift is superluminal for $\rho > \ln(1+\sqrt{2}) \simeq 0.881$. The components of the unit normal with respect to the adapted coordinates are obtained via formulas (5.36) and (5.37):

$$n_\alpha = (-\cosh\rho, 0, 0, 0) \quad \text{and} \quad n^\alpha = \left(\frac{1}{\cosh\rho},\ \frac{1}{b}\tanh\rho,\ 0,\ 0\right). \tag{5.61}$$

Note that these components are different from (4.15) and (4.16), for the latter are the components $n_{\hat{\alpha}}$ and $n^{\hat{\alpha}}$ with respect to the Minkowskian coordinates.

5.2.4 Choice of Coordinates via the Lapse and the Shift

We have seen above that giving a coordinate system (x^α) on $\mathscr{M}$ such that the hypersurfaces $x^0 = \text{const.}$ are spacelike determines uniquely a lapse function N and a shift vector $\boldsymbol{\beta}$. The converse is true in the following sense: setting on some hypersurface Σ_0 a scalar field N, a vector field $\boldsymbol{\beta}$ and a coordinate system (x^i) uniquely specifies a coordinate system (x^α) in some neighbourhood of Σ_0, such that the hypersurface $x^0 = 0$ is Σ_0. Indeed, the knowledge of the lapse function a each point of Σ_0 determines a unique vector $\boldsymbol{m} = N\boldsymbol{n}$ and consequently the location of the "next" hypersurface $\Sigma_{\delta t}$ by Lie transport along $\boldsymbol{m}$ (cf. Sect. 4.3.2). Graphically, we may also say that for each point of Σ_0 the lapse function specifies how far is the point of $\Sigma_{\delta t}$ located "above" it ("above" meaning perpendicularly to Σ_0, cf. Fig. 4.2). Then the shift vector tells how to propagate the coordinate system (x^i) from Σ_0 to $\Sigma_{\delta t}$ (cf. Fig. 5.1).

This way of choosing coordinates via the lapse function and the shift vector is one of the main topics in 3+1 numerical relativity and will be discussed in detail in Chap. 10.

5.3 3+1 Einstein Equation as a PDE System

5.3.1 Lie Derivatives Along m as Partial Derivatives

Let $\boldsymbol{T}$ be any tensor field tangent to Σ_t. Thanks to Eq. (5.27), we can write

$$\mathcal{L}_{\boldsymbol{m}}\boldsymbol{T} = \mathcal{L}_{\boldsymbol{\partial}_t}\boldsymbol{T} - \mathcal{L}_{\boldsymbol{\beta}}\boldsymbol{T}. \tag{5.62}$$

This implies that $\mathcal{L}_{\boldsymbol{\partial}_t}\boldsymbol{T}$ is a tensor field tangent to Σ_t, since both $\mathcal{L}_{\boldsymbol{m}}\boldsymbol{T}$ and $\mathcal{L}_{\boldsymbol{\beta}}\boldsymbol{T}$ are tangent to Σ_t, the former by the property (4.36) and the latter because $\boldsymbol{\beta}$ and $\boldsymbol{T}$ are tangent to Σ_t. Moreover, if one uses tensor components with respect to a coordinate system $(x^\alpha) = (t, x^i)$ adapted to the foliation, the Lie derivative along $\boldsymbol{\partial}_t$ reduces simply to the partial derivative with respect to t [cf. Eq. (2.86)]:

$$\mathcal{L}_{\boldsymbol{\partial}_t} T^{i\ldots}{}_{j\ldots} = \frac{\partial}{\partial t} T^{i\ldots}{}_{j\ldots}.$$

Accordingly, in terms of components with respect to adapted coordinates, Eq. (5.62) can be written as

$$\boxed{\mathcal{L}_{\boldsymbol{m}} T^{i\ldots}{}_{j\ldots} = \left(\frac{\partial}{\partial t} - \mathcal{L}_{\boldsymbol{\beta}}\right) T^{i\ldots}{}_{j\ldots}}. \tag{5.63}$$

We may apply this formula to the term $\mathcal{L}_{\boldsymbol{m}}\boldsymbol{K}$ which occurs in the 3+1 Einstein equation (5.17):

$$\mathcal{L}_{\boldsymbol{m}} K_{ij} = \left(\frac{\partial}{\partial t} - \mathcal{L}_{\beta}\right) K_{ij}. \tag{5.64}$$

By means of formula (2.90), one can express $\mathcal{L}_{\boldsymbol{\beta}}\boldsymbol{K}$ in terms of partial derivatives:

$$\mathcal{L}_{\boldsymbol{\beta}} K_{ij} = \beta^k \frac{\partial K_{ij}}{\partial x^k} + K_{kj}\frac{\partial \beta^K}{\partial x^i} + K_{ik}\frac{\partial \beta^K}{\partial x^j}. \tag{5.65}$$

Similarly, thanks to (5.63), the relation (4.30) between $\mathcal{L}_{\boldsymbol{m}}\boldsymbol{\gamma}$ and $\boldsymbol{K}$ becomes

$$\left(\frac{\partial}{\partial t} - \mathcal{L}_{\boldsymbol{\beta}}\right) \gamma_{ij} = -2NK_{ij}. \tag{5.66}$$

In this relation, one may evaluate the Lie derivative with the connection $\boldsymbol{D}$ instead of partial derivatives [cf. Eq. (2.92)]:

$$\mathcal{L}_{\boldsymbol{\beta}} \gamma_{ij} = \beta^k \underbrace{D_k \gamma_{ij}}_{0} + \gamma_{kj} D_i \beta^k + \gamma_{ik} D_j \beta^k,$$

i.e.

$$\mathcal{L}_{\boldsymbol{\beta}} \gamma_{ij} = D_i \beta_j + D_j \beta_i. \tag{5.67}$$

5.3.2 3+1 Einstein System

Using Eqs. (5.64) and (5.66), we rewrite the 3+1 Einstein system (5.19), (5.20) and (5.22) as

$$\boxed{\left(\frac{\partial}{\partial t} - \mathcal{L}_{\boldsymbol{\beta}}\right) \gamma_{ij} = -2NK_{ij}} \tag{5.68}$$

$$\boxed{\left(\frac{\partial}{\partial t} - \mathcal{L}_{\boldsymbol{\beta}}\right) K_{ij} = -D_i D_j N + N\left\{R_{ij} + KK_{ij} - 2K_{ik}K^k_{\ j} + 4\pi\left[(S-E)\gamma_{ij} - 2S_{ij}\right]\right\}} \tag{5.69}$$

$$\boxed{R + K^2 - K_{ij}K^{ij} = 16\pi E} \tag{5.70}$$

$$\boxed{D_j K^j_{\ i} - D_i K = 8\pi p_i}\,. \tag{5.71}$$

In this system, the covariant derivatives D_i can be expressed in terms of partial derivatives with respect to the spatial coordinates (x^i) by means of the Christoffel symbols $\Gamma^i_{\ jk}$ of $\boldsymbol{D}$ associated with (x^i):

$$D_i D_j N = \frac{\partial^2 N}{\partial x^i \partial x^j} - \Gamma^k_{\ ij}\frac{\partial N}{\partial x^k}, \tag{5.72}$$

$$D_j K^j_{\ i} = \frac{\partial K^j_{\ i}}{\partial x^j} + \Gamma^j_{\ jk}K^k_{\ i} - \Gamma^k_{\ ji}K^j_{\ k}, \tag{5.73}$$

$$D_i K = \frac{\partial K}{\partial x^i}. \tag{5.74}$$

The Lie derivatives along $\boldsymbol{\beta}$ can be expressed in terms of partial derivatives with respect to the spatial coordinates (x^i), via Eqs. (5.65) and (5.67):

$$\mathcal{L}_{\boldsymbol{\beta}}\gamma_{ij} = \frac{\partial \beta_i}{\partial x^j} + \frac{\partial \beta_j}{\partial x^i} - 2\Gamma^k_{\ ij}\beta_k \tag{5.75}$$

$$\mathcal{L}_{\boldsymbol{\beta}}K_{ij} = \beta^k\frac{\partial K_{ij}}{\partial x^k} + K_{kj}\frac{\partial \beta^k}{\partial x^i} + K_{ik}\frac{\partial \beta^K}{\partial x^j}. \tag{5.76}$$

Finally, the Ricci tensor and scalar curvature of $\boldsymbol{\gamma}$ are expressible according to Eqs. (2.76), (2.69) and (2.78):

$$R_{ij} = \frac{\partial \Gamma^k_{\ ij}}{\partial x^k} - \frac{\partial \Gamma^k_{\ ik}}{\partial x^j} + \Gamma^k_{\ ij}\Gamma^l_{\ kl} - \Gamma^l_{\ ik}\Gamma^k_{\ lj} \tag{5.77}$$

$$R = \gamma^{ij}R_{ij}. \tag{5.78}$$

For completeness, let us recall the expression of the Christoffel symbols in terms of partial derivatives of the metric [cf. Eq. (2.62)]:

$$\Gamma^k{}_{ij} = \frac{1}{2}\gamma^{kl}\left(\frac{\partial \gamma_{lj}}{\partial x^i} + \frac{\partial \gamma_{il}}{\partial x^j} - \frac{\partial \gamma_{ij}}{\partial x^l}\right). \tag{5.79}$$

Assuming that matter "source terms" (E, p_i, S_{ij}) are given, the system (5.68)–(5.71), with all the terms made explicit according to Eqs. (5.72)–(5.79), constitutes a second-order non-linear PDE system for the unknowns $(\gamma_{ij}, K_{ij}, N, \beta^i)$. It has been first derived by Darmois, as early as 1927 [3], in the special case $N = 1$ and $\boldsymbol{\beta} = 0$ (Gaussian normal coordinates, to be discussed in Sect. 5.4.2). The case $N \neq 1$, but still with $\boldsymbol{\beta} = 0$, has been obtained by Lichnerowicz in 1939 [4, 5] and the general case (arbitrary lapse and shift) by Choquet-Bruhat in 1948 [6, 2]. A slightly different form, with K_{ij} replaced by the "momentum conjugate to γ_{ij}", namely $\pi^{ij} := \sqrt{\gamma}(K\gamma^{ij} - K^{ij})$, has been derived by Arnowitt, Deser and Misner [7] from their Hamiltonian formulation of general relativity (to be discussed in Sect. 5.5).

Remark 5.8 In the numerical relativity literature, the 3+1 Einstein equations (5.68)–(5.71) are sometimes called the "*ADM equations*", in reference of the above mentioned work by Arnowitt, Deser and Misner [7]. However, the major contribution of ADM is a Hamiltonian formulation of general relativity (which we will discuss succinctly in Sect. 5.5). This Hamiltonian approach is not used in numerical relativity, which proceeds by integrating the system (5.68)–(5.71). The latter was known before ADM work. In particular, the recognition of the extrinsic curvature $\boldsymbol{K}$ as a fundamental 3+1 variable was already achieved by Darmois in 1927 [3]. Moreover, as stressed by York [8] (see also Ref. [9]), Eq. (5.69) is the spatial projection of the spacetime *Ricci tensor* [i.e. is derived from the Einstein equation in the form (5.2), cf. Sect. 5.1.3] whereas the dynamical equation in the ADM work [7] is instead the spatial projection of the *Einstein tensor* [i.e. is derived from the Einstein equation in the form (5.1)].

5.4 The Cauchy Problem

5.4.1 General Relativity as a Three-Dimensional Dynamical System

The system (5.68)–(5.79) involves only three-dimensional quantities, i.e. tensor fields defined on the hypersurface Σ_t, and their time derivatives. Consequently one may forget about the four-dimensional origin of the system and consider that (5.68)–(5.79) describes *time evolving* tensor fields on a *single* three-dimensional manifold Σ, without any reference to some ambient four-dimensional spacetime. This constitutes the *geometrodynamics* point of view developed by Wheeler [1] (see also Fischer and Marsden [10, 11] for a more formal treatment).

It is to be noticed that the system (5.68)–(5.79) does not contain any time derivative of the lapse function N nor of the shift vector $\boldsymbol{\beta}$. This means that N and $\boldsymbol{\beta}$ are not dynamical variables. This should not be surprising if one remembers that they are associated with the choice of coordinates (t, x^i) (cf. Sect. 5.2.4). Actually the coordinate freedom of general relativity implies that we may choose the lapse and shift freely, without changing the physical solution $\boldsymbol{g}$ of the Einstein equation. The only things to avoid are coordinate singularities, to which a arbitrary choice of lapse and shift might lead.

5.4.2 Analysis Within Gaussian Normal Coordinates

To gain some insight in the nature of the system (5.68)–(5.79), let us simplify it by using the freedom in the choice of lapse and shift: we set

$$N = 1 \tag{5.80}$$

$$\boldsymbol{\beta} = 0, \tag{5.81}$$

in some neighbourhood of a given hypersurface Σ_0 where the coordinates (x^i) are specified arbitrarily. This means that the lines of constant spatial coordinates are orthogonal to the hypersurfaces Σ_t (see Fig. 5.1). Moreover, with $N = 1$, the coordinate time t coincides with the proper time measured by the Eulerian observers between neighbouring hypersurfaces Σ_t [cf. Eq. (4.17)]. Such coordinates are called ***Gaussian normal coordinates***. The foliation away from Σ_0 selected by the choice (5.80) of the lapse function is called a ***geodesic slicing***. This name stems from the fact that the worldlines of the Eulerian observers are geodesics, the parameter t being then an affine parameter along them. This is immediate from Eq. (4.19), which, for $N = 1$, implies the vanishing of the 4-accelerations of the Eulerian observers (free fall).

In Gaussian normal coordinates, the spacetime metric tensor takes a simple form [cf. Eq. (5.50)]:

$$g_{\mu\nu}\mathrm{d}x^\mu \mathrm{d}x^\nu = -\mathrm{d}t^2 + \gamma_{ij}\mathrm{d}x^i \mathrm{d}x^j. \tag{5.82}$$

In general it is not possible to get a Gaussian normal coordinate system that covers all $\mathscr{M}$. This results from the well known tendencies of timelike geodesics without vorticity (such as the worldlines of the Eulerian observers) to focus and eventually cross. This reflects the attractive nature of gravity and is best seen on the Raychaudhuri equation (cf. Lemma 9.2.1 in [12]). However, for the purpose of the present discussion it is sufficient to consider Gaussian normal coordinates in some neighbourhood of the hypersurface Σ_0; provided that the neighbourhood is small enough, this is always possible. The 3+1 Einstein system (5.68)–(5.71) reduces then to

$$\frac{\partial \gamma_{ij}}{\partial t} = -2K_{ij} \tag{5.83}$$

$$\frac{\partial K_{ij}}{\partial t} = R_{ij} + KK_{ij} - 2K_{ik}K^k{}_j + 4\pi \left[(S-E)\gamma_{ij} - 2S_{ij}\right] \tag{5.84}$$

$$R + K^2 - K_{ij}K^{ij} = 16\pi E \tag{5.85}$$

$$D_j K^j{}_i - D_i K = 8\pi p_i. \tag{5.86}$$

Using the short-hand notation

$$\dot{\gamma}_{ij} := \frac{\partial \gamma_{ij}}{\partial t} \tag{5.87}$$

and replacing everywhere K_{ij} thanks to Eq. (5.83), we get

$$-\frac{\partial^2 \gamma_{ij}}{\partial t^2} = 2R_{ij} + \frac{1}{2}\gamma^{kl}\dot{\gamma}_{kl}\dot{\gamma}_{ij} - \gamma^{kl}\dot{\gamma}_{ik}\dot{\gamma}_{lj} + 8\pi \left[(S-E)\gamma_{ij} - 2S_{ij}\right] \tag{5.88}$$

$$R + \frac{1}{4}(\gamma^{ij}\dot{\gamma}_{ij})^2 - \frac{1}{4}\gamma^{ik}\gamma^{jl}\dot{\gamma}_{ij}\dot{\gamma}_{kl} = 16\pi E \tag{5.89}$$

$$D_j(\gamma^{jk}\dot{\gamma}_{ki}) - \frac{\partial}{\partial x^i}\left(\gamma^{kl}\dot{\gamma}_{kl}\right) = -16\pi p_i. \tag{5.90}$$

As far as the gravitational field is concerned, this equation contains only the 3-metric $\boldsymbol{\gamma}$. In particular the Ricci tensor can be made explicit by plugging Eq. (5.79) into Eq. (5.77). We need only the ***principal part*** for our analysis, that is the part containing the derivative of γ_{ij} of highest degree (two in the present case). We get, denoting by "$\cdots$" everything but a second order derivative of γ_{ij}:

$$\begin{aligned} R_{ij} &= \frac{\partial \Gamma^k{}_{ij}}{\partial x^k} - \frac{\partial \Gamma^k{}_{ik}}{\partial x^j} + \cdots \\ &= \frac{1}{2}\frac{\partial}{\partial x^k}\left[\gamma^{kl}\left(\frac{\partial \gamma_{lj}}{\partial x^i} + \frac{\partial \gamma_{il}}{\partial x^j} - \frac{\partial \gamma_{ij}}{\partial x^l}\right)\right] - \frac{1}{2}\frac{\partial}{\partial x^j}\left[\gamma^{kl}\left(\frac{\partial \gamma_{lk}}{\partial x^i} + \frac{\partial \gamma_{il}}{\partial x^k} - \frac{\partial \gamma_{ik}}{\partial x^l}\right)\right] + \cdots \\ &= \frac{1}{2}\gamma^{kl}\left(\frac{\partial^2 \gamma_{lj}}{\partial x^k \partial x^i} + \frac{\partial^2 \gamma_{il}}{\partial x^k \partial x^j} - \frac{\partial^2 \gamma_{ij}}{\partial x^k \partial x^l} - \frac{\partial^2 \gamma_{lk}}{\partial x^j \partial x^i} - \frac{\partial^2 \gamma_{il}}{\partial x^j \partial x^k} + \frac{\partial^2 \gamma_{ik}}{\partial x^j \partial x^l}\right) + \cdots \\ R_{ij} &= -\frac{1}{2}\gamma^{kl}\left(\frac{\partial^2 \gamma_{ij}}{\partial x^k \partial x^l} + \frac{\partial^2 \gamma_{kl}}{\partial x^i \partial x^j} - \frac{\partial^2 \gamma_{lj}}{\partial x^i \partial x^k} - \frac{\partial^2 \gamma_{il}}{\partial x^j \partial x^k}\right) + \mathcal{Q}_{ij}\left(\gamma_{kl}, \frac{\partial \gamma_{kl}}{\partial x^m}\right), \end{aligned} \tag{5.91}$$

where $\mathcal{Q}_{ij}(\gamma_{kl}, \partial\gamma_{kl}/\partial x^m)$ is a (non-linear) expression containing the components γ_{kl} and their first spatial derivatives only. Taking the trace of (5.91) (i.e. contracting with γ^{ij}), we get

$$R = \gamma^{ik}\gamma^{jl}\frac{\partial^2 \gamma_{ij}}{\partial x^k \partial x^l} - \gamma^{ij}\gamma^{kl}\frac{\partial^2 \gamma_{ij}}{\partial x^k \partial x^l} + \mathcal{Q}\left(\gamma_{kl}, \frac{\partial \gamma_{kl}}{\partial x^m}\right). \tag{5.92}$$

Besides

$$D_j(\gamma^{jk}\dot{\gamma}_{ki}) = \gamma^{jk}D_j\dot{\gamma}_{ki} = \gamma^{jk}\left(\frac{\partial\dot{\gamma}_{ki}}{\partial x^j} - \Gamma^l{}_{jk}\dot{\gamma}_{li} - \Gamma^l{}_{ji}\dot{\gamma}_{kl}\right)$$
$$= \gamma^{jk}\frac{\partial^2\gamma_{ki}}{\partial x^j\partial t} + \mathscr{Q}_i\left(\gamma_{kl}, \frac{\partial\gamma_{kl}}{\partial x^m}, \frac{\partial\gamma_{kl}}{\partial t}\right), \tag{5.93}$$

where $\mathscr{Q}_i(\gamma_{kl}, \partial\gamma_{kl}/\partial x^m, \partial\gamma_{kl}/\partial t)$ is some expression that does not contain any second order derivative of γ_{kl}. Substituting Eqs. (5.91), (5.92) and (5.93) in Eqs. (5.88)–(5.90) gives

$$-\frac{\partial^2\gamma_{ij}}{\partial t^2} + \gamma^{kl}\left(\frac{\partial^2\gamma_{ij}}{\partial x^k\partial x^l} + \frac{\partial^2\gamma_{kl}}{\partial x^i\partial x^j} - \frac{\partial^2\gamma_{lj}}{\partial x^i\partial x^k} - \frac{\partial^2\gamma_{il}}{\partial x^j\partial x^k}\right)$$
$$= 8\pi\left[(S-E)\gamma_{ij} - 2S_{ij}\right] + \mathscr{Q}_{ij}\left(\gamma_{kl}, \frac{\partial\gamma_{kl}}{\partial x^m}, \frac{\partial\gamma_{kl}}{\partial t}\right) \tag{5.94}$$

$$\gamma^{ik}\gamma^{jl}\frac{\partial^2\gamma_{ij}}{\partial x^k\partial x^l} - \gamma^{ij}\gamma^{kl}\frac{\partial^2\gamma_{ij}}{\partial x^k\partial x^l} = 16\pi E + \mathscr{Q}\left(\gamma_{kl}, \frac{\partial\gamma_{kl}}{\partial x^m}, \frac{\partial\gamma_{kl}}{\partial t}\right) \tag{5.95}$$

$$\gamma^{jk}\frac{\partial^2\gamma_{ki}}{\partial x^j\partial t} - \gamma^{kl}\frac{\partial^2\gamma_{kl}}{\partial x^i\partial t} = -16\pi p_i + \mathscr{Q}_i\left(\gamma_{kl}, \frac{\partial\gamma_{kl}}{\partial x^m}, \frac{\partial\gamma_{kl}}{\partial t}\right). \tag{5.96}$$

Notice that we have incorporated the first order time derivatives into the $\mathscr{Q}$ terms.

Equations (5.94)–(5.96) constitute a system of PDEs for the unknowns γ_{ij}. This system is of second order and non linear, but ***quasi-linear***, i.e. linear with respect to all the second order derivatives. Let us recall that, in this system, the γ^{ij}'s are to be considered as functions of the γ_{ij}'s, these functions being given by expressing the matrix (γ_{ij}) as the inverse of the matrix (γ_{ij}) (e.g. via Cramer's rule).

A key feature of the system (5.94)–(5.96) is that it contains 6+1+3 = 10 equations for the 6 unknowns γ_{ij}. Hence it is an over-determined system. Among the three sub-systems (5.94), (5.95) and (5.96), only the first one involves second-order time derivatives. Moreover the sub-system (5.94) contains the same numbers of equations than unknowns (six) and it is in a form tractable as a *Cauchy problem*, namely one could search for a solution, given some initial data. More precisely, the sub-system (5.94) being of second order and in the form

$$\frac{\partial^2\gamma_{ij}}{\partial t^2} = F_{ij}\left(\gamma_{kl}, \frac{\partial\gamma_{kl}}{\partial x^m}, \frac{\partial\gamma_{kl}}{\partial t}, \frac{\partial^2\gamma_{kl}}{\partial x^m\partial x^n}\right), \tag{5.97}$$

the Cauchy problem amounts to finding a solution γ_{ij} for $t > 0$ given the knowledge of γ_{ij} and $\partial\gamma_{ij}/\partial t$ at $t = 0$, i.e. the values of γ_{ij} and $\partial\gamma_{ij}/\partial t$ on the hypersurface Σ_0. Since F_{ij} is a analytical function,[1] we can invoke the Cauchy–Kovalevskaya

[1] It is polynomial in the derivatives of γ_{kl} and involves at most rational fractions in γ_{kl} (to get the inverse metric γ^{kl}).

theorem (see e.g. [13]) to guarantee the existence and uniqueness of a solution γ_{ij} in a neighbourhood of Σ_0, for any initial data $(\gamma_{ij}, \partial\gamma_{ij}/\partial t)$ on Σ_0 that are analytical functions of the coordinates (x^i).

The complication arises because of the extra equations (5.95) and (5.96), which must be fulfilled to ensure that the metric $\boldsymbol{g}$ reconstructed from γ_{ij} via Eq. (5.82) is indeed a solution of Einstein equation. Equations (5.95) and (5.96), which cannot be put in the form such that the Cauchy–Kovalevskaya theorem applies, constitute ***constraints*** for the Cauchy problem (5.94). In particular one has to make sure that the initial data $(\gamma_{ij}, \partial\gamma_{ij}/\partial t)$ on Σ_0 satisfy these constraints. A natural question which arises is then: suppose that we prepare initial data $(\gamma_{ij}, \partial\gamma_{ij}/\partial t)$ which satisfy the constraints (5.95)–(5.96) and that we get a solution of the Cauchy problem (5.94) from these initial data, are the constraints satisfied by the solution for $t > 0$? The answer is yes, thanks to the Bianchi identities, as we shall see in Sect. 11.3.2.

5.4.3 Constraint Equations

The main conclusions of the above discussion remain valid for the general 3+1 Einstein system as given by Eqs. (5.68)–(5.71): Eqs. (5.68) and (5.69) constitute a time evolution system tractable as a Cauchy problem, whereas Eqs. (5.70) and (5.71) constitute constraints. This partly justifies the names *Hamiltonian constraint* and *momentum constraint* given respectively to Eq. (5.70) and to Eq. (5.71).

The existence of constraints is not specific to general relativity. For instance the Maxwell equations for the electromagnetic field can be treated as a Cauchy problem subject to the constraints $\boldsymbol{D}\cdot\boldsymbol{B} = 0$ and $\boldsymbol{D}\cdot\boldsymbol{E} = \rho/\varepsilon_0$ [Eqs. (6.96)–(6.98) below]. We refer the reader to [14] or Sect. 2.3 of [15] for details of the electromagnetic analogy.

5.4.4 Existence and Uniqueness of Solutions to the Cauchy Problem

In the general case of arbitrary lapse and shift, the time derivative $\dot{\gamma}_{ij}$ introduced in Sect. 5.4.2 has to be replaced by the extrinsic curvature K_{ij}, so that the initial data on a given hypersurface Σ_0 is $(\boldsymbol{\gamma}, \boldsymbol{K})$. The couple $(\boldsymbol{\gamma}, \boldsymbol{K})$ has to satisfy the constraint equations (5.70) and (5.71) on Σ_0. One may then ask the question: given a set $(\Sigma_0, \boldsymbol{\gamma}, \boldsymbol{K}, E, \boldsymbol{p})$, where Σ_0 is a three-dimensional manifold, $\boldsymbol{\gamma}$ a Riemannian metric on Σ_0, $\boldsymbol{K}$ a symmetric bilinear form field on Σ_0, E a scalar field on Σ_0 and $\boldsymbol{p}$ a vector field on Σ_0, which obeys the constraint equations (5.70) and (5.71):

$$R + K^2 - K_{ij}K^{ij} = 16\pi E \tag{5.98}$$

$$D_j K^j{}_i - D_i K = 8\pi p_i, \tag{5.99}$$

does there exist a spacetime $(\mathscr{M}, \boldsymbol{g}, \boldsymbol{T})$ such that $(\boldsymbol{g}, \boldsymbol{T})$ fulfills the Einstein equation and Σ_0 can be embedded as an hypersurface of $\mathscr{M}$ with induced metric $\boldsymbol{\gamma}$ and extrinsic curvature $\boldsymbol{K}$?

Darmois [3] and Lichnerowicz [4] have shown that the answer is yes for the vacuum case ($E = 0$ and $p_i = 0$), when the initial data $(\boldsymbol{\gamma}, \boldsymbol{K})$ are *analytical* functions of the coordinates (x^i) on Σ_0. Their analysis is based on the Cauchy–Kovalevskaya theorem mentioned in Sect. 5.4.2 (cf. Chap. 10 of Wald's textbook [12] for details). However, on physical grounds, the analytical case is too restricted. One would like to deal instead with *smooth* (*i.e. differentiable*) initial data. There are at least two reasons for this:

- The smooth manifold structure of $\mathscr{M}$ imposes only that the change of coordinates are differentiable, not necessarily analytical. Consequently if $(\boldsymbol{\gamma}, \boldsymbol{K})$ are analytical functions of the coordinates, they might not be analytical functions of another coordinate system (x'^i).
- An analytical function is fully determined by its value and those of all its derivatives at a single point. Equivalently an analytical function is fully determined by its value in some small open domain D. This fits badly with causality requirements, because a small change to the initial data, localized in a small region, should not change the whole solution at all points of $\mathscr{M}$. The change should take place only in the so-called *domain of dependence* of D.

This is why the major breakthrough in the Cauchy problem of general relativity has been achieved by Choquet-Bruhat in 1952 [16] when she showed existence and uniqueness of the solution in a small neighbourhood of Σ_0 for *smooth* (at least C^5) initial data $(\boldsymbol{\gamma}, \boldsymbol{K})$. We shall not give any sketch on the proof (beside the original publication [16], see the review articles [17] and [18]) but simply mentioned that it is based on *harmonic coordinates*.

A major improvement has been then the *global* existence and uniqueness theorem by Choquet-Bruhat and Geroch [19]. The latter tells that among all the spacetimes $(\mathscr{M}, \boldsymbol{g})$ solution of the Einstein equation and such that $(\Sigma_0, \boldsymbol{\gamma}, \boldsymbol{K})$ is an embedded Cauchy surface, there exists a maximal spacetime $(\mathscr{M}^*, \boldsymbol{g}^*)$ and it is unique. ***Maximal*** means that any spacetime $(\mathscr{M}, \boldsymbol{g})$ solution of the Cauchy problem is isometric to a subpart of $(\mathscr{M}^*, \boldsymbol{g}^*)$. For more details about the existence and uniqueness of solutions to the Cauchy problem, see the reviews by Choquet-Bruhat and York [18], Klainerman and Nicolò [20], Andersson [21] and Rendall [22], as well as Choquet-Bruhat textbook [23].

5.5 ADM Hamiltonian Formulation

Further insight in the 3+1 Einstein equations is provided by the Hamiltonian formulation of general relativity. Indeed the latter makes use of the 3+1 formalism, since any Hamiltonian approach involves the concept of a physical state "at a certain time", which is translated in general relativity by the state on a spacelike hypersurface

Σ_t. The Hamiltonian formulation of general relativity has been developed notably by Dirac in the late fifties [24, 25] (see also Ref. [26]), by Arnowitt, Deser and Misner (ADM) in the early sixties [7] and by Regge and Teitelboim in the seventies [27]. Pedagogical presentations are given in Chap. 21 of MTW [28], in Chap. 4 of Poisson's book [29], in M. Henneaux's lectures [30] and in G. Schäfer's ones [31]. Here we focus on the ADM approach, which makes a direct use of the lapse function and shift vector (contrary to Dirac's one). For simplicity, we consider only the vacuum Einstein equation in this section. Also we shall disregard any boundary term in the action integrals. Such terms will be restored in Chap. 8 in order to discuss total energy and momentum.

5.5.1 3+1 Form of the Hilbert Action

Let us consider the standard ***Hilbert action*** for general relativity (see e.g. [32, 12]):

$$S = \int_{\mathscr{V}} {}^4\!R \sqrt{-g}\, \mathrm{d}^4x, \tag{5.100}$$

where $\mathscr{V}$ is a part of $\mathscr{M}$ delimited by two hypersurfaces Σ_{t_1} and Σ_{t_2} ($t_1 < t_2$) of the foliation $(\Sigma_t)_{t\in\mathbb{R}}$:

$$\mathscr{V} := \bigcup_{t=t_1}^{t_2} \Sigma_t. \tag{5.101}$$

Thanks to the 3+1 decomposition of ${}^4\!R$ provided by Eq. (4.49) and to the relation $\sqrt{-g} = N\sqrt{\gamma}$ [Eq. (5.55)] we can write

$$S = \int_{\mathscr{V}} \left[N\left(R + K^2 + K_{ij}K^{ij}\right) - 2\mathscr{L}_{\boldsymbol{m}} K - 2D_iD^iN \right] \sqrt{\gamma}\, \mathrm{d}^4x. \tag{5.102}$$

Now

$$\begin{aligned}\mathscr{L}_{\boldsymbol{m}} K &= m^\mu \nabla_\mu K = N n^\mu \nabla_\mu K = N\big[\nabla_\mu (K n^\mu) - K \underbrace{\nabla_\mu n^\mu}_{-K}\big] \\ &= N\big[\nabla_\mu (K n^\mu) + K^2\big].\end{aligned}$$

Hence Eq. (5.102) becomes

$$S = \int_{\mathscr{V}} \left[N\left(R + K_{ij}K^{ij} - K^2\right) - 2N\nabla_\mu (K n^\mu) - 2D_iD^iN \right] \sqrt{\gamma}\, \mathrm{d}^4x.$$

But

$$\int_{\mathscr{V}} N\nabla_\mu(Kn^\mu)\sqrt{\gamma}\,\mathrm{d}^4x = \int_{\mathscr{V}} \nabla_\mu(Kn^\mu)\sqrt{-g}\,\mathrm{d}^4x = \int_{\mathscr{V}} \frac{\partial}{\partial x^\mu}\left(\sqrt{-g}Kn^\mu\right)\,\mathrm{d}^4x$$

is the integral of a pure divergence and we can disregard this term in the action. Accordingly, the latter becomes

$$S = \int_{t_1}^{t_2}\left\{\int_{\Sigma_t}\left[N\left(R + K_{ij}K^{ij} - K^2\right) - 2D_iD^iN\right]\sqrt{\gamma}\,\mathrm{d}^3x\right\}\mathrm{d}t,$$

where we have used (5.101) to split the four-dimensional integral into a time integral and a three-dimensional one. Again we have a divergence term:

$$\int_{\Sigma_t} D_iD^iN\sqrt{\gamma}\,\mathrm{d}^3x = \int_{\Sigma_t}\frac{\partial}{\partial x^i}\left(\sqrt{\gamma}D^iN\right)\mathrm{d}^3x,$$

which we can disregard. Hence the 3+1 writing of the Hilbert action is

$$\boxed{S = \int_{t_1}^{t_2}\left\{\int_{\Sigma_t} N\left(R + K_{ij}K^{ij} - K^2\right)\sqrt{\gamma}\,\mathrm{d}^3x\right\}\mathrm{d}t}\,. \tag{5.103}$$

5.5.2 Hamiltonian Approach

The action (5.103) is to be considered as a functional of the "configuration" variables $q = (\gamma_{ij}, N, \beta^i)$ [which describe the full spacetime metric components $g_{\alpha\beta}$, cf. Eq. (5.49)] and their time derivatives[2] $\dot{q} = (\dot{\gamma}_{ij}, \dot{N}, \dot{\beta}^i)$: $S = S[q, \dot{q}]$. In particular K_{ij} in Eq. (5.103) is the function of $\dot{\gamma}_{ij}$, γ_{ij}, N and β^i given by Eqs. (5.68) and (5.67):

$$K_{ij} = \frac{1}{2N}\left(\gamma_{ik}D_j\beta^k + \gamma_{jk}D_i\beta^k - \dot{\gamma}_{ij}\right). \tag{5.104}$$

From Eq. (5.103), we read that the gravitational field Lagrangian density is

$$\boxed{L(q,\dot{q}) = N\sqrt{\gamma}(R + K_{ij}K^{ij} - K^2) = N\sqrt{\gamma}\left[R + (\gamma^{ik}\gamma^{jl} - \gamma^{ij}\gamma^{kl})K_{ij}K_{kl}\right]}\,, \tag{5.105}$$

with K_{ij} and K_{kl} expressed as (5.104). Notice that this Lagrangian does not depend upon the time derivatives of N and β^i: this shows that the lapse function and the shift vector are not dynamical variables. Consequently the only dynamical variable is γ_{ij}. The momentum canonically conjugate to it is

$$\pi^{ij} := \frac{\partial L}{\partial \dot{\gamma}_{ij}}. \tag{5.106}$$

---

2 we use the same notation as that defined by Eq. (5.87).

From Eqs. (5.105) and (5.104), we get

$$\pi^{ij} = N\sqrt{\gamma}\left[(\gamma^{ik}\gamma^{jl} - \gamma^{ij}\gamma^{kl})K_{kl} + (\gamma^{ki}\gamma^{lj} - \gamma^{kl}\gamma^{ij})K_{kl}\right] \times \left(-\frac{1}{2N}\right),$$

i.e.

$$\boxed{\pi^{ij} = \sqrt{\gamma}\left(K\gamma^{ij} - K^{ij}\right)}. \tag{5.107}$$

The Hamiltonian density is given by the Legendre transform

$$\mathscr{H} = \pi^{ij}\dot{\gamma}_{ij} - L. \tag{5.108}$$

Using Eqs. (5.104), (5.107) and (5.105), we have

$$\begin{aligned}
\mathscr{H} &= \sqrt{\gamma}\left(K\gamma^{ij} - K^{ij}\right)\left(-2NK_{ij} + D_i\beta_j + D_j\beta_i\right) - N\sqrt{\gamma}(R + K_{ij}K^{ij} - K^2) \\
&= \sqrt{\gamma}\left[-N(R + K^2 - K_{ij}K^{ij}) + 2\left(K\gamma^j{}_i - K^j{}_i\right)D_j\beta^i\right] \\
&= -\sqrt{\gamma}\left[N(R + K^2 - K_{ij}K^{ij}) + 2\beta^i\left(D_iK - D_jK^j{}_i\right)\right] \\
&\quad + 2\sqrt{\gamma}D_j\left(K\beta^j - K^j{}_i\beta^i\right).
\end{aligned} \tag{5.109}$$

The corresponding Hamiltonian is

$$H = \int_{\Sigma_t} \mathscr{H}\, \mathrm{d}^3x. \tag{5.110}$$

Noticing that the last term in Eq. (5.109) is a divergence and therefore does not contribute to the integral, we get

$$\boxed{H = -\int_{\Sigma_t}\left(NC_0 - 2\beta^iC_i\right)\sqrt{\gamma}\,\mathrm{d}^3x}, \tag{5.111}$$

where

$$C_0 := R + K^2 - K_{ij}K^{ij}, \tag{5.112}$$

$$C_i := D_jK^j{}_i - D_iK \tag{5.113}$$

are the left-hand sides of the constraint equations (5.70) and (5.71) respectively.

The quantity H defined by (5.111) is called the ***ADM Hamiltonian***; it is a functional of the configuration variables $(\gamma_{ij}, N, \beta^i)$ and their conjugate momenta $(\pi^{ij}, \pi^N, \pi^\beta_i)$, the last two ones being identically zero since

$$\pi^N := \frac{\partial L}{\partial \dot{N}} = 0 \quad \text{and} \quad \pi_i^\beta := \frac{\partial L}{\partial \dot{\beta}^i} = 0. \tag{5.114}$$

The scalar curvature R which appears in H via C_0 is a function of γ_{ij} and its spatial derivatives, via Eqs. (5.77)–(5.79), whereas K_{ij} which appears in both C_0 and C_i is a function of γ_{ij} and π^{ij}, obtained by "inverting" relation (5.107):

$$K_{ij} = K_{ij}[\boldsymbol{\gamma}, \boldsymbol{\pi}] = \frac{1}{\sqrt{\gamma}} \left(\frac{1}{2} \gamma_{kl} \pi^{kl} \gamma_{ij} - \gamma_{ik} \gamma_{jl} \pi^{kl} \right). \tag{5.115}$$

The minimization of the Hilbert action is equivalent to the Hamilton equations

$$\frac{\delta H}{\delta \pi^{ij}} = \dot{\gamma}_{ij} \tag{5.116}$$

$$\frac{\delta H}{\delta \gamma_{ij}} = -\dot{\pi}^{ij} \tag{5.117}$$

$$\frac{\delta H}{\delta N} = -\dot{\pi}^N = 0 \tag{5.118}$$

$$\frac{\delta H}{\delta \beta^i} = -\dot{\pi}_i^\beta = 0. \tag{5.119}$$

Computing the functional derivatives from the expression (5.111) of H leads to the equations

$$\frac{\delta H}{\delta \pi^{ij}} = -2NK_{ij} + D_i\beta_j + D_j\beta_i = \dot{\gamma}_{ij} \tag{5.120}$$

$$\frac{\delta H}{\delta \gamma_{ij}} = -\dot{\pi}^{ij} \tag{5.121}$$

$$\frac{\delta H}{\delta N} = -C_0 = 0 \tag{5.122}$$

$$\frac{\delta H}{\delta \beta^i} = 2C_i = 0. \tag{5.123}$$

Equation (5.120) is nothing but the first equation of the 3+1 Einstein system (5.68)–(5.71). We do not perform the computation of the variation (5.121) but the explicit calculation (see e.g. Sect. 4.2.7 of Ref. [29]) yields an equation which is equivalent to the dynamical Einstein equation (5.69). Finally, Eq. (5.122) is the Hamiltonian constraint (5.70) with $E = 0$ (vacuum) and Eq. (5.123) is the momentum constraint (5.71) with $p_i = 0$.

Equations (5.122) and (5.123) show that in the ADM Hamiltonian approach, the lapse function and the shift vector turn out to be Lagrange multipliers to enforce respectively the Hamiltonian constraint and the momentum constraint, the true dynamical variables being γ_{ij} and π^{ij}.

References

1. Wheeler, J.A.: Geometrodynamics and the issue of the final state. In: DeWitt, C., DeWitt, B.S. (eds) Relativity Groups and Topology, pp. 316. Gordon and Breach, New York (1964)
2. Fourés-Bruhat, Y., Choquet-Bruhat, Y.: Sur l'Intégration des Équations de la Relativité Générale. J. Ration. Mech. Anal. **5**, 951 (1956)
3. Darmois, G.: Les équations de la gravitation einsteinienne, Mémorial des Sciences Mathématiques **25**, Gauthier-Villars, Paris (1927)
4. Lichnerowicz, A.: Sur certains problèmes globaux relatifs au système des équations d'Einstein. Hermann, Paris; Actual. Sci. Ind. **833**, (1939)
5. Lichnerowicz, A.: L'intégration des équations de la gravitation relativiste et le problème des n corps. J. Math. Pures Appl. **23**, 37 (1944); reprinted in Lichnerowicz, A.: Choix d'œuvres mathématiques, p. 4. Hermann, Paris (1982)
6. Fourés-Bruhat, Y., Choquet-Bruhat, Y.: Sur l'intégration des équations d'Einstein. C. R. Acad. Sci. Paris **226**, 1071 (1948)
7. Arnowitt, R., Deser, S., Misner, C.W.: The dynamics of general relativity. In: Witten, L. (ed.) Gravitation: an introduction to current research, p. 227. Wiley, New York (1962) http://arxiv.org/abs/gr-qc/0405109
8. York, J.W.: Velocities and momenta in an extended elliptic form of the initial value conditions. Nuovo Cim. **B119**, 823 (2004)
9. Anderson, A., York, J.W.: Hamiltonian time evolution for general relativity. Phys. Rev. Lett. **81**, 1154 (1998)
10. Fischer, A.E., Marsden, J.: The Einstein equation of evolution—a geometric approach. J. Math. Phys. **13**, 546 (1972)
11. Fischer, A., Marsden, J.: The initial value problem and the dynamical formulation of general relativity. In: Hawking, S.W., Israel, W. (eds) General Relativity: An Einstein Centenary Survey, pp. 138. Cambridge University Press, Cambridge (1979)
12. Wald, R.M.: General relativity. University of Chicago Press, Chicago (1984)
13. Courant, R., Hilbert, D.: Methods of mathematical physics; vol. II: partial differential equations. Interscience, New York (1962)
14. Knapp, A.M., Walker, E.J., Baumgarte, T.W.: Illustrating stability properties of numerical relativity in electrodynamics. Phys. Rev. D **65**, 064031 (2002)
15. Baumgarte, T.W., Shapiro, S.L.: Numerical relativity and compact binaries. Phys. Rep. **376**, 41 (2003)
16. Fourés-Bruhat, Y., (Choquet-Bruhat, Y).: Théorème d'existence pour certains systèmes d'équations aux dérivées partielles non linéaires. Acta Mathematica **88**, 141 (1952) http://fanfreluche.math.univ-tours.fr/
17. Bartnik, R., Isenberg, J.: The Constraint equations, in Ref. [33], p. 1
18. Choquet-Bruhat, Y., York, J.W.: The Cauchy problem. In: Held, A. (eds) General Relativity and Gravitation one hundred Years after the Birth of Albert Einstein, pp. 99. Plenum Press, New York (1980)
19. Choquet-Bruhat, Y., Geroch, R.: Global aspects of the Cauchy problem in general relativity. Commun. Math. Phys. **14**, 329 (1969)
20. Klainerman, S., Nicoló, F.: On the local and global aspects of the Cauchy problem in general relativity. Class. Quantum Grav. **16**, R73 (1999)
21. Andersson, L.: The global existence problem in general relativity, in Ref. [33], p. 71
22. Rendall, A.D.: Theorems on existence and global dynamics for the Einstein equations. Living Rev. Relativ. **8**, 6 (2005) http://www.livingreviews.org/lrr-2005-6
23. Choquet-Bruhat, Y.: General relativity and Einstein's equations. Oxford University Press, New York (2009)
24. Dirac, P.A.M.: The theory of gravitation in Hamiltonian form. Proc. Roy. Soc. Lond. A **246**, 333 (1958)

25. Dirac, P.A.M.: Fixation of coordinates in the Hamiltonian theory of gravitation. Phys. Rev. **114**, 924 (1959)
26. Deser, S.: Some remarks on Dirac's contributions to general relativity. Int. J. Mod. Phys. A **19S1**, 99 (2004)
27. Regge, T., Teitelboim, C.: Role of surface integrals in the Hamiltonian formulation of general relativity. Ann. Phys. (N.Y.) **88**, 286 (1974)
28. Misner, C.W., Thorne, K.S., Wheeler, J.A.: Gravitation. Freeman, New York (1973)
29. Poisson, E.: A relativist's toolkit, the mathematics of black-hole mechanics. Cambridge University Press, Cambridge (2004)
30. Henneaux, M.: Hamiltonian formalism of general relativity, lectures at Institut Henri Poincaré, Paris (2006), http://www.luth.obspm.fr/IHP06/
31. Schäfer, G.: Equations of motion in the ADM formalism, lectures at Institut Henri Poincaré, Paris (2006), http://www.luth.obspm.fr/IHP06/
32. Deruelle, N.: General relativity: a primer, lectures at Institut Henri Poincaré, Paris (2006), http://www.luth.obspm.fr/IHP06/
33. Chruściel, P.T., Friedrich, H. (eds): The Einstein equations and the large scale behavior of gravitational fields—50 years of the Cauchy problem in general relativity. Birkhäuser Verlag, Basel (2004)

Chapter 6
3+1 Equations for Matter and Electromagnetic Field

Abstract We present the 3+1 treatment of fields other than the gravitational one, describing the matter content of spacetime or the electromagnetic field. After deriving the general laws of energy conservation and momentum conservation from the vanishing of the divergence of the stress-energy tensor, we focus on the particular case of a perfect fluid. We introduce the basic quantities describing the fluid from the point of view of the Eulerian observer and derive the 3+1 versions of the laws of conservation of baryon number and of energy, as well as the relativistic Euler equation for the fluid velocity with respect to the Eulerian observer. We also present the flux-conservative form of these laws, which is at the basis of the so-called high-resolution shock-capturing schemes in numerical relativity. The case of the electromagnetic field is then contemplated. The Maxwell equations are written in 3+1 form, in terms of the electric and magnetic fields, both measured by the Eulerian observer. The final section deals with ideal magnetohydrodynamics. We present both the 3+1 MHD-Euler equation and the system of MHD equations in flux-conservative form.

6.1 Introduction

After having considered mostly the left-hand side of Einstein equation, in this chapter we focus on the right-hand side, namely on the matter represented by its stress-energy tensor $\boldsymbol{T}$. By "matter", we actually mean any kind of non-gravitational field, which is minimally coupled to gravity. This includes the electromagnetic field, which we shall treat in Sect. 6.4. The matter obeys two types of equations. The first one is the vanishing of the spacetime divergence of the stress-energy tensor[1]:

$$\boxed{\boldsymbol{\nabla} \cdot \vec{\boldsymbol{T}} = 0}\,, \tag{6.1}$$

[1] Let us recall that the covariant divergence operator $\boldsymbol{\nabla}\cdot$ has been defined in Sect. 2.4.1 and that $\vec{\boldsymbol{T}}$ stands for the type-(1,1) tensor associated by metric duality to $\boldsymbol{T}$, via Eq. (2.39). Accordingly, in index notation, Eq. (6.1) would be written $\nabla_\mu T^\mu{}_\alpha = 0$.

É. Gourgoulhon, *3+1 Formalism in General Relativity*, Lecture Notes in Physics 846, DOI: 10.1007/978-3-642-24525-1_6, © Springer-Verlag Berlin Heidelberg 2012

which, thanks to the contracted Bianchi identity (2.79), is a consequence of Einstein equation (5.1). The second type of equations are the field equations that must be satisfied independently of the Einstein equation, for instance the Maxwell equations for the electromagnetic field or the conservation of the baryon number.

After having established the general energy and momentum conservation laws from the 3+1 decomposition of Eq. (6.1) (Sect. 6.2), we focus on the specific cases of a perfect fluid (Sect. 6.3), an electromagnetic field (Sect. 6.4) and finally a conducting fluid in the magnetohydrodynamics regime (Sect. 6.5). Other models of matter are discussed in Refs. [1] (collisionless matter and scalar fields), [2] (scalar fields) or [3] (collisionless matter, dissipative fluids).

6.2 Energy and Momentum Conservation

6.2.1 3+1 Decomposition of the 4-Dimensional Equation

Let us replace $\boldsymbol{T}$ in Eq. (6.1) by its 3+1 expression (5.14) in terms of the energy density E, the momentum density $\boldsymbol{p}$ and the stress tensor $\boldsymbol{S}$, all of them as measured by the Eulerian observer. We get successively

$$
\begin{aligned}
&\nabla_\mu T^\mu{}_\alpha = 0,\\
&\nabla_\mu \left(S^\mu{}_\alpha + n^\mu p_\alpha + p^\mu n_\alpha + E n^\mu n_\alpha\right) = 0,\\
&\nabla_\mu S^\mu{}_\alpha - K p_\alpha + n^\mu \nabla_\mu p_\alpha + \nabla_\mu p^\mu n_\alpha - p^\mu K_{\mu\alpha} - K E n_\alpha + E D_\alpha \ln N\\
&\qquad + n^\mu \nabla_\mu E n_\alpha = 0,
\end{aligned}
\tag{6.2}
$$

where we have used Eq. (4.24) to express $\boldsymbol{\nabla}\underline{\boldsymbol{n}}$ in terms of $\boldsymbol{K}$ and $\boldsymbol{D}\ln N$.

6.2.2 Energy Conservation

Let us project Eq. (6.2) along the normal to the hypersurfaces Σ_t, i.e. contract Eq. (6.2) with n^α. We get, since $\boldsymbol{p}$, $\boldsymbol{K}$ and $\boldsymbol{D}\ln N$ are all orthogonal to $\boldsymbol{n}$:

$$
n^\nu \nabla_\mu S^\mu{}_\nu + n^\mu n^\nu \nabla_\mu p_\nu - \nabla_\mu p^\mu + K E - n^\mu \nabla_\mu E = 0. \tag{6.3}
$$

Now, since $\boldsymbol{n}\cdot\boldsymbol{S} = 0$,

$$
n^\nu \nabla_\mu S^\mu{}_\nu = -S^\mu{}_\nu \nabla_\mu n^\nu = S^\mu{}_\nu (K^\nu{}_\mu + D^\nu \ln N n_\mu) = K_{\mu\nu} S^{\mu\nu}, \tag{6.4}
$$

where we have used again Eq. (4.24). Similarly

$$
n^\mu n^\nu \nabla_\mu p_\nu = -p_\nu n^\mu \nabla_\mu n^\nu = -p_\nu D^\nu \ln N. \tag{6.5}
$$

Besides, let us express the 4-dimensional divergence $\nabla_\mu p^\mu$ is terms of the 3-dimensional one, $D_\mu p^\mu$. For any vector $\boldsymbol{v}$ tangent to Σ_t, like $\overrightarrow{\boldsymbol{p}}$, Eq. (3.68) gives

$$\begin{aligned} D_\mu v^\mu &= \gamma^\rho{}_\mu \gamma^\mu{}_\sigma \nabla_\rho v^\sigma = \gamma^\rho{}_\sigma \nabla_\rho v^\sigma = (\delta^\rho{}_\sigma + n^\rho n_\sigma)\nabla_\rho v^\sigma = \nabla_\rho v^\rho - v^\sigma n^\rho \nabla_\rho n_\sigma \\ &= \nabla_\rho v^\rho - v^\sigma D_\sigma \ln N. \end{aligned}$$

Hence the useful relation between the two divergences

$$\boxed{\forall \boldsymbol{v} \in \mathscr{T}(\Sigma_t), \quad \boldsymbol{\nabla} \cdot \boldsymbol{v} = \boldsymbol{D} \cdot \boldsymbol{v} + \boldsymbol{v} \cdot \boldsymbol{D} \ln N}, \tag{6.6}$$

or in terms of components,

$$\forall \boldsymbol{v} \in \mathscr{T}(\Sigma_t), \quad \nabla_\mu v^\mu = D_i v^i + v^i D_i \ln N. \tag{6.7}$$

Applying this relation to $\boldsymbol{v} = \boldsymbol{p}$ and taking into account Eqs. (6.4) and (6.5), Eq. (6.3) becomes

$$\mathscr{L}_{\boldsymbol{n}} E + \boldsymbol{D} \cdot \overrightarrow{\boldsymbol{p}} + 2\overrightarrow{\boldsymbol{p}} \cdot \boldsymbol{D} \ln N - KE - K_{ij} S^{ij} = 0. \tag{6.8}$$

Remark 6.1 We have written the derivative of E along $\boldsymbol{n}$ as a Lie derivative. E being a scalar field, we have of course the alternative expressions

$$\mathscr{L}_{\boldsymbol{n}} E = \nabla_{\boldsymbol{n}} E = \boldsymbol{n} \cdot \boldsymbol{\nabla} E = n^\mu \nabla_\mu E = n^\mu \frac{\partial E}{\partial x^\mu} = \langle \boldsymbol{\nabla} E, \boldsymbol{n} \rangle. \tag{6.9}$$

$\mathscr{L}_{\boldsymbol{n}} E$ is the derivative of E with respect to the proper time of the Eulerian observers: $\mathscr{L}_{\boldsymbol{n}} E = dE/d\tau$, for $\boldsymbol{n}$ is the 4-velocity of these observers. It is easy to let appear the derivative with respect to the coordinate time t instead, thanks to the relation $\boldsymbol{n} = N^{-1}(\boldsymbol{\partial}_t - \boldsymbol{\beta})$ [cf. Eq. (5.29)]:

$$\mathscr{L}_{\boldsymbol{n}} E = \frac{1}{N}\left(\frac{\partial}{\partial t} - \mathscr{L}_{\boldsymbol{\beta}}\right) E.$$

Then

$$\boxed{\left(\frac{\partial}{\partial t} - \mathscr{L}_{\boldsymbol{\beta}}\right) E + N\left(\boldsymbol{D} \cdot \overrightarrow{\boldsymbol{p}} - KE - K_{ij} S^{ij}\right) + 2\overrightarrow{\boldsymbol{p}} \cdot \boldsymbol{D} N = 0}, \tag{6.10}$$

in components:

$$\left(\frac{\partial}{\partial t} - \beta^i \frac{\partial}{\partial x^i}\right) E + N\left(D_i p^i - KE - K_{ij} S^{ij}\right) + 2 p^i D_i N = 0. \tag{6.11}$$

This equation has been obtained by York in the seminal article [4].

6.2.3 Newtonian Limit

As a check, let us consider the Newtonian limit of Eq. (6.10). For this purpose let us assume that the gravitational field is weak and static. It is then always possible to find a coordinate system $(x^\alpha) = (x^0 = ct, x^i)$ such that the metric components take the form (cf. Deruelle's lectures [5])

$$g_{\mu\nu}\mathrm{d}x^\mu \mathrm{d}x^\nu = -(1+2\Phi)\,\mathrm{d}t^2 + (1-2\Phi)f_{ij}\mathrm{d}x^i\mathrm{d}x^j, \tag{6.12}$$

where Φ is the Newtonian gravitational potential (solution of Poisson equation $\Delta\Phi = 4\pi G\rho$) and f_{ij} are the components the flat Euclidean metric $\boldsymbol{f}$ in the 3-dimensional space. For a weak gravitational field (Newtonian limit), $|\Phi| \ll 1$ (in units where the velocity of light is not one, this should read $|\Phi|/c^2 \ll 1$). Comparing Eq. (6.12) with (5.50), we get $N = \sqrt{1+2\Phi} \simeq 1+\Phi$, $\boldsymbol{\beta} = 0$ and $\boldsymbol{\gamma} = (1-2\Phi)\boldsymbol{f}$. From Eq. (5.68), we then obtain immediately that $\boldsymbol{K} = 0$. To summarize:

$$\text{Newtonian limit}: \quad N = 1+\Phi, \quad \boldsymbol{\beta} = 0, \quad \boldsymbol{\gamma} = (1-2\Phi)\boldsymbol{f}, \quad \boldsymbol{K} = 0, \quad |\Phi| \ll 1. \tag{6.13}$$

Notice that the Eulerian observer becomes a Galilean (inertial) observer for he is non-rotating (cf. Remark 4.4, p. 62).

Taking into account the limits (6.13), Eq. (6.10) reduces to

$$\frac{\partial E}{\partial t} + \boldsymbol{D}\cdot\vec{\boldsymbol{p}} = -2\vec{\boldsymbol{p}}\cdot\boldsymbol{D}\Phi. \tag{6.14}$$

Let us denote by $\bar{\boldsymbol{D}}$ the Levi–Civita connection associated with the flat metric $\boldsymbol{f}$. Obviously $\boldsymbol{D}\Phi = \bar{\boldsymbol{D}}\Phi$. On the other side, let us express the divergence $\boldsymbol{D}\cdot\vec{\boldsymbol{p}}$ in terms of the divergence $\bar{\boldsymbol{D}}\cdot\vec{\boldsymbol{p}}$. From Eq. (6.13), we have $\gamma^{ij} = (1-2\Phi)^{-1}f^{ij} \simeq (1+2\Phi)f^{ij}$ as well as the relation $\sqrt{\gamma} = \sqrt{(1-2\Phi)^3 f} \simeq (1-3\Phi)\sqrt{f}$ between the determinants γ and f of respectively (γ_{ij}) and (f_{ij}). Therefore, evaluating the divergence by means of formula (2.65), we get

$$\begin{aligned}
\boldsymbol{D}\cdot\vec{\boldsymbol{p}} &= \frac{1}{\sqrt{\gamma}}\frac{\partial}{\partial x^i}\left(\sqrt{\gamma}p^i\right) = \frac{1}{\sqrt{\gamma}}\frac{\partial}{\partial x^i}\left(\sqrt{\gamma}\gamma^{ij}p_j\right) \\
&\simeq \frac{1}{(1-3\Phi)\sqrt{f}}\frac{\partial}{\partial x^i}\left[(1-3\Phi)\sqrt{f}(1+2\Phi)f^{ij}p_j\right] \\
&\simeq \frac{1}{\sqrt{f}}\frac{\partial}{\partial x^i}\left[(1-\Phi)\sqrt{f}f^{ij}p_j\right] \\
&\simeq \frac{1}{\sqrt{f}}\frac{\partial}{\partial x^i}\left(\sqrt{f}f^{ij}p_j\right) - f^{ij}p_j\frac{\partial\Phi}{\partial x^i} \\
&\simeq \bar{\boldsymbol{D}}\cdot\vec{\boldsymbol{p}} - \vec{\boldsymbol{p}}\cdot\bar{\boldsymbol{D}}\Phi.
\end{aligned} \tag{6.15}$$

Consequently Eq. (6.14) becomes

$$\frac{\partial E}{\partial t} + \bar{\boldsymbol{D}} \cdot \overrightarrow{\boldsymbol{p}} = -\overrightarrow{\boldsymbol{p}} \cdot \bar{\boldsymbol{D}}\Phi. \tag{6.16}$$

This is the standard energy conservation relation in a Galilean frame with the source term $-\overrightarrow{\boldsymbol{p}} \cdot \bar{\boldsymbol{D}}\Phi$. The latter constitutes the density of power provided to the system by the gravitational field; this will be clear in the perfect fluid case, to be discussed below.

Remark 6.2 In the left-hand side of Eq. (6.16), the quantity $\boldsymbol{p}$ plays the role of an *energy flux*, whereas it had been defined in Sect. 5.12 as a *momentum density*. But we have seen that both aspects are equivalent [Eq. (5.10)].

6.2.4 Momentum Conservation

Let us now project Eq. (6.2) onto Σ_t:

$$\gamma^\nu{}_\alpha \nabla_\mu S^\mu{}_\nu - K p_\alpha + \gamma^\nu{}_\alpha n^\mu \nabla_\mu p_\nu - K_{\alpha\mu} p^\mu + E D_\alpha \ln N = 0. \tag{6.17}$$

Now, from relation (3.68),

$$\begin{aligned} D_\mu S^\mu{}_\alpha &= \gamma^\rho{}_\mu \gamma^\mu{}_\sigma \gamma^\nu{}_\alpha \nabla_\rho S^\sigma{}_\nu = \gamma^\rho{}_\sigma \gamma^\nu{}_\alpha \nabla_\rho S^\sigma{}_\nu \\ &= \gamma^\nu{}_\alpha (\delta^\rho{}_\sigma + n^\rho n_\sigma) \nabla_\rho S^\sigma{}_\nu = \gamma^\nu{}_\alpha \big(\nabla_\rho S^\rho{}_\nu - S^\sigma{}_\nu \underbrace{n^\rho \nabla_\rho n_\sigma}_{D_\sigma \ln N}\big) \\ &= \gamma^\nu{}_\alpha \nabla_\mu S^\mu{}_\nu - S^\mu{}_\alpha D_\mu \ln N. \end{aligned} \tag{6.18}$$

Besides

$$\begin{aligned} \gamma^\nu{}_\alpha n^\mu \nabla_\mu p_\nu &= N^{-1} \gamma^\nu{}_\alpha m^\mu \nabla_\mu p_\nu = N^{-1} \gamma^\nu{}_\alpha \left(\mathcal{L}_{\boldsymbol{m}} p_\nu - p_\mu \nabla_\nu m^\mu \right) \\ &= N^{-1} \mathcal{L}_{\boldsymbol{m}} p_\alpha + K_{\alpha\mu} p^\mu, \end{aligned} \tag{6.19}$$

where use has been made of Eqs. (4.36) and (4.26) to get the second line. In view of Eqs. (6.18) and (6.19), Eq. (6.17) becomes

$$\frac{1}{N} \mathcal{L}_{\boldsymbol{m}} p_\alpha + D_\mu S^\mu{}_\alpha + S^\mu{}_\alpha D_\mu \ln N - K p_\alpha + E D_\alpha \ln N = 0$$

Using the property (5.63): $\mathcal{L}_{\boldsymbol{m}} = \partial/\partial t - \mathcal{L}_{\boldsymbol{\beta}}$, we obtain

$$\boxed{\left(\frac{\partial}{\partial t} - \mathcal{L}_{\boldsymbol{\beta}}\right) \boldsymbol{p} + N \boldsymbol{D} \cdot \overrightarrow{\boldsymbol{S}} + \boldsymbol{S} \cdot \overrightarrow{\boldsymbol{D}} N - N K \boldsymbol{p} + E \boldsymbol{D} N = 0}, \tag{6.20}$$

or in components

$$\left(\frac{\partial}{\partial t} - \mathcal{L}_{\boldsymbol{\beta}}\right) p_i + N D_j S^j{}_i + S_{ij} D^j N - N K p_i + E D_i N = 0. \tag{6.21}$$

Again, this equation appears in York's article [4]. Actually York's version [his Eq. (41)] contains an additional term, for it is written for the vector $\vec{p}$ dual to the linear form $\boldsymbol{p}$, and since $\mathscr{L}_{\boldsymbol{m}}\gamma^{ij} \neq 0$, this generates the extra term $p_j \mathscr{L}_{\boldsymbol{m}}\gamma^{ij} = 2NK^{ij}p_j$.

To take the Newtonian limit of Eq. (6.20), we shall consider not only Eq. (6.13), which provides the Newtonian limit of the gravitational field, but in addition the property

$$\text{Newtonian limit :} \quad |S^i{}_j| \ll E, \tag{6.22}$$

which expresses that the matter is not relativistic. Then the Newtonian limit of (6.20) is

$$\frac{\partial \boldsymbol{p}}{\partial t} + \bar{\boldsymbol{D}} \cdot \vec{S} = -E\bar{\boldsymbol{D}}\Phi. \tag{6.23}$$

Note that in relating $\boldsymbol{D} \cdot \vec{S}$ to $\bar{\boldsymbol{D}} \cdot \vec{S}$, there should appear derivatives of Φ, as in Eq. (6.15), but thanks to property (6.22), these terms are negligible in front of $E\bar{\boldsymbol{D}}\Phi$. Equation (6.23) is the standard momentum conservation law, with $-E\bar{\boldsymbol{D}}\Phi$ being the gravitational force density.

6.3 Perfect Fluid

6.3.1 Kinematics

The ***perfect fluid*** model of matter is based on a vector field $\boldsymbol{u}$ which is timelike and unitary: $\boldsymbol{u} \cdot \boldsymbol{u} = -1$. It gives the 4-velocity of the so-called ***fluid particles***. In addition the perfect fluid is characterized by an isotropic pressure in the fluid frame. More precisely, the perfect fluid model is entirely defined by the following stress-energy tensor:

$$\boxed{\boldsymbol{T} = (\rho + P)\underline{\boldsymbol{u}} \otimes \underline{\boldsymbol{u}} + P\boldsymbol{g}}, \tag{6.24}$$

where ρ and P are two scalar fields, representing respectively the matter energy density and the pressure, both measured in the fluid frame (i.e. by an observer who is comoving with the fluid), and $\underline{\boldsymbol{u}}$ is the 1-form associated to the 4-velocity $\boldsymbol{u}$ by the metric tensor $\boldsymbol{g}$ [cf. Eq. (2.35)].

Let us consider a fluid element at point $p \in \Sigma_t$ (cf. Fig. 6.1). Let τ be the Eulerian observer's proper time at p. At the coordinate time $t+\mathrm{d}t$, the fluid element has moved to the point $q \in \Sigma_{t+\mathrm{d}t}$. The date $\tau + \mathrm{d}\tau$ attributed to the event q by the Eulerian observer moving through p is given by the orthogonal projection q' of q onto the worldline of that observer. Indeed, let us recall that the space of simultaneous events (local rest frame) for the Eulerian observer is the space orthogonal to his 4-velocity

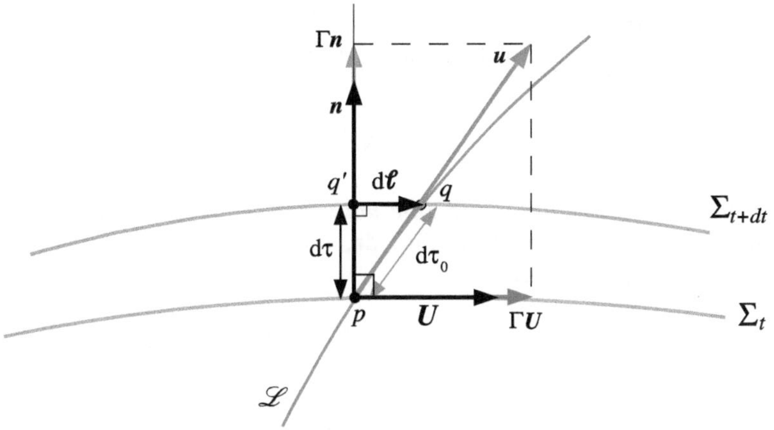

Fig. 6.1 Worldline $\mathscr{L}$ of a fluid element crossing the spacetime foliation $(\Sigma_t)_{t\in\mathbb{R}}$ · $\boldsymbol{u}$ is the fluid 4-velocity and $\boldsymbol{U} = \mathrm{d}\boldsymbol{\ell}/\mathrm{d}\tau$ the relative velocity of the fluid with respect to the Eulerian observer, whose 4-velocity is $\boldsymbol{n}$. The vector $\boldsymbol{U}$ is tangent to Σ_t and enters in the orthogonal decomposition of $\boldsymbol{u}$ with respect to Σ_t, via $\boldsymbol{u} = \Gamma(\boldsymbol{n} + \boldsymbol{U})$. *NB*: contrary to what the figure might suggest, $\mathrm{d}\tau > \mathrm{d}\tau_0$ (do not interpret the figure with the Euclidean metric but with the Lorentzian one !)

$\boldsymbol{u}$, i.e. locally Σ_t (cf. Sect. 4.3.3). Let $\mathrm{d}\boldsymbol{\ell}$ be the infinitesimal vector connecting q' to q. Let $\mathrm{d}\tau_0$ be the increment of the fluid proper time between the events p and q. The ***Lorentz factor*** of the fluid with respect to the Eulerian observer is defined as being the proportionality factor Γ between the proper times $\mathrm{d}\tau_0$ and $\mathrm{d}\tau$:

$$\boxed{\mathrm{d}\tau =: \Gamma \mathrm{d}\tau_0}. \tag{6.25}$$

One has the triangle identity (cf. Fig. 6.1):

$$\mathrm{d}\tau_0 \boldsymbol{u} = \mathrm{d}\tau \boldsymbol{n} + \mathrm{d}\boldsymbol{\ell}. \tag{6.26}$$

Taking the scalar product with $\boldsymbol{n}$ yields

$$\mathrm{d}\tau_0 \boldsymbol{n}\cdot\boldsymbol{u} = \mathrm{d}\tau \underbrace{\boldsymbol{n}\cdot\boldsymbol{n}}_{-1} + \underbrace{\boldsymbol{n}\cdot\mathrm{d}\boldsymbol{\ell}}_{0}, \tag{6.27}$$

hence, using relation (6.25),

$$\boxed{\Gamma = -\boldsymbol{n}\cdot\boldsymbol{u}}. \tag{6.28}$$

From a pure geometrical point of view, the Lorentz factor is thus nothing but minus the scalar product of the two 4-velocities, the fluid's one and the Eulerian observer's one.

Remark 6.3 In the numerical relativity literature, Γ is often denoted W (cf. e.g. [1, 2, 6, 7]).

Using the components n_α of $\underline{\boldsymbol{n}}$ given by Eq. (5.37), Eq. (6.28) gives an expression of the Lorentz factor in terms of the component u^0 of $\boldsymbol{u}$ with respect to the coordinates (t, x^i):

$$\Gamma = N u^0. \tag{6.29}$$

The fluid ***velocity relative to the Eulerian observer*** is defined as the quotient of the displacement $\mathrm{d}\boldsymbol{\ell}$ by the proper time $\mathrm{d}\tau$, both quantities being relative to the Eulerian observer (cf. Fig. 6.1):

$$\boxed{\boldsymbol{U} := \frac{\mathrm{d}\boldsymbol{\ell}}{\mathrm{d}\tau}}. \tag{6.30}$$

Notice that by construction, $\boldsymbol{U}$ is tangent to Σ_t. Dividing the identity (6.26) by $\mathrm{d}\tau$ and making use of Eq. (6.25) results in

$$\boxed{\boldsymbol{u} = \Gamma(\boldsymbol{n} + \boldsymbol{U})}. \tag{6.31}$$

Since $\boldsymbol{n} \cdot \boldsymbol{U} = 0$, the above writing constitutes the orthogonal 3+1 decomposition of the fluid 4-velocity $\boldsymbol{u}$. The normalization relation of the fluid 4-velocity, i.e. $\boldsymbol{u} \cdot \boldsymbol{u} = -1$, combined with Eq. (6.31), results in

$$-1 = \Gamma^2(\underbrace{\boldsymbol{n} \cdot \boldsymbol{n}}_{-1} + 2\underbrace{\boldsymbol{n} \cdot \boldsymbol{U}}_{0} + \boldsymbol{U} \cdot \boldsymbol{U}), \tag{6.32}$$

hence

$$\boxed{\Gamma = (1 - \boldsymbol{U} \cdot \boldsymbol{U})^{-1/2}}. \tag{6.33}$$

Thus, in terms of the velocity $\boldsymbol{U}$, the Lorentz factor is expressed by a formula identical of that of special relativity, except of course that the scalar product in Eq. (6.33) is to be taken with the (curved) metric $\boldsymbol{\gamma}$, whereas in special relativity it is taken with a flat metric.

It is worth to introduce another type of fluid velocity, namely the ***fluid coordinate velocity*** defined by

$$\boxed{\boldsymbol{v} := \frac{\mathrm{d}\boldsymbol{x}}{\mathrm{d}t}}, \tag{6.34}$$

where $\mathrm{d}\boldsymbol{x}$ is the displacement of the fluid worldline with respect to the line of constant spatial coordinates (cf. Fig. 6.2). More precisely, if the fluid moves from the point p of coordinates (t, x^i) to the point q of coordinates $(t + \mathrm{d}t, x^i + \mathrm{d}x^i)$, the fluid coordinate velocity is defined as the vector tangent to Σ_t, the components of which are

$$V^i = \frac{\mathrm{d}x^i}{\mathrm{d}t}. \tag{6.35}$$

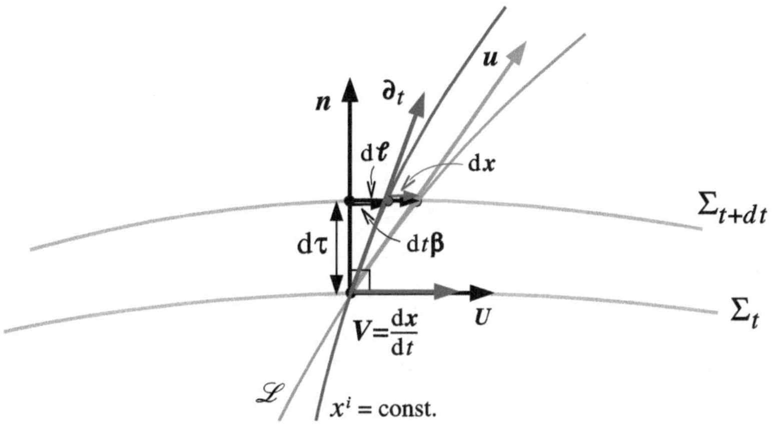

Fig. 6.2 Coordinate velocity $\boldsymbol{v}$ of the fluid defined as the ratio of the fluid displacement with respect to the line of constant spatial coordinates to the coordinate time increment dt

Noticing that the components of the fluid 4-velocity are $u^\alpha = dx^\alpha/d\tau_0$, the above formula can be written

$$V^i = \frac{u^i}{u^0}. \tag{6.36}$$

From the very definition of the shift vector (cf. Sect. 5.2.2), the drift of the coordinate line $x^i = $ const from the Eulerian observer worldline between t and $t+\mathrm{d}t$ is the vector $\mathrm{d}t\boldsymbol{\beta}$. Hence we have (cf. Fig. 6.2)

$$\mathrm{d}\boldsymbol{\ell} = \mathrm{d}t\boldsymbol{\beta} + \mathrm{d}\boldsymbol{x}. \tag{6.37}$$

Dividing this relation by $\mathrm{d}\tau$, using Eqs. (6.30), (4.17) and (6.34) yields

$$\boxed{\boldsymbol{U} = \frac{1}{N}(\boldsymbol{v} + \boldsymbol{\beta})}. \tag{6.38}$$

On this expression, it is clear that at the Newtonian limit as given by (6.13), $\boldsymbol{U} = \boldsymbol{v}$.

6.3.2 Baryon Number Conservation

In addition to $\nabla \cdot \boldsymbol{T} = 0$, the perfect fluid must obey to the law of baryon number conservation:

$$\boxed{\nabla \cdot \boldsymbol{j}_\mathrm{b} = 0}, \tag{6.39}$$

where $\boldsymbol{j}_\mathrm{b}$ is the ***baryon number 4-current***, expressible in terms of the fluid 4-velocity and the fluid ***proper baryon number density*** n_b as

$$\boxed{\boldsymbol{j}_{\rm b} = n_{\rm b}\boldsymbol{u}}. \tag{6.40}$$

The ***baryon number density measured by the Eulerian observer*** is

$$\mathscr{N}_{\rm b} := -\boldsymbol{j}_{\rm b} \cdot \boldsymbol{n}. \tag{6.41}$$

Combining Eqs. (6.28) and (6.40), we get

$$\boxed{\mathscr{N}_{\rm b} = \Gamma n_{\rm b}}. \tag{6.42}$$

This relation is easily interpretable by remembering that $\mathscr{N}_{\rm b}$ and $n_{\rm b}$ are volume densities and invoking the Lorentz–FitzGerald "length contraction" in the direction of motion.

The ***baryon number current measured by the Eulerian observer*** is given by the orthogonal projection of $\boldsymbol{j}_{\rm b}$ onto Σ_t:

$$\boldsymbol{J}_{\rm b} := \vec{\boldsymbol{\gamma}}(\boldsymbol{j}_{\rm b}). \tag{6.43}$$

Taking into account that $\vec{\boldsymbol{\gamma}}(\boldsymbol{u}) = \Gamma \boldsymbol{U}$ [Eq. (6.31)], we get the simple relation

$$\boxed{\boldsymbol{J}_{\rm b} = \mathscr{N}_{\rm b}\boldsymbol{U}}. \tag{6.44}$$

Using the above formulæ, as well as the orthogonal decomposition (6.31) of $\boldsymbol{u}$, the baryon number conservation law (6.39) can be written

$$\begin{aligned}
&\boldsymbol{\nabla} \cdot (n_{\rm b}\boldsymbol{u}) = 0 \\
\Rightarrow &\boldsymbol{\nabla} \cdot [n_{\rm b}\Gamma(\boldsymbol{n} + \boldsymbol{U})] = 0 \\
\Rightarrow &\boldsymbol{\nabla} \cdot [\mathscr{N}_{\rm b}\boldsymbol{n} + \mathscr{N}_{\rm b}\boldsymbol{U}] = 0 \\
\Rightarrow &\boldsymbol{n} \cdot \boldsymbol{\nabla}\mathscr{N}_{\rm b} + \mathscr{N}_{\rm b}\underbrace{\boldsymbol{\nabla} \cdot \boldsymbol{n}}_{-K} + \boldsymbol{\nabla} \cdot (\mathscr{N}_{\rm b}\boldsymbol{U}) = 0
\end{aligned} \tag{6.45}$$

Since $\mathscr{N}_{\rm b}\boldsymbol{U} \in \mathscr{T}(\Sigma_t)$, we may use the divergence formula (6.6) and obtain

$$\mathscr{L}_{\boldsymbol{n}}\mathscr{N}_{\rm b} - K\mathscr{N}_{\rm b} + \boldsymbol{D} \cdot (\mathscr{N}_{\rm b}\boldsymbol{U}) + \mathscr{N}_{\rm b}\boldsymbol{U} \cdot \boldsymbol{D}\ln N = 0, \tag{6.46}$$

where we have written $\boldsymbol{n} \cdot \boldsymbol{\nabla}\mathscr{N}_{\rm b} = \mathscr{L}_{\boldsymbol{n}}\mathscr{N}_{\rm b}$. Since $\boldsymbol{n} = N^{-1}(\boldsymbol{\partial}_t - \boldsymbol{\beta})$ [Eq. (5.29)], we may rewrite the above equation as

$$\boxed{\left(\frac{\partial}{\partial t} - \mathscr{L}_{\boldsymbol{\beta}}\right)\mathscr{N}_{\rm b} + \boldsymbol{D} \cdot (N\mathscr{N}_{\rm b}\boldsymbol{U}) - NK\mathscr{N}_{\rm b} = 0}. \tag{6.47}$$

Using Eq. (6.38), we can put this equation in an alternative form

$$\frac{\partial}{\partial t}\mathscr{N}_{\rm b} + \boldsymbol{D} \cdot (\mathscr{N}_{\rm b}\boldsymbol{v}) + \mathscr{N}_{\rm b}(\boldsymbol{D} \cdot \boldsymbol{\beta} - NK) = 0. \tag{6.48}$$

6.3.3 Dynamical Quantities

The fluid energy density as measured by the Eulerian observer is given by formula (5.4): $E = \boldsymbol{T}(\boldsymbol{n}, \boldsymbol{n})$, with the stress-energy tensor (6.24). Hence $E = (\rho + P)(\boldsymbol{u} \cdot \boldsymbol{n})^2 + P\boldsymbol{g}(\boldsymbol{n}, \boldsymbol{n})$. Since $\boldsymbol{u} \cdot \boldsymbol{n} = -\Gamma$ [Eq. (6.28)] and $\boldsymbol{g}(\boldsymbol{n}, \boldsymbol{n}) = -1$, we get

$$\boxed{E = \Gamma^2(\rho + P) - P}. \tag{6.49}$$

Remark 6.4 For pressureless matter (dust), the above formula reduces to $E = \Gamma^2\rho$. The reader familiar with the formula $E = \Gamma mc^2$ may then be puzzled by the Γ^2 factor in (6.49). However he should remind that E is not an energy, but an energy per unit volume: the extra Γ factor arises from "length contraction" in the direction of motion.

Introducing the proper baryon density n_b, one may decompose the proper energy density ρ in terms of a ***proper rest-mass energy density*** ρ_0 and an ***proper internal energy*** ε_{int} as

$$\rho = \rho_0 + \varepsilon_{\text{int}}, \quad \text{with} \quad \rho_0 := m_b n_b, \tag{6.50}$$

m_b being a constant, namely the mean baryon rest mass ($m_b \simeq 1.66 \times 10^{-27}$ kg). Inserting the above relation into Eq. (6.49) and writing $\Gamma^2\rho = \Gamma\rho + (\Gamma - 1)\Gamma\rho$ leads to the following decomposition of E:

$$E = E_0 + E_{\text{kin}} + E_{\text{int}}, \tag{6.51}$$

with the rest-mass energy density

$$E_0 := m_b \mathcal{N}_b, \tag{6.52}$$

the kinetic energy density

$$E_{\text{kin}} := (\Gamma - 1)E_0 = (\Gamma - 1)m_b\mathcal{N}_b, \tag{6.53}$$

the internal energy density

$$E_{\text{int}} := \Gamma^2(\varepsilon_{\text{int}} + P) - P. \tag{6.54}$$

The three quantities E_0, E_{kin} and E_{int} are relative to the Eulerian observer.

At the Newtonian limit, we shall suppose that the fluid is not relativistic [cf. (6.22)]:

$$P \ll \rho_0, \quad |\varepsilon_{\text{int}}| \ll \rho_0, \quad U^2 := \boldsymbol{U} \cdot \boldsymbol{U} \ll 1. \tag{6.55}$$

Then we get

$$\text{Newtonian limit}: \quad \Gamma \simeq 1+\frac{U^2}{2}, \quad E \simeq E+P \simeq E_0 \simeq \rho_0, \quad E-E_0 \simeq \frac{1}{2}\rho_0 U^2+\varepsilon_{\rm int}. \quad (6.56)$$

The fluid momentum density as measured by the Eulerian observer is obtained by applying formula (5.5):

$$\begin{aligned} \boldsymbol{p} &= -\boldsymbol{T}(\boldsymbol{n}, \vec{\boldsymbol{\gamma}}(.)) = -(\rho + P) \underbrace{\langle \underline{\boldsymbol{u}}, \boldsymbol{n}\rangle}_{-\Gamma} \underbrace{\langle \underline{\boldsymbol{u}}, \vec{\boldsymbol{\gamma}}(.)\rangle}_{\Gamma \underline{U}} - P \underbrace{\boldsymbol{g}(\boldsymbol{n}, \vec{\boldsymbol{\gamma}}(.))}_{0} \\ &= \Gamma^2(\rho + P)\underline{\boldsymbol{U}}, \end{aligned}$$

where Eqs. (6.28) and (6.31) have been used to get the second line. Taking into account Eq. (6.49), the above relation becomes

$$\boxed{\boldsymbol{p} = (E + P)\underline{\boldsymbol{U}}}. \quad (6.57)$$

Finally, by applying formula (5.11), we get the fluid stress tensor with respect to the Eulerian observer:

$$\begin{aligned} \boldsymbol{S} = \vec{\boldsymbol{\gamma}}^*\boldsymbol{T} &= (\rho + P) \underbrace{\vec{\boldsymbol{\gamma}}^* \underline{\boldsymbol{u}}}_{\Gamma \underline{U}} \otimes \underbrace{\vec{\boldsymbol{\gamma}}^* \underline{\boldsymbol{u}}}_{\Gamma \underline{U}} + P \underbrace{\vec{\boldsymbol{\gamma}}^* \boldsymbol{g}}_{\gamma} \\ &= P\boldsymbol{\gamma} + \Gamma^2(\rho + P)\underline{\boldsymbol{U}} \otimes \underline{\boldsymbol{U}}, \end{aligned}$$

or, taking into account Eq. (6.49),

$$\boxed{\boldsymbol{S} = P\boldsymbol{\gamma} + (E + P)\underline{\boldsymbol{U}} \otimes \underline{\boldsymbol{U}}}. \quad (6.58)$$

6.3.4 Energy Conservation Law

By means of Eqs. (6.57) and (6.58), the energy conservation law (6.10) becomes

$$\boxed{\begin{aligned} &\left(\frac{\partial}{\partial t} - \mathscr{L}_{\boldsymbol{\beta}}\right) E + N\{\boldsymbol{D} \cdot [(E + P)\boldsymbol{U}] - (E + P)(K + K_{ij}U^iU^j)\} \\ &+ 2(E + P)\boldsymbol{U} \cdot \boldsymbol{D}N = 0. \end{aligned}} \quad (6.59)$$

To take the Newtonian limit, we may combine the Newtonian limit of the baryon number conservation law (6.47) with Eq. (6.16) to get

$$\frac{\partial E'}{\partial t} + \bar{\boldsymbol{D}} \cdot [(E' + P)\boldsymbol{U}] = -\boldsymbol{U} \cdot (\rho_0 \bar{\boldsymbol{D}}\Phi), \quad (6.60)$$

where $E' := E - E_0 = E_{\rm kin} + E_{\rm int}$ and we clearly recognize in the right-hand side the power provided to a unit volume fluid element by the gravitational force.

6.3.5 Relativistic Euler Equation

Injecting the expressions (6.57) and (6.58) into the momentum conservation law (6.20), we get

$$\left(\frac{\partial}{\partial t} - \mathcal{L}_{\boldsymbol{\beta}}\right) [(E+P)U_i] + ND_j \left[P\delta^j{}_i + (E+P)U^j U_i\right]$$
$$+ \left[P\gamma_{ij} + (E+P)U_i U_j\right] D^j N - NK(E+P)U_i + ED_i N = 0.$$

Expanding and making use of Eq. (6.59) yields

$$\left(\frac{\partial}{\partial t} - \mathcal{L}_{\boldsymbol{\beta}}\right) U_i + NU^j D_j U_i - U^j D_j N U_i + D_i N + NK_{kl} U^k U^l U_i$$
$$+ \frac{1}{E+P}\left[ND_i P + U_i \left(\frac{\partial}{\partial t} - \mathcal{L}_{\boldsymbol{\beta}}\right) P\right] = 0.$$

Now, from Eq. (6.38), $NU^j D_j U_i = V^j D_j U_i + \beta^j D_j U_i$, so that $-\mathcal{L}_{\boldsymbol{\beta}} U_i + NU^j D_j U_i = V^j D_j U_i - U_j D_i \beta^j$ [cf. Eq. (2.91)]. Hence the above equation can be written

$$\boxed{\begin{aligned}\frac{\partial U_i}{\partial t} + V^j D_j U_i = &- \frac{1}{E+P}\left[ND_i P + U_i \left(\frac{\partial P}{\partial t} - \beta^j \frac{\partial P}{\partial x^j}\right)\right] + U_j D_i \beta^j \\ &- D_i N + U_i U^j \left(D_j N - NK_{jk} U^k\right).\end{aligned}} \qquad (6.61)$$

The Newtonian limit of this equation is [cf. Eqs. (6.13) and (6.56)]

$$\frac{\partial U_i}{\partial t} + U^j \bar{D}_j U_i = -\frac{1}{\rho_0} \bar{D}_i P - \bar{D}_i \Phi, \qquad (6.62)$$

i.e. the standard Euler equation in presence of a gravitational field of potential Φ.

Remark 6.5 Usually the general relativistic Euler equation is written in terms of the momentum density $\boldsymbol{p}$, i.e. as (6.20) with $\boldsymbol{p} = (E+P)\boldsymbol{U}$ [Eq. (6.57)]. We have written it here in terms of the fluid velocity with respect to the Eulerian observer, $\boldsymbol{U}$, appealing to the energy conservation law (6.59) to simplify some terms. In this way, the relativistic Euler equation takes a shape which is closer to its Newtonian counterpart (6.62). The version (6.61) has been exhibited in [8] for the particular case of spherically symmetric spacetimes. A form equivalent to (6.61) has been obtained in a tetrad formalism by Salgado [9]. See also [10] for the special relativistic limit.

6.3.6 Flux-Conservative Form

The ***flux-conservative form*** of the relativistic hydrodynamics equations is [6, 11]

$$\boxed{\frac{\partial \mathscr{U}_A}{\partial t} + \frac{\partial}{\partial x^j}\mathscr{F}^j_A = \mathscr{S}_A}, \quad 0 \le A \le 4, \tag{6.63}$$

where (i) $\mathscr{U}_A$ is the ***state vector***, defined by [cf. Eqs. (6.42), (6.57), (6.49) and (6.52)]

$$\mathscr{U}_0 := \sqrt{\gamma}\,\mathscr{N}_\mathrm{b} = \sqrt{\gamma}\,\Gamma n_\mathrm{b} \tag{6.64a}$$

$$\mathscr{U}_i := \sqrt{\gamma}\,p_i = \sqrt{\gamma}(E+P)U_i = \sqrt{\gamma}\,\Gamma^2(\rho+P)U_i, \quad 1 \le i \le 3 \tag{6.64b}$$

$$\mathscr{U}_4 := \sqrt{\gamma}(E - E_0) = \sqrt{\gamma}\left[\Gamma^2(\rho+P) - P - \Gamma m_\mathrm{b} n_\mathrm{b}\right], \tag{6.64c}$$

(ii) $(\mathscr{F}^j_A)_{1\le j\le 3}$ are the components of the Ath ***flux vector*** and (iii) $\mathscr{S}_A$ is the ***source vector***. $\mathscr{F}^i_A$ and $\mathscr{S}_A$ will be made explicit below. The form (6.63) is clearly that of a *conservation law* in flat spacetime with Cartesian coordinates. This form is at the basis of the so-called *high-resolution shock-capturing* (*HRSC*) *schemes* introduced in numerical relativity by Martí et al. in 1991 [12] (spherically symmetric case) and Banuyls et al. in 1997 [13] (3D case). HRSC schemes are the most powerful numerical methods to date (see [6] for a review). They are devised from Godunov's idea [14] to reduce the problem of numerical integration of the hydrodynamics equation to local Riemann shock tube problems.

To show that the 3+1 relativistic hydrodynamics equations presented above can indeed be recast as (6.63), let us start from the baryon number conservation law (6.48). We shall first transform the term $\boldsymbol{D}\cdot\boldsymbol{\beta} - NK$ which appears in this equation as follows. By taking the trace of Eq. (5.68) with $\mathscr{L}_{\boldsymbol{\beta}}\gamma_{ij}$ expressed via Eq. (5.67), we obtain

$$\gamma^{ij}\left(\frac{\partial \gamma_{ij}}{\partial t} - D_i\beta_j - D_j\beta_i\right) = -2N\underbrace{\gamma^{ij}K_{ij}}_{K},$$

i.e.

$$D_j\beta^j - NK = \frac{1}{2}\gamma^{ij}\frac{\partial \gamma_{ij}}{\partial t}. \tag{6.65}$$

Recalling that (γ^{ij}) is the inverse matrix of (γ_{ij}), the general rule (2.64) for the variation of a determinant yields the useful relation

$$\boxed{\boldsymbol{D}\cdot\boldsymbol{\beta} - NK = \frac{1}{\sqrt{\gamma}}\frac{\partial\sqrt{\gamma}}{\partial t}}. \tag{6.66}$$

On the other side, by means of the divergence formula (2.65), we have

$$\boldsymbol{D} \cdot (\mathcal{N}_{\mathrm{b}} \boldsymbol{v}) = \frac{1}{\sqrt{\gamma}} \frac{\partial}{\partial x^j} \left(\sqrt{\gamma} \mathcal{N}_{\mathrm{b}} V^j \right).$$

Accordingly, the baryon number conservation law (6.48) can be written

$$\boxed{\frac{\partial \mathcal{D}}{\partial t} + \frac{\partial}{\partial x^j} \left(\mathcal{D} V^j \right) = 0}, \tag{6.67}$$

with

$$\mathcal{D} := \sqrt{\gamma} \mathcal{N}_{\mathrm{b}} = \sqrt{\gamma} \Gamma n_{\mathrm{b}}. \tag{6.68}$$

The form (6.67), although fully relativistic, is identical to the continuity equation in Newtonian physics. The price to pay is that this equation is not covariant, for $\mathcal{D}$ is not a scalar field independent of the coordinates (x^i), contrary to $\mathcal{N}_{\mathrm{b}}$ (recall that the determinant γ which relates $\mathcal{D}$ to $\mathcal{N}_{\mathrm{b}}$ does depend on the coordinates). Equation (6.67) was first exhibited by Wilson [15, 16].

Let us now turn to the momentum conservation equation. For our purpose, it is more convenient to start from the general form (6.21) than from the Euler form (6.61). Expressing the Lie derivative in (6.21) according to $\mathcal{L}_{\beta} p_i = \beta^j D_j p_i + p_j D_i \beta^j$ [formula (2.92) with $(k, \ell) = (0, 1)$ and $\nabla = \boldsymbol{D}$], we get

$$\frac{\partial p_i}{\partial t} + D_j (N S^j{}_i - \beta^j p_i) + (D_j \beta^j - NK) p_i = p_j D_i \beta^j - E D_i N. \tag{6.69}$$

Again, we can use (6.66) to replace $D_j \beta^j - NK$. In addition, for any type-(1, 1) tensor $\boldsymbol{Q}$, formulas (2.54) and (2.63) lead to:

$$D_j Q^j{}_i = \frac{\partial}{\partial x^j} Q^j{}_i + \underbrace{\Gamma^j{}_{kj}}_{\frac{\partial}{\partial x^k} \ln \sqrt{\gamma}} Q^k_i - \Gamma^k{}_{ij} Q^j{}_k = \frac{1}{\sqrt{\gamma}} \frac{\partial}{\partial x^j} \left(\sqrt{\gamma} Q^j{}_i \right) - \Gamma^k{}_{ij} Q^j{}_k.$$

Accordingly, Eq. (6.69) becomes

$$\boxed{\begin{aligned} \frac{\partial}{\partial t} (\sqrt{\gamma} p_i) + \frac{\partial}{\partial x^j} \left[\sqrt{\gamma} \left(N S^j{}_i - \beta^j p_i \right) \right] = \sqrt{\gamma} \Big[& p_j D_i \beta^j - E D_i N \\ & + \Gamma^k{}_{ij} \left(N S^j{}_k - \beta^j p_k \right) \Big]. \end{aligned}} \tag{6.70}$$

In the case of a perfect fluid, we have from Eqs. (6.58), (6.57) and (6.38)

$$N S^j{}_i - \beta^j p_i = N P \delta^j{}_i + N U^j p_i - \beta^j p_i = N P \delta^j{}_i + V^j p_i.$$

Hence the general law (6.70) becomes

$$\boxed{\begin{aligned} \frac{\partial}{\partial t} (\sqrt{\gamma} p_i) + \frac{\partial}{\partial x^j} \left[\sqrt{\gamma} \left(N P \delta^j{}_i + V^j p_i \right) \right] = \sqrt{\gamma} \Big[& p_j D_i \beta^j - E D_i N \\ & + \Gamma^k{}_{ij} \left(N P \delta^j{}_k + V^j p_k \right) \Big]. \end{aligned}} \tag{6.71}$$

Finally let us consider the energy conservation law, in the general form (6.10). Writing $\mathcal{L}_{\boldsymbol{\beta}} E = \beta^j D_j E = D_j(E\beta^j) - ED_j\beta^j$, we get

$$\frac{\partial E}{\partial t} + D_j(Np^j - E\beta^j) + E(D_j\beta^j - NK) = NK_{jk}S^{jk} - p^j D_j N.$$

Using the identities (2.65) and (6.66), this equation can be recast as

$$\boxed{\frac{\partial}{\partial t}\left(\sqrt{\gamma}E\right) + \frac{\partial}{\partial x^j}\left[\sqrt{\gamma}(Np^j - E\beta^j)\right] = \sqrt{\gamma}\left(NK_{jk}S^{jk} - p^j D_j N\right)}. \tag{6.72}$$

In the special case of a perfect fluid, Eqs. (6.57), (6.38) and (6.58) yield

$$Np^j - E\beta^j = EV^j + NPU^j$$
$$NK_{jk}S^{jk} - p^j D_j N = N\left[KP + p^j(K_{jk}U^k - D_j \mathrm{ln} N)\right].$$

Moreover, noticing that $\sqrt{\gamma}E_0 = m_b\mathcal{D}$ [cf. Eqs. (6.52) and (6.68)], we may combine Eq. (6.72) with (6.67) to write

$$\boxed{\begin{aligned}\frac{\partial}{\partial t}\left[\sqrt{\gamma}(E - E_0)\right] + \frac{\partial}{\partial x^j}\left\{\sqrt{\gamma}\left[(E - E_0)V^j + NPU^j\right]\right\} \\ = \sqrt{\gamma}N\left[KP + p^j(K_{jk}U^k - D_j \ln N)\right].\end{aligned}} \tag{6.73}$$

Collecting Eqs. (6.67), (6.71) and (6.73), we have proved that the equations of 3+1 hydrodynamics can be put in the form (6.63) with the state vector defined by (6.64), the flux vectors

$$\mathcal{F}^j_0 = \mathcal{D}V^j \tag{6.74a}$$

$$\mathcal{F}^j_i = \sqrt{\gamma}\left[NP\delta^j{}_i + p_i V^j\right], \quad 1 \le i \le 3 \tag{6.74b}$$

$$\mathcal{F}^j_4 = \sqrt{\gamma}\left[(E - E_0)V^j + NPU^j\right] \tag{6.74c}$$

and the source vector

$$\mathcal{S}_0 = 0 \tag{6.75a}$$

$$\mathcal{S}_i = \sqrt{\gamma}\left\{p_j D_i\beta^j - ED_iN + \Gamma^k{}_{ij}\left[NP\delta^j{}_k + p_k V^j\right]\right\}, \quad 1 \le i \le 3 \tag{6.75b}$$

$$\mathcal{S}_4 = \sqrt{\gamma}N\left[KP + p^j(K_{jk}U^k - D_j \ln N)\right]. \tag{6.75c}$$

Note that the source vector does not involve any derivative of the matter quantities.

Remark 6.6 The source vector $\mathscr{S}_A$ is very often presented in terms of 4-dimensional quantities, i.e. the components $T_{\alpha\beta}$ of the stress-energy tensor and the Christoffel symbols ${}^4\Gamma^{\alpha}_{\beta\gamma}$ of the 4-metric $\boldsymbol{g}$ (e.g. [1, 6, 7, 13]; an exception is [2]). We have written it in terms of 3+1 quantities only; in particular in Eq. (6.75b), the $\Gamma^k{}_{ij}$'s are the Christoffel symbols of the 3-metric $\boldsymbol{\gamma}$ [cf. Eq. (5.79)].

All modern codes for general relativistic hydrodynamics are using HRSC schemes and hence are based on the flux-conservative form (6.63); this regards the codes Whisky [17–19], SACRA [20] (see [21] for a comparison between Whisky and SACRA), Shibata's code [22], CoCoNuT [23], SpEC [24, 25] and BAM-GRHD [26] (see [6] or [27] for a more complete list of codes).

6.3.7 Further Developments

For further developments in 3+1 relativistic hydrodynamics, we refer to the review article by Font [6]. One may recommend also the reading of Chap. 7 of Alcubierre's book [7] for the hyperbolicity analysis and the determination of the speed of sound. As a final remark, let us point out that the 3+1 decomposition presented above is definitively well adapted to the numerical integration but is not very convenient for discussing conservation laws, such as the relativistic generalizations of Bernoulli's theorem or Kelvin's circulation theorem. For this purpose the Carter–Lichnerowicz approach, which is based on exterior calculus (cf. Remark 2.16), is much more powerful, as discussed in Ref. [28].

6.4 Electromagnetism

6.4.1 Electromagnetic Field

The ***electromagnetic field*** is represented by a 2-form $\boldsymbol{F}$ in the spacetime $(\mathscr{M}, \boldsymbol{g})$, i.e. by a valence-2 tensor field that is twice covariant and antisymmetric (cf. Sect. 2.2.5). In other words, at every point $p \in \mathscr{M}$, $\boldsymbol{F}$ is an antisymmetric bilinear form on the tangent space $\mathscr{T}_p(\mathscr{M})$. $\boldsymbol{F}$ is sometimes called the ***Faraday tensor*** [29]. The physical interpretation of $\boldsymbol{F}$ relies on the ***Lorentz force***: a particle of rest mass m, electric charge q and 4-velocity $\boldsymbol{u}$ is subject to the 4-acceleration $\boldsymbol{\nabla}_{\boldsymbol{u}}\boldsymbol{u}$ obeying

$$m\boldsymbol{\nabla}_{\boldsymbol{u}}\underline{\boldsymbol{u}} = q\boldsymbol{F}(., \boldsymbol{u}). \tag{6.76}$$

Of course, if $q = 0$, this equation reduces to $\boldsymbol{\nabla}_{\boldsymbol{u}}\underline{\boldsymbol{u}} = 0$, which implies that the particle follows a spacetime geodesic.

The ***electric field*** and the ***magnetic field*** measured by the Eulerian observer are respectively the vector field $\boldsymbol{E}$ and the vector field $\boldsymbol{B}$ defined in terms of $\boldsymbol{F}$ and the

4-velocity $\boldsymbol{n}$ of the Eulerian observer according to the formulas

$$\underline{\boldsymbol{E}} = \boldsymbol{F}(.,\boldsymbol{n}) \tag{6.77a}$$
$$\underline{\boldsymbol{B}} = \star\boldsymbol{F}(\boldsymbol{n},.). \tag{6.77b}$$

In the second formula, $\star\boldsymbol{F}$ is the ***Hodge dual*** of $\boldsymbol{F}$, i.e. the 2-form defined by

$$\star F_{\alpha\beta} := \frac{1}{2}{}^4\varepsilon^{\mu\nu}{}_{\alpha\beta} F_{\mu\nu}, \tag{6.78}$$

where ${}^4\boldsymbol{\varepsilon}$ is the spacetime Levi–Civita tensor (cf. Sect. 2.3.4). We assume indeed that the spacetime $\mathscr{M}$ is an orientable manifold and that we have chosen some orientation (cf. Sect. 2.3.4). The Levi–Civita tensor ${}^4\boldsymbol{\varepsilon}$ is such that for any $\boldsymbol{g}$-orthonormal basis $(\boldsymbol{e}_\alpha)$,

$${}^4\boldsymbol{\varepsilon}(\boldsymbol{e}_0, \boldsymbol{e}_1, \boldsymbol{e}_2, \boldsymbol{e}_3) = \pm 1 \tag{6.79}$$

[cf. Eq. (2.43)].

By construction, since $\boldsymbol{F}(\boldsymbol{n},\boldsymbol{n}) = 0$ and $\star\boldsymbol{F}(\boldsymbol{n},\boldsymbol{n}) = 0$,

$$\boldsymbol{n}\cdot\boldsymbol{E} = 0 \quad \text{and} \quad \boldsymbol{n}\cdot\boldsymbol{B} = 0, \tag{6.80}$$

i.e. the vectors $\boldsymbol{E}$ and $\boldsymbol{B}$ are tangent to the hypersurfaces Σ_t.

From the knowledge of $\boldsymbol{n}$, $\boldsymbol{E}$ and $\boldsymbol{B}$, we can reconstruct the electromagnetic field tensor according to

$$\boxed{\boldsymbol{F} = \underline{\boldsymbol{n}} \otimes \underline{\boldsymbol{E}} - \underline{\boldsymbol{E}} \otimes \underline{\boldsymbol{n}} + {}^4\boldsymbol{\varepsilon}(\boldsymbol{n}, \boldsymbol{B}, ., .)}, \tag{6.81}$$

or in index notation,

$$F_{\alpha\beta} = n_\alpha E_\beta - E_\alpha n_\beta + {}^4\varepsilon_{\mu\nu\alpha\beta} n^\mu B^\nu. \tag{6.82}$$

Actually (6.81) is the unique 3+1 decomposition of an *antisymmetric* rank-2 tensor, as (5.14) was the unique 3+1 decomposition of a *symmetric* rank-2 tensor.

Remark 6.7 The electric field $\boldsymbol{E}$ has been defined here as a vector, whereas it would have been more natural to define it as a linear form: this explains why an underbar (metric duality) appears under $\boldsymbol{E}$ in Eq. (6.81). On the contrary, the magnetic field is naturally a vector, hence the absence of underbar on $\boldsymbol{B}$ in Eq. (6.81). In the context of the 3+1 formalism, we have however preferred the "vector" definition for $\boldsymbol{E}$. Among other things, it puts $\boldsymbol{E}$ on the same footing as $\boldsymbol{B}$.

The 3+1 decomposition of the Hodge dual $\star\boldsymbol{F}$ is

$$\star\boldsymbol{F} = -\underline{\boldsymbol{n}} \otimes \underline{\boldsymbol{B}} + \underline{\boldsymbol{B}} \otimes \underline{\boldsymbol{n}} + {}^4\boldsymbol{\varepsilon}(\boldsymbol{n}, \boldsymbol{E}, ., .). \tag{6.83}$$

Note that it can deduced from (6.81) by replacing $\boldsymbol{E}$ by $-\boldsymbol{B}$ and $\boldsymbol{B}$ by $\boldsymbol{E}$. It is easy to establish (6.83) by plugging (6.82) into the definition (6.78) and using the identity (see e.g. Appendix B of [30])

$${}^4\varepsilon^{\mu\nu\alpha\beta}\,{}^4\varepsilon_{\mu\nu\kappa\lambda} = -2\left(\delta^\alpha{}_\kappa \delta^\beta{}_\lambda - \delta^\beta{}_\kappa \delta^\alpha{}_\lambda\right).$$

6.4.2 3+1 Maxwell Equations

The electromagnetic field $\boldsymbol{F}$ is ruled by Maxwell equations; they are best expressed in terms of the exterior derivative operator **d** mentioned in Remark 2.16 (see e.g. [10]). However, for the purpose of the 3+1 formalism, we shall express them in terms of divergence with respect to the spacetime connection $\boldsymbol{\nabla}$ [cf. Eq. (2.58)]. The ***Maxwell equations*** are then

$$\boxed{\boldsymbol{\nabla}\cdot\star\overrightarrow{\boldsymbol{F}}=0} \tag{6.84a}$$

$$\boxed{\boldsymbol{\nabla}\cdot\overrightarrow{\boldsymbol{F}}=\mu_0\boldsymbol{j}}, \tag{6.84b}$$

where $\star\overrightarrow{\boldsymbol{F}}$ and $\overrightarrow{\boldsymbol{F}}$ are the twice-contravariant tensor fields associated to $\star\boldsymbol{F}$ and $\boldsymbol{F}$, respectively, by metric duality [cf. Eq. (2.40)], the constant μ_0 is the magnetic permeability of vacuum and the vector field $\boldsymbol{j}$ is the ***electric 4-current***. In term of components, Eqs. (6.84a) and (6.84b) are written (cf. Remark 2.12)

$$\nabla_\mu \star F^{\alpha\mu}=0 \tag{6.85a}$$

$$\nabla_\mu F^{\alpha\mu}=\mu_0 j^\alpha. \tag{6.85b}$$

The 3+1 decomposition of $\boldsymbol{j}$ is

$$\boldsymbol{j}=\rho_{\rm e}\boldsymbol{n}+\boldsymbol{J}, \qquad \text{with} \quad \boldsymbol{n}\cdot\boldsymbol{J}=0. \tag{6.86}$$

The scalar field $\rho_{\rm e}$ is the ***electric charge density*** with respect to the Eulerian observer and is given by

$$\rho_{\rm e}=-\boldsymbol{n}\cdot\boldsymbol{j}. \tag{6.87}$$

The vector $\boldsymbol{J}$, tangent to Σ_t, is the ***electric current*** measured by the Eulerian observer; it is nothing but the orthogonal projection of $\boldsymbol{j}$ onto Σ_t:

$$\boldsymbol{J}=\overrightarrow{\boldsymbol{\gamma}}(\boldsymbol{j}). \tag{6.88}$$

Let us perform the 3+1 split of the source-free Maxwell equation (6.84a). Replacing $\star\boldsymbol{F}$ by its 3+1 expression (6.83), we may write it as

$$\nabla_\mu\left(-n^\alpha B^\mu+B^\alpha n^\mu+{}^4\!\varepsilon^{\rho\sigma\alpha\mu}n_\rho E_\sigma\right)=0. \tag{6.89}$$

Let us compute separately the $\boldsymbol{B}$ and $\boldsymbol{E}$ terms. We have

$$\nabla_\mu\left(-n^\alpha B^\mu+B^\alpha n^\mu\right)=\underbrace{n^\mu\nabla_\mu B^\alpha-B^\mu\nabla_\mu n^\alpha}_{\mathcal{L}_{\boldsymbol{n}}B^\alpha}-\nabla_\mu B^\mu n^\alpha+\underbrace{\nabla_\mu n^\mu}_{-K}B^\alpha,$$

where we have let appear the Lie derivative of $\boldsymbol{B}$ along $\boldsymbol{n}$ [cf. Eq. (2.89)] and have expressed the divergence of $\boldsymbol{n}$ in terms of the trace of the extrinsic curvature tensor [Eq. (3.66)]. Introducing the normal evolution vector $\boldsymbol{m} = N\boldsymbol{n}$ [Eq. (4.9)], we can write

$$\mathcal{L}_{\boldsymbol{n}} B^\alpha = \frac{1}{N}\left(\mathcal{L}_{\boldsymbol{m}} B^\alpha + B^\mu \nabla_\mu N n^\alpha\right),$$

so that

$$\begin{aligned}\nabla_\mu\left(-n^\alpha B^\mu + B^\alpha n^\mu\right) &= \frac{1}{N}\mathcal{L}_{\boldsymbol{m}} B^\alpha - K B^\alpha - (\nabla_\mu B^\mu - B^\mu D_\mu \ln N) n^\alpha \\ &= \frac{1}{N}\mathcal{L}_{\boldsymbol{m}} B^\alpha - K B^\alpha - D_\mu B^\mu n^\alpha, \end{aligned} \qquad (6.90)$$

where we have used the identity $B^\mu \nabla_\mu N = B^\mu D_\mu N$ ($\boldsymbol{B}$ being tangent to Σ_t) and the property (6.6). Let us now turn to the second part in (6.89). Thanks to the vanishing of the covariant derivative of the Levi–Civita tensor [property (2.61)], it can be written

$$\begin{aligned}\nabla_\mu\left({}^4\varepsilon^{\rho\sigma\alpha\mu} n_\rho E_\sigma\right) &= {}^4\varepsilon^{\rho\sigma\alpha\mu}\nabla_\mu n_\rho E_\sigma + {}^4\varepsilon^{\rho\sigma\alpha\mu} n_\rho \nabla_\mu E_\sigma \\ &= -(\underbrace{{}^4\varepsilon^{\rho\sigma\alpha\mu} K_{\mu\rho}}_{0} + {}^4\varepsilon^{\rho\sigma\alpha\mu} D_\rho \ln N n_\mu) E_\sigma + {}^4\varepsilon^{\rho\sigma\alpha\mu} n_\rho \nabla_\mu E_\sigma \\ &= -{}^4\varepsilon^{\rho\sigma\alpha\mu} D_\rho \ln N n_\mu E_\sigma + {}^4\varepsilon^{\rho\sigma\alpha\mu} n_\rho \nabla_\mu E_\sigma, \end{aligned} \qquad (6.91)$$

where use has been made of (4.24) and the vanishing of ${}^4\varepsilon^{\rho\sigma\alpha\mu} K_{\mu\rho}$ stems from the fact that ${}^4\boldsymbol{\varepsilon}$ is antisymmetric whereas $\boldsymbol{K}$ is symmetric. Now

$${}^4\varepsilon^{\rho\sigma\alpha\mu} n_\rho \nabla_\mu E_\sigma = {}^4\varepsilon^{\rho\sigma\alpha\mu} n_\rho (\nabla \underline{E})_{\sigma\mu} = {}^4\varepsilon^{\rho\sigma\alpha\mu} n_\rho (\vec{\boldsymbol{\gamma}}^* \nabla \underline{E})_{\sigma\mu},$$

since any part of $\nabla \underline{E}$ along $\underline{\boldsymbol{n}}$ is annihilated by ${}^4\varepsilon^{\rho\sigma\alpha\mu} n_\rho$, due to the alternate character of ${}^4\varepsilon^{\rho\sigma\alpha\mu}$. Thanks to (3.67), $\vec{\boldsymbol{\gamma}}^* \nabla \underline{E} = \boldsymbol{D}\underline{E}$, so that

$${}^4\varepsilon^{\rho\sigma\alpha\mu} n_\rho \nabla_\mu E_\sigma = {}^4\varepsilon^{\rho\sigma\alpha\mu} n_\rho D_\mu E_\sigma = {}^4\varepsilon^{\mu\sigma\alpha\rho} n_\mu D_\rho E_\sigma = {}^4\varepsilon^{\mu\alpha\rho\sigma} n_\mu D_\rho E_\sigma.$$

Accordingly, Eq.(6.91) becomes (noticing that $-{}^4\varepsilon^{\rho\sigma\alpha\mu} = {}^4\varepsilon^{\mu\alpha\rho\sigma}$)

$$\nabla_\mu\left({}^4\varepsilon^{\rho\sigma\alpha\mu} n_\rho E_\sigma\right) = \frac{1}{N} n_\mu {}^4\varepsilon^{\mu\alpha\rho\sigma} D_\rho (N E_\sigma). \qquad (6.92)$$

In this relation appears the Levi–Civita tensor $\boldsymbol{\varepsilon}$ of the Riemannian manifold $(\Sigma_t, \boldsymbol{\gamma})$, with the orientation induced by that of $\mathcal{M}$:

$$\boxed{\boldsymbol{\varepsilon} = {}^4\boldsymbol{\varepsilon}(\boldsymbol{n}, ., ., .).} \qquad (6.93)$$

Indeed the above relation defines a 3-form and if $(\boldsymbol{e}_i)$ is an orthonormal basis of $(\Sigma_t, \boldsymbol{\gamma})$, then $(\boldsymbol{n}, \boldsymbol{e}_i)$ is an orthonormal basis of $(\mathcal{M}, \boldsymbol{g})$, so that (6.79) implies

$$\boldsymbol{\varepsilon}(\boldsymbol{e}_1, \boldsymbol{e}_2, \boldsymbol{e}_3) = {}^4\boldsymbol{\varepsilon}(\boldsymbol{n}, \boldsymbol{e}_1, \boldsymbol{e}_2, \boldsymbol{e}_3) = \pm 1.$$

The above relation shows that $\boldsymbol{\varepsilon}$ is indeed the Levi–Civita tensor of $(\Sigma_t, \boldsymbol{\gamma})$ (cf. Sect. 2.3.4). We thus rewrite Eq. (6.92) as

$$\nabla_\mu \left({}^4\varepsilon^{\rho\sigma\alpha\mu} n_\rho E_\sigma \right) = \frac{1}{N} \varepsilon^{\alpha\rho\sigma} D_\rho (N E_\sigma). \tag{6.94}$$

Collecting the results (6.90) and (6.94), the first Maxwell equation (6.89) becomes

$$\mathscr{L}_{\boldsymbol{m}} B^\alpha - NKB^\alpha + \varepsilon^{\alpha\rho\sigma} D_\rho (NE_\sigma) - ND_\mu B^\mu n^\alpha = 0. \tag{6.95}$$

It is easy to decompose this vectorial equation into a part along $\boldsymbol{n}$ and a part tangent to Σ_t. Indeed, $\boldsymbol{B}$ being tangent to Σ_t, so is $\mathscr{L}_{\boldsymbol{m}}\boldsymbol{B}$ thanks to the property (4.12). We thus conclude that (i) the part of the first Maxwell equation along $\boldsymbol{n}$ is simply $-ND_\mu B^\mu n^\alpha = 0$, i.e.

$$\boxed{D_i B^i = 0} \tag{6.96}$$

and (ii) the orthogonal projection of the first Maxwell equation onto Σ_t is

$$\mathscr{L}_{\boldsymbol{m}} B^\alpha - NKB^\alpha + \varepsilon^{\alpha\rho\sigma} D_\rho (NE_\sigma) = 0.$$

This equation involving only space tensors, we may replace the Greek indices by Latin ones. Moreover, we may use (5.63) to replace $\mathscr{L}_{\boldsymbol{m}}$ by $\partial/\partial t - \mathscr{L}_{\boldsymbol{\beta}}$, thereby getting

$$\boxed{\left(\frac{\partial}{\partial t} - \mathscr{L}_{\boldsymbol{\beta}} \right) B^i - NKB^i + \varepsilon^{ijk} D_j (NE_k) = 0}. \tag{6.97}$$

This is the ***3+1 Maxwell-Faraday equation***.

We may now proceed straightforwardly to the *3+1* decomposition of the Maxwell equation with source, Eq. (6.84b), since (i) it merely differs from the Maxwell equation (6.84a) by the non-vanishing right-hand side and the replacement of $\star\boldsymbol{F}$ by $\boldsymbol{F}$ and (ii) we have already noticed that one goes from $\star\boldsymbol{F}$ to $\boldsymbol{F}$ by replacing $\boldsymbol{E}$ by $\boldsymbol{B}$ and $\boldsymbol{B}$ by $-\boldsymbol{E}$ [cf. Eqs. (6.82) and (6.83)]. Performing these substitutions in Eq. (6.95) and making use of (6.86) we obtain the *3+1* writing of the Maxwell equation (6.84b):

$$-\mathscr{L}_{\boldsymbol{m}} E^\alpha + NKE^\alpha + \varepsilon^{\alpha\rho\sigma} D_\rho (NB_\sigma) + ND_\mu E^\mu n^\alpha = \mu_0 N (\rho_{\rm e} n^\alpha + J^\alpha).$$

This equation can be split into a part along $\boldsymbol{n}$:

$$\boxed{D_i E^i = \mu_0 \rho_{\rm e}} \tag{6.98}$$

and a part tangent to Σ_t:

$$\boxed{\left(\frac{\partial}{\partial t} - \mathcal{L}_{\boldsymbol{\beta}}\right) E^i - NKE^i - \varepsilon^{ijk} D_j(NB_k) = -\mu_0 N J^i}. \tag{6.99}$$

Equations (6.98) and (6.99) are called respectively the ***3+1 Maxwell–Gauss*** and ***3+1 Maxwell–Ampère equations***. The 3+1 Maxwell equations (6.96)–(6.99) have been exhibited by Thorne and Macdonald in 1982 [31] (see also [32] and [33]).

In the flat spacetime limit, with the Eulerian observer chosen to be inertial ($N = 1$, $\boldsymbol{\beta} = 0$ and $K = 0$), Eqs. (6.96)–(6.99) reduce to the standard Maxwell equations:

$$\begin{cases} D_i B^i = 0 \\ \varepsilon^{ijk} D_j E_k = -\frac{\partial B^i}{\partial t} \\ D_i E^i = \mu_0 \rho_{\mathrm{e}} \\ \varepsilon^{ijk} D_j B_k = \mu_0 J^i + \frac{\partial E^i}{\partial t}. \end{cases} \qquad \begin{pmatrix} \text{inertial frame in} \\ \text{Minkowski spacetime} \end{pmatrix}$$

6.4.3 Electromagnetic Energy, Momentum and Stress

The stress-energy tensor of the electromagnetic field is given by the formula [34]

$$T_{\alpha\beta} = \frac{1}{\mu_0}\left(F_{\mu\alpha} F^{\mu}{}_{\beta} - \frac{1}{4} F_{\mu\nu} F^{\mu\nu}\, g_{\alpha\beta}\right). \tag{6.100}$$

Substituting the 3+1 expression (6.82) for $F_{\alpha\beta}$, expanding and making use of the identity (see e.g. Appendix B of [30])

$$\begin{aligned} {}^4\varepsilon_{\mu\alpha\beta\gamma}\, {}^4\varepsilon^{\mu\rho\sigma\tau} = & -\delta^{\rho}{}_{\alpha}\delta^{\sigma}{}_{\beta}\delta^{\tau}{}_{\gamma} + \delta^{\rho}{}_{\alpha}\delta^{\tau}{}_{\beta}\delta^{\sigma}{}_{\gamma} - \delta^{\sigma}{}_{\alpha}\delta^{\tau}{}_{\beta}\delta^{\rho}{}_{\gamma} + \delta^{\sigma}{}_{\alpha}\delta^{\rho}{}_{\beta}\delta^{\tau}{}_{\gamma} \\ & - \delta^{\tau}{}_{\alpha}\delta^{\rho}{}_{\beta}\delta^{\sigma}{}_{\gamma} + \delta^{\tau}{}_{\alpha}\delta^{\sigma}{}_{\beta}\delta^{\rho}{}_{\gamma}, \end{aligned}$$

we obtain the 3+1 form (5.14) for $\boldsymbol{T}$ with

- the energy density[2]

$$\boxed{E = \frac{1}{2\mu_0}(\boldsymbol{E}\cdot\boldsymbol{E} + \boldsymbol{B}\cdot\boldsymbol{B})} \tag{6.101}$$

- the momentum density

$$\boxed{\boldsymbol{p} = \frac{1}{\mu_0}\boldsymbol{\varepsilon}(.,\boldsymbol{E},\boldsymbol{B})} \tag{6.102}$$

- the stress tensor

[2] Do not confuse E (the energy density measured by the Eulerian observer) with $\boldsymbol{E}$ (the electric field measured by the same observer).

$$\boxed{S = \frac{1}{\mu_0}\left[\frac{1}{2}(E \cdot E + B \cdot B)\gamma - \underline{E} \otimes \underline{E} - \underline{B} \otimes \underline{B}\right]}. \qquad (6.103)$$

All the above quantities are relative to the Eulerian observer. Remembering that the energy flux 1-form is equal to the momentum density [Eq. (5.10)], we recognize in Eq. (6.102) the classical expression for the ***Poynting vector***.[3]

Remark 6.8 Evaluating the trace of T via formula (5.15), we get

$$T = S - E = \frac{1}{\mu_0}\left[\frac{1}{2}(E \cdot E + B \cdot B) \times 3 - E \cdot E - B \cdot B - \frac{1}{2}(E \cdot E + B \cdot B)\right] = 0.$$

We recover a well-known property of the electromagnetic stress-energy tensor, which is obvious on the 4-dimensional expression (6.100): its trace is vanishing.

Remark 6.9 Expressed in terms of (E, B), the energy density (6.101), the Poynting vector (6.102) and the stress tensor (6.103) of the electromagnetic field have exactly the same expressions than in special relativity. There is no additional terms from spacetime curvature as in the Maxwell equations (6.96)–(6.99).

6.5 3+1 Ideal Magnetohydrodynamics

6.5.1 Basic Settings

The ***ideal magnetohydrodynamics*** (ideal MHD) model consists of an electromagnetic field and a perfect fluid of infinite electric conductivity. Via Ohm's law, the latter property implies that the electric field vanishes in the fluid frame. The electric and magnetic fields in the fluid frame are defined by formulas analogous to (6.77a) and (6.77b), by replacing n by the fluid 4-velocity u:

$$\underline{e} = F(., u) \quad \text{and} \quad \underline{b} = \star F(u, .). \qquad (6.104)$$

The ideal MHD condition is thus

$$e = 0. \qquad (6.105)$$

Consequently the formulas equivalent to (6.81) and (6.83) simplify to

$$F = {}^4\varepsilon(u, b, ., .) \qquad (6.106a)$$

$$\star F = \underline{b} \otimes \underline{u} - \underline{u} \otimes \underline{b}. \qquad (6.106b)$$

In other words, in the ideal MHD approximation, the electromagnetic field tensor F is entirely determined by the fluid 4-velocity u and a vector field b that is orthogonal to u. Physically b is the magnetic field in the fluid frame.

[3] More precisely, the Poynting vector is $\vec{p}$, i.e. the metric dual of the 1-form p.

The magnetic field with respect to the Eulerian observer is obtained by substituting expression (6.106b) for $\star\boldsymbol{F}$ into Eq. (6.77b):

$$\underline{\boldsymbol{B}} = \star\boldsymbol{F}(\boldsymbol{n}, .) = (\boldsymbol{b}\cdot\boldsymbol{n})\underline{\boldsymbol{u}} - \underbrace{(\boldsymbol{u}\cdot\boldsymbol{n})}_{-\Gamma}\boldsymbol{b},$$

where we have let appear the Lorentz factor of the fluid with respect to the Eulerian observer [cf. Eq. (6.28)]. Hence

$$\boldsymbol{B} = \Gamma\boldsymbol{b} + (\boldsymbol{n}\cdot\boldsymbol{b})\boldsymbol{u}. \tag{6.107}$$

Let us take the scalar product of this relation with $\boldsymbol{u}$ and use the property $\boldsymbol{u}\cdot\boldsymbol{b} = 0$, as well as the 3+1 decomposition of $\boldsymbol{u}$, Eq. (6.31). We get

$$\underbrace{\boldsymbol{u}\cdot\boldsymbol{B}}_{\Gamma\boldsymbol{U}\cdot\boldsymbol{B}} = \Gamma\underbrace{\boldsymbol{u}\cdot\boldsymbol{b}}_{0} + (\boldsymbol{n}\cdot\boldsymbol{b})\underbrace{\boldsymbol{u}\cdot\boldsymbol{u}}_{-1}.$$

Hence

$$\boldsymbol{n}\cdot\boldsymbol{b} = -\Gamma\boldsymbol{U}\cdot\boldsymbol{B}.$$

Using again Eq. (6.31), we may then rewrite (6.107) as

$$\boldsymbol{b} = \Gamma(\boldsymbol{U}\cdot\boldsymbol{B})\boldsymbol{n} + \Gamma^{-1}\boldsymbol{B} + \Gamma(\boldsymbol{U}\cdot\boldsymbol{B})\boldsymbol{U}. \tag{6.108}$$

This constitutes the 3+1 decomposition of $\boldsymbol{b}$, since the vectors $\boldsymbol{B}$ and $\boldsymbol{U}$ are tangent to Σ_t.

The electric field with respect to the Eulerian observer is obtained by substituting expression (6.106a) for $\boldsymbol{F}$ into Eq. (6.77a):

$$\begin{aligned}\underline{\boldsymbol{E}} &= \boldsymbol{F}(., \boldsymbol{n}) = {}^4\boldsymbol{\varepsilon}(\boldsymbol{u}, \boldsymbol{b}, ., \boldsymbol{n}) = {}^4\boldsymbol{\varepsilon}(\boldsymbol{n}, \boldsymbol{b}, \boldsymbol{u}, .) = {}^4\boldsymbol{\varepsilon}\left(\boldsymbol{n}, \Gamma^{-1}\boldsymbol{B} - \Gamma^{-1}(\boldsymbol{n}\cdot\boldsymbol{b})\boldsymbol{u}, \boldsymbol{u}, .\right) \\ &= (\Gamma^{-1})^4\boldsymbol{\varepsilon}(\boldsymbol{n}, \boldsymbol{B}, \boldsymbol{u}, .) = {}^4\boldsymbol{\varepsilon}(\boldsymbol{n}, \boldsymbol{B}, \boldsymbol{n} + \boldsymbol{U}, .) = {}^4\boldsymbol{\varepsilon}(\boldsymbol{n}, \boldsymbol{B}, \boldsymbol{U}, .),\end{aligned} \tag{6.109}$$

where we have used Eqs. (6.107) and (6.31), as well as the alternate character of ${}^4\boldsymbol{\varepsilon}$. We recognize in ${}^4\boldsymbol{\varepsilon}(\boldsymbol{n}, ., ., .)$ the Levi–Civita tensor of $(\Sigma_t, \boldsymbol{\gamma})$, $\boldsymbol{\varepsilon}$ [cf. Eq. (6.93)], so we can write

$$\boxed{\underline{\boldsymbol{E}} = \boldsymbol{\varepsilon}(., \boldsymbol{B}, \boldsymbol{U})}, \tag{6.110}$$

or, in terms of components,

$$E_i = \varepsilon_{ijk}B^j U^k. \tag{6.111}$$

Hence in ideal MHD and with respect to the Eulerian observer, the electric field is fully determined by the magnetic field and the fluid velocity.

Remark 6.10 The ***cross product*** of two vectors $\boldsymbol{a}$ and $\boldsymbol{b}$ tangent to Σ_t is defined via the Levi–Civita tensor $\boldsymbol{\varepsilon}$ as

$$\boldsymbol{a} \times \boldsymbol{b} := \boldsymbol{c} \quad \text{with} \quad \underline{\boldsymbol{c}} = \boldsymbol{\varepsilon}(., \boldsymbol{a}, \boldsymbol{b}). \tag{6.112}$$

Accordingly, Eq. (6.110) can be rewritten in the more familiar form

$$\boldsymbol{E} = -\boldsymbol{U} \times \boldsymbol{B}. \tag{6.113}$$

6.5.2 Maxwell Equations

Let us substitute expression (6.110) for $\boldsymbol{E}$ in the Maxwell–Faraday equation (6.97):

$$\left(\frac{\partial}{\partial t} - \mathcal{L}_{\boldsymbol{\beta}}\right) B^i - NKB^i + \varepsilon^{ijk} D_j (N \varepsilon_{klm} B^l U^m) = 0.$$

Since $\boldsymbol{\varepsilon}$ and $\boldsymbol{D}$ are respectively the Levi–Civita tensor and the covariant derivative associated with the metric $\boldsymbol{\gamma}$, we have $D_j \varepsilon_{klm} = 0$ [cf. Eq. (2.61)], so that we may write

$$\left(\frac{\partial}{\partial t} - \mathcal{L}_{\boldsymbol{\beta}}\right) B^i - NKB^i + \varepsilon^{ijk} \varepsilon_{klm} D_j (NB^l U^m) = 0.$$

Using the identity

$$\varepsilon^{kij} \varepsilon_{klm} = \delta^i{}_l \delta^j{}_m - \delta^j{}_l \delta^i{}_m, \tag{6.114}$$

as well as the expression of the Lie derivative $\mathcal{L}_{\boldsymbol{\beta}} B^i$ in terms of the connection $\boldsymbol{D}$ [cf. formula (2.89)], we get

$$\frac{\partial B^i}{\partial t} - \beta^j D_j B^i + B^j D_j \beta^i - NKB^i + D_j (NB^i U^j) - D_j (NB^j U^i) = 0.$$

Now, from Eq. (6.38), $NU^i = V^i + \beta^i$, where V^i are the components of the fluid coordinate velocity. Accordingly, the above equation becomes, after simplification,

$$\frac{\partial B^i}{\partial t} + (D_j \beta^j - NK) B^i + D_j (B^i V^j - V^i B^j) = 0. \tag{6.115}$$

Note that we have used the Maxwell equation (6.96) to get rid of the term $D_j B^j \beta^i$. Equation (6.115) is the MHD version of the Maxwell–Faraday equation (6.97) written in covariant form, in terms of the Levi–Civita connection $\boldsymbol{D}$ of $(\Sigma_t, \boldsymbol{\gamma})$. It can be given a simpler form in terms of the partial derivatives with respect to the coordinates (x^i)

on Σ_t: let us express the last term in Eq. (6.115) via the general formula (2.66) for the divergence of an antisymmetric tensor:

$$D_j(B^i V^j - V^i B^j) = \frac{1}{\sqrt{\gamma}} \frac{\partial}{\partial x^j} \left[\sqrt{\gamma}(B^i V^j - V^i B^j) \right]$$

and make use of (6.66) to replace $D_j\beta^j - NK$. We hence put Eq. (6.115) in the form

$$\frac{\partial}{\partial t}\left(\sqrt{\gamma}B^i\right) + \frac{\partial}{\partial x^j}\left[\sqrt{\gamma}(B^i V^j - V^i B^j)\right] = 0.$$

This suggests to introduce the quantity

$$\boxed{\mathcal{B}^i := \sqrt{\gamma}B^i}, \tag{6.116}$$

to write

$$\boxed{\frac{\partial \mathcal{B}^i}{\partial t} + \frac{\partial}{\partial x^j}\left(\mathcal{B}^i V^j - V^i \mathcal{B}^j\right) = 0}. \tag{6.117}$$

Moreover, thanks to the divergence formula (2.65) for a vector field, we may also rewrite the Maxwell equation (6.96) in terms of $\mathcal{B}^i$ as

$$\boxed{\frac{\partial}{\partial x^i}\mathcal{B}^i = 0}. \tag{6.118}$$

Since the determinant γ is not a scalar field (it depends on the choice of the coordinates (x^i) on Σ_t), the $\mathcal{B}^i$'s defined by (6.116) are not the components of a tensor field, but rather the components a *tensor density*. We shall discuss tensor densities in more details in Sect. 7.2.1.

Remark 6.11 One could arrive at Eqs. (6.117), (6.118) directly from the 4-dimensional Maxwell equation (6.84a) with the divergence of $\star\boldsymbol{F}$ expressed according to formula (2.66):

$$\frac{\partial}{\partial x^\mu}\left(\sqrt{-g}\,\star F^{\alpha\mu}\right) = 0.$$

Substituting (6.106b) for $\star\boldsymbol{F}$ and using $\sqrt{-g} = N\sqrt{\gamma}$ [Eq. (5.55)], we get

$$\frac{\partial}{\partial x^\mu}\left[N\sqrt{\gamma}\left(b^\alpha u^\mu - u^\alpha b^\mu\right)\right] = 0.$$

Now thanks to (6.107), $b^\alpha u^\mu - u^\alpha b^\mu = \Gamma^{-1}(B^\alpha u^\mu - u^\alpha B^\mu)$. Since $\Gamma = Nu^0$ [Eq. (6.29)], we obtain

$$\frac{\partial}{\partial x^\mu}\left[\sqrt{\gamma}\left(B^\alpha \frac{u^\mu}{u^0} - \frac{u^\alpha}{u^0}B^\mu\right)\right] = 0.$$

Since $B^0 = 0$ (for $\boldsymbol{B}$ is tangent to Σ_t), we may write the above formula as

$$\frac{\partial}{\partial t}\left(\sqrt{\gamma}B^\alpha\right) + \frac{\partial}{\partial x^j}\left[\sqrt{\gamma}\left(B^\alpha \frac{u^j}{u^0} - \frac{u^\alpha}{u^0}B^j\right)\right] = 0.$$

Using $V^j = u^j/u^0$ [Eq. (6.36)] and $B^0 = 0$, we conclude that the above equation gives (6.118) for $\alpha = 0$ and (6.117) for $\alpha = i \in \{1, 2, 3\}$.

Since for ideal MHD the electric field $\boldsymbol{E}$ is entirely determined by $\boldsymbol{B}$ and $\boldsymbol{U}$ [cf. Eq. (6.110)], there is no need to re-express the 3+1 Maxwell equations (6.98) and (6.99). We are instead taking the point of view that these equations give the electric charge density ρ_e and the electric current $\boldsymbol{J}$ once $\boldsymbol{B}$ (and hence $\boldsymbol{E}$) is known.

6.5.3 Electromagnetic Energy, Momentum and Stress

Let us evaluate the 3+1 components $E_{\rm em}$, $\boldsymbol{p}_{\rm em}$ and $\boldsymbol{S}_{\rm em}$ of the electromagnetic stress-energy tensor, as given by Eqs. (6.101)–(6.103) for the specific case of ideal MHD. This amounts to replacing $\boldsymbol{E}$ in these expressions by the value (6.110). Using the identity (6.114), as well as $\varepsilon_{ikl}\varepsilon^{jnm} = 6\delta^{[j}_{\ i}\delta^{m}_{\ k}\delta^{n]}_{\ l}$, we obtain

$$E_{\rm em} = \frac{1}{2\mu_0}\left[(1 + \boldsymbol{U}\cdot\boldsymbol{U})\boldsymbol{B}\cdot\boldsymbol{B} - (\boldsymbol{U}\cdot\boldsymbol{B})^2\right] \tag{6.119}$$

$$\boldsymbol{p}_{\rm em} = \frac{1}{\mu_0}\left[(\boldsymbol{B}\cdot\boldsymbol{B})\underline{\boldsymbol{U}} - (\boldsymbol{U}\cdot\boldsymbol{B})\underline{\boldsymbol{B}}\right] \tag{6.120}$$

$$\boldsymbol{S}_{\rm em} = \frac{1}{\mu_0}\left\{\frac{1}{2}\left[\frac{1}{\Gamma^2}\boldsymbol{B}\cdot\boldsymbol{B} + (\boldsymbol{U}\cdot\boldsymbol{B})^2\right]\boldsymbol{\gamma} - \frac{1}{\Gamma^2}\underline{\boldsymbol{B}}\otimes\underline{\boldsymbol{B}} + (\boldsymbol{B}\cdot\boldsymbol{B})\underline{\boldsymbol{U}}\otimes\underline{\boldsymbol{U}}\right.$$
$$\left. -(\boldsymbol{U}\cdot\boldsymbol{B})\left(\underline{\boldsymbol{U}}\otimes\underline{\boldsymbol{B}} + \underline{\boldsymbol{B}}\otimes\underline{\boldsymbol{U}}\right)\right\}. \tag{6.121}$$

6.5.4 MHD-Euler Equation

The MHD-Euler equation is derived from the energy-momentum conservation equation (6.1) in the form

$$\nabla\cdot(\vec{\boldsymbol{T}}_{\rm fl} + \vec{\boldsymbol{T}}_{\rm em}) = 0, \tag{6.122}$$

where $\boldsymbol{T}_{\rm fl}$ is the fluid stress-energy tensor as given by (6.100) and $\boldsymbol{T}_{\rm em}$ is the electromagnetic field stress-energy tensor as given by (6.24). Now it is a standard result[4] that

$$\nabla\cdot\vec{\boldsymbol{T}}_{\rm em} = -\boldsymbol{F}(.,\boldsymbol{j}). \tag{6.123}$$

[4] It is easy to establish it starting from expression (6.100) for $\boldsymbol{T}_{\rm em}$ and using the Maxwell equation (6.84b) to let appear $\boldsymbol{j}$ and the Maxwell equation (6.84a) in the equivalent form $\nabla_\alpha F_{\beta\gamma} + \nabla_\beta F_{\gamma\alpha} + \nabla_\gamma F_{\alpha\beta} = 0$.

Therefore (6.122) becomes

$$\nabla \cdot \vec{\boldsymbol{T}}_{\text{fl}} = \boldsymbol{F}(.,\boldsymbol{j}). \tag{6.124}$$

Let us evaluate $\boldsymbol{F}(.,\boldsymbol{j})$. By plugging (6.110) into (6.81), we get the 3+1 expression of the electromagnetic field tensor in ideal MHD:

$$\boldsymbol{F} = \underline{\boldsymbol{n}} \otimes \boldsymbol{\varepsilon}(.,\boldsymbol{B},\boldsymbol{U}) - \boldsymbol{\varepsilon}(.,\boldsymbol{B},\boldsymbol{U}) \otimes \underline{\boldsymbol{n}} + \boldsymbol{\varepsilon}(\boldsymbol{B},.,.), \tag{6.125}$$

hence

$$\boldsymbol{F}(.,\boldsymbol{j}) = \boldsymbol{\varepsilon}(\boldsymbol{j},\boldsymbol{B},\boldsymbol{U})\underline{\boldsymbol{n}} - (\boldsymbol{n}\cdot\boldsymbol{j})\boldsymbol{\varepsilon}(.,\boldsymbol{B},\boldsymbol{U}) + \boldsymbol{\varepsilon}(\boldsymbol{B},.,\boldsymbol{j}).$$

From the 3+1 decomposition (6.86) of $\boldsymbol{j}$, we have $\boldsymbol{\varepsilon}(\boldsymbol{j},\boldsymbol{B},\boldsymbol{U}) = \boldsymbol{\varepsilon}(\boldsymbol{J},\boldsymbol{B},\boldsymbol{U})$, $\boldsymbol{n}\cdot\boldsymbol{j} = -\rho_{\text{e}}$ and $\boldsymbol{\varepsilon}(\boldsymbol{B},.,\boldsymbol{j}) = \boldsymbol{\varepsilon}(\boldsymbol{B},.,\boldsymbol{J}) = \boldsymbol{\varepsilon}(.,\boldsymbol{J},\boldsymbol{B})$. Consequently, we arrive at the following 3+1 decomposition of $\boldsymbol{F}(.,\boldsymbol{j})$:

$$\boldsymbol{F}(.,\boldsymbol{j}) = \boldsymbol{\varepsilon}(\boldsymbol{J},\boldsymbol{B},\boldsymbol{U})\underline{\boldsymbol{n}} + \boldsymbol{\varepsilon}(.,\ \boldsymbol{J} - \rho_{\text{e}}\boldsymbol{U},\ \boldsymbol{B}). \tag{6.126}$$

Accordingly the 3+1 decomposition of Eq. (6.124) is

$$\boldsymbol{n}\cdot(\nabla\cdot\vec{\boldsymbol{T}}_{\text{fl}}) = -\boldsymbol{\varepsilon}(\boldsymbol{J},\boldsymbol{B},\boldsymbol{U}) \tag{6.127a}$$
$$\vec{\boldsymbol{\gamma}}^{*}\nabla\cdot\vec{\boldsymbol{T}}_{\text{fl}} = \boldsymbol{\varepsilon}(.,\ \boldsymbol{J} - \rho_{\text{e}}\boldsymbol{U},\ \boldsymbol{B}). \tag{6.127b}$$

The terms $\boldsymbol{n}\cdot(\nabla\cdot\vec{\boldsymbol{T}}_{\text{fl}})$ and $\vec{\boldsymbol{\gamma}}^{*}\nabla\cdot\vec{\boldsymbol{T}}_{\text{fl}}$ have been computed in Sect. 6.3. Noticing that (i) a multiplication by $-N$ has been performed to go from $\boldsymbol{n}\cdot(\nabla\cdot\vec{\boldsymbol{T}}_{\text{fl}}) = 0$ to Eq. (6.59), and (ii) a multiplication by $N/(E+P)$ to go from $\vec{\boldsymbol{\gamma}}^{*}\nabla\cdot\vec{\boldsymbol{T}}_{\text{fl}} = 0$ to Eq. (6.61), we conclude that the energy conservation law for ideal MHD is

$$\boxed{\begin{aligned}&\frac{\partial E}{\partial t} - \beta^i D_i E + N\left\{D_i\left[(E+P)U^i\right] - (E+P)(K + K_{ij}U^iU^j)\right\}\\ &\quad + 2(E+P)U^iD_iN = N\varepsilon_{ijk}J^iB^jU^k\end{aligned}} \tag{6.128}$$

and that the ***MHD-Euler equation*** is

$$\boxed{\begin{aligned}\tfrac{\partial U_i}{\partial t} + V^jD_jU_i = &-\frac{N}{E+P}\left[D_iP + \frac{U_i}{N}\left(\frac{\partial P}{\partial t} - \beta^j\frac{\partial P}{\partial x^j}\right) - \varepsilon_{ijk}(J^j - \rho_{\text{e}}U^j)B^k\right]\\ &-D_iN + U_iU^j\left(D_jN - NK_{jk}U^k\right) + U_jD_i\beta^j.\end{aligned}} \tag{6.129}$$

In these equations, ρ_{e} and $\boldsymbol{J}$ are to be considered as functions of $(\boldsymbol{B},\boldsymbol{U})$ expressed by Eqs. (6.98) and (6.99), with $\boldsymbol{E}$ given by (6.110).

At the non-relativistic limit, Eqs. (6.128) and (6.129) reduce to respectively [cf. Eqs. (6.60) and (6.62)]

$$\frac{\partial E'}{\partial t} + \bar{D}_i[(E'+P)U^i] = -U^i(\rho_0 \bar{D}_i \Phi) + U^i \varepsilon_{ijk} J^j B^k, \tag{6.130}$$

$$\frac{\partial U_i}{\partial t} + U^j \bar{D}_j U_i = -\frac{1}{\rho_0}\bar{D}_i P - \bar{D}_i \Phi + \frac{1}{\rho_0}\varepsilon_{ijk} J^j B^k. \tag{6.131}$$

We recognize in $\varepsilon_{ijk} J^j B^k$ the *Lorentz force* term of non-relativistic MHD and in $U^i \varepsilon_{ijk} J^j B^k$ the power transferred to the fluid by that force.

6.5.5 MHD in Flux-Conservative Form

As for the pure hydrodynamical case (cf. Sect. 6.3.6), the relativistic MHD equations are amenable to a flux-conservative form:

$$\boxed{\frac{\partial \mathscr{U}_A}{\partial t} + \frac{\partial}{\partial x^j}\mathscr{F}^j_A = \mathscr{S}_A}, \quad 0 \le A \le 7, \tag{6.132}$$

with the state vector

$$\mathscr{U}_0 := \mathscr{D} = \sqrt{\gamma}\,\mathscr{N}_\mathrm{b} = \sqrt{\gamma}\,\Gamma n_\mathrm{b} \tag{6.133a}$$

$$\mathscr{U}_i := \sqrt{\gamma}\,p_i = \sqrt{\gamma}\left[(E+P)U_i + \frac{1}{\mu_0}\left(B_j B^j U_i - U_j B^j B_i\right)\right], \quad 1 \le i \le 3 \tag{6.133b}$$

$$\mathscr{U}_4 := \sqrt{\gamma}\left\{E - E_0 + \frac{1}{2\mu_0}\left[(1 + U_j U^j) B_j B^j - (U_j B^j)^2\right]\right\} \tag{6.133c}$$

$$\mathscr{U}_{4+i} := \mathscr{B}^i = \sqrt{\gamma}\,B^i, \quad 1 \le i \le 3, \tag{6.133d}$$

the flux vectors

$$\mathscr{F}^j_0 := \mathscr{D} V^j \tag{6.134a}$$

$$\begin{aligned}\mathscr{F}^j_i := \sqrt{\gamma}\Bigg\{ & N\left[P + \frac{1}{\mu_0}\left(\frac{B_k B^k}{\Gamma^2} + (U_k B^k)^2\right)\right]\delta^j{}_i + \frac{B_k B^k}{\mu_0} V_i U^j \\ & + \left[(E+P)U_i - \frac{U_K B^k}{\mu_0} B_i\right] V^j - \frac{N}{\mu_0}\left[\frac{B_i}{\Gamma^2} - U_k B^k U_i\right] B^j \Bigg\}, \\ & \qquad 1 \le i \le 3 \end{aligned} \tag{6.134b}$$

$$\mathscr{F}^j_4 := \sqrt{\gamma}\left[(E - E_0)V^j + NPU^j + \frac{N}{\mu_0}(B_k B^k U^j - U_k B^k B^j) - E_\mathrm{em}\beta^j\right] \tag{6.134c}$$

$$\mathscr{F}^j_{4+i} := \mathscr{B}^i V^j - V^i \mathscr{B}^j, \quad 1 \le i \le 3 \tag{6.134d}$$

and the source vector

$$\mathscr{S}_0 := 0 \tag{6.135a}$$

$$\mathscr{S}_i := \sqrt{\gamma}\left[p_j D_i \beta^j - (E + E_{\rm em}) D_i N + \Gamma^k{}_{ij}\left(N S^j{}_k - \beta^j p_k\right)\right] \tag{6.135b}$$

$$\mathscr{S}_4 := \sqrt{\gamma}\left(N K_{jk} S^{jk} - p^j D_j N\right) \tag{6.135c}$$

$$\mathscr{S}_{4+i} := 0, \quad 1 \le i \le 3. \tag{6.135d}$$

In the above relations, E stands for the energy density (with respect to the Eulerian observer) of the fluid only, i.e. the quantity given by (6.49), whereas p_i is the total momentum density with respect to the Eulerian observer: $\boldsymbol{p} = \boldsymbol{p}_{\rm fl} + \boldsymbol{p}_{\rm em}$ with $\boldsymbol{p}_{\rm fl}$ given by (6.57) and $\boldsymbol{p}_{\rm em}$ by (6.120). Similarly, S_{ij} is the total stress tensor with respect to the Eulerian observer: $\boldsymbol{S} = \boldsymbol{S}_{\rm fl} + \boldsymbol{S}_{\rm em}$ with $\boldsymbol{S}_{\rm fl}$ given by (6.58) and $\boldsymbol{S}_{\rm em}$ by (6.121). The system (6.132)–(6.135a), (6.135b), (6.135c), (6.135d) follows from the baryon number conservation (6.67) (index $A = 0$), the momentum conservation (6.70) ($A = 1, 2, 3$), the energy conservation (6.72) (with $E \to E + E_{\rm em}$) ($A = 4$) and the Maxwell–Faraday equation (6.117) ($A = 5, 6, 7$). Note that, in addition to the system (6.132)–(6.135a), (6.135b), (6.135c), (6.135d), the constraint (6.118) has to be satisfied.

The flux-conservative form of the MHD equations has been developed by Gammie, McKinney and Tóth [35], Komissarov [36], Duez et al. [37], Shibata and Sekiguchi [38] and Antón et al. [39]. Most modern codes for general relativistic MHD employ such a formulation, via HRSC schemes. This concerns, among others, WhiskyMHD [19, 40, 41], HAD GRMHD [42], X-ECHO [43], the code of Kiuchi et al. [44, 45], the code of Cerdá-Durán, Font, Antón and Müller [46] and the code of Etienne et al. [47]. Let us mention that modern special relativistic MHD codes, such as the AMRVAC code [48], are also based on the flux-conservative form.

References

1. Baumgarte, T.W., Shapiro, S.L.: Numerical relativity solving einstein's equations on the computer. Cambridge University Press, Cambridge (2010)
2. Bona, C., Palenzuela-Luque, C., Bona-Casas, C.: Elements of numerical relativity and relativistic hydrodynamics: from Einstein's equations to astrophysical simulations . 2nd edn. Springer, Berlin (2009)
3. Choquet-Bruhat, Y.: General relativity and einstein's equations. Oxford University Press, New York (2009)
4. York, J.W.: Kinematics and dynamics of general relativity. In: Smarr, L.L. (ed.) Sources of gravitational radiation, p. 83. Cambridge University Press, Cambridge (1979)
5. Deruelle, N.: General relativity: a primer. Lectures at Institut Henri Poincaré, Paris. http://www.luth.obspm.fr/IHP06/ (2006)
6. Font, J.A.: Numerical hydrodynamics and magnetohydrodynamics in general relativity. Living Rev. Relat. **11**, 7. http://www.livingreviews.org/lrr-2008-7 (2008)

7. Alcubierre, M.: Introduction to 3+1 numerical relativity. Oxford University Press, Oxford (2008)
8. Gourgoulhon, E.: Simple equations for general relativistic hydrodynamics in spherical symmetry applied to neutron star collapse. Astron. Astrophys. **252**, 651 (1991)
9. Salgado, M.: General relativistic hydrodynamics: a new approach. Rev. Mex. Fís. **44**, 0001 (1998)
10. Gourgoulhon, E.: Relativité restreinte, des particules à l'astrophysique. EDP Sciences, Les Ulis / CNRS Éditions, Paris (2010)
11. Font, J.A.: An introduction to relativistic hydrodynamics. J. Phys. Conf. Ser. **91**, 012002 (2007)
12. Martí, J.M., Ibáñez, J.M., Miralles, J.A.: Numerical relativistic hydrodynamics: local characteristic approach. Phys. Rev. D **43**, 3794 (1991)
13. Banyuls, F., Font, J.A., Ibáñez, J.M., Martí, J.M., Miralles, J.A.: Numerical 3+1 general relativistic hydrodynamics: a local characteristic approach. Astrophys. J. **476**, 221 (1997)
14. Godunov, S.K.: A finite difference method for the numerical computation and discontinuous solutions of the equations of fluid dynamics (in Russian). Math. Sbornik **47**, 271 (1959)
15. Wilson, J.R.: Numerical study of fluid flow in a Kerr space. Astrophys. J. **173**, 431 (1972)
16. Wilson, J.R., Mathews, G.J.: Relativistic hydrodynamics. In: Evans, C.R., Finn, L.S., Hobill, D.W. (eds.) Frontiers in numerical relativity, p. 306. Cambridge University Press, Cambridge (1989)
17. Baiotti, L., Hawke, I., Montero, P.J., Rezzolla, L.: A new three-dimensional general-relativistic hydrodynamics code. Mem. S.A.It. Suppl. **1**, 210 (2003)
18. Baiotti, L., Hawke, I., Montero, P.J., Löffler, F., Rezzolla, L., Stergioulas, N., Font, J.A., Seidel, E.: Three-dimensional relativistic simulations of rotating neutron-star collapse to a Kerr black hole. Phys. Rev. D **71**, 024035 (2005)
19. Whisky code: http://www.whiskycode.org/
20. Yamamoto, T., Shibata, M., Taniguchi, K.: Simulating coalescing compact binaries by a new code. Phys. Rev. D **78**, 064054 (2008)
21. Baiotti, L., Shibata, S., Yamamoto, T.: Binary neutron-star mergers with Whisky and SACRA: First quantitative comparison of results from independent general-relativistic hydrodynamics codes. Phys. Rev. D **82**, 064015 (2010)
22. Shibata, M., Taniguchi, K., Uryu, K.: Merger of binary neutron stars with realistic equations of state in full general relativity. Phys. Rev. D **71**, 084021 (2005)
23. Dimmelmeier, H., Novak, J., Font, J.A., Ibáñez, J.M., Müller, E.: Combining spectral and shock-capturing methods: a new numerical approach for 3D relativistic core collapse simulations. Phys. Rev. D **71**, 064023 (2005)
24. Duez, M.D., Foucart, F., Kidder, L.E., Pfeiffer, H.P., Scheel, M.A., Teukolsky, S.A.: Evolving black hole-neutron star binaries in general relativity using pseudospectral and finite difference methods. Phys. Rev. D **78**, 104015 (2008)
25. Spectral Einstein code: http://www.black-holes.org/SpEC.html
26. Thierfelder, M., Bernuzzi, S., Brügmann, B.: Numerical relativity simulations of binary neutron stars. Phys. Rev. D **84**, 044012 (2011)
27. Duez, M.D.: Numerical relativity confronts compact neutron star binaries: a review and status report. Class. Quant. Grav. **27**, 114002 (2010)
28. Gourgoulhon, E.: An introduction to relativistic hydrodynamics. In: Rieutord, M., Dubrulle, B. (eds.) Stellar fluid dynamics and numerical simulations: from the Sun to neutron stars. EAS Publications Series, vol. 21, p. 43. EDP Sciences, Les Ulis. http://arxiv.org/abs/gr-qc/0603009 (2006)
29. Misner, C.W., Thorne, K.S., Wheeler, J.A.: Gravitation. Freeman, New York (1973)
30. Wald, R.M.: General relativity. University of Chicago Press, Chicago (1984)
31. Thorne, K.S., Macdonald, D.: Electrodynamics in curved spacetime: 3+1 formulation. Mon. Not. R. Astron. Soc. **198**, 339 (1982)
32. Baumgarte, T.W., Shapiro, S.L.: General relativistic magnetohydrodynamics for the numerical construction of dynamical spacetimes. Astrophys. J. **585**, 921 (2003)

33. Alcubierre, M., Degollado, J.C., Salgado, M.: Einstein–Maxwell system in 3+1 form and initial data for multiple charged black holes. Phys. Rev. D **80**, 104022 (2009)
34. Jackson, J.D.: Classical electrodynamics. 2nd edn. Wiley, New york (1975)
35. Gammie, C.F., McKinney, J.C., Tóth, G.: HARM: a numerical scheme for general relativistic magnetohydrodynamics. Astrophys. J. **589**, 444 (2003)
36. Komissarov, S.S.: Observations of the Blandford–Znajek process and the magnetohydrodynamic Penrose process in computer simulations of black hole magnetospheres. Mon. Not. Roy. Astron. Soc. **359**, 801 (2005)
37. Duez, M.D., Liu, Y.T., Shapiro, S.L., Stephens, B.C.: Relativistic magnetohydrodynamics in dynamical spacetimes: numerical methods and tests. Phys. Rev. D **72**, 024028 (2005)
38. Shibata, M., Sekiguchi, Y.: Magnetohydrodynamics in full general relativity: formulation and tests. Phys. Rev. D **72**, 044014 (2005)
39. Antón, L., Zanotti, O., Miralles, J.A., Martí, J.M., Ibáñez, J.M., Font, J.A., Pons, J.A.: Numerical 3+1 general relativistic magnetohydrodynamics: a local characteristic approach. Astrophys. J. **637**, 296 (2006)
40. Giacomazzo, B., Rezzolla, L.: WhiskyMHD: a new numerical code for general relativistic magnetohydrodynamics. Class. Quant. Grav. **24**, S235 (2007)
41. Giacomazzo, B., Rezzolla, L., Baioti, L.: Accurate evolutions of inspiralling and magnetized neutron stars: equal-mass binaries. Phys. Rev. D **83**, 044014 (2011)
42. Liebling, S.L., Lehner, L., Neilsen, D., Palenzuela, C.: Evolutions of magnetized and rotating neutron stars. Phys. Rev. D **81**, 124023 (2010)
43. Bucciantini, N., Del Zanna, L.: General relativistic magnetohydrodynamics in axisymmetric dynamical spacetimes: the X-ECHO code. Astron. Astrophys. **528**, A101 (2011)
44. Kiuchi, K., Shibata, M., Yoshida, S.: Evolution of neutron stars with toroidal magnetic fields: axisymmetric simulation in full general relativity. Phys. Rev. D **78**, 024029 (2008)
45. Kiuchi, K., Yoshida, S., Shibata, M.: Non-axisymmetric instabilities of neutron star with toroidal magnetic fields. Astron. Astrophys. **532**, A30 (2011)
46. Cerdá-Durán, P., Font, J.A., Antón, L., Müller, E.: A new general relativistic magnetohydrodynamics code for dynamical spacetimes. Astron. Astrophys. **492**, 937 (2008)
47. Etienne, Z.B., Liu, Y.T., Shapiro, S.L.: Relativistic magnetohydrodynamics in dynamical spacetimes: a new adaptive mesh refinement implementation. Phys. Rev. D **82**, 084031 (2010)
48. van der Holst, B., Keppens, R., Meliani, Z.: A multidimensional grid-adaptive relativistic magnetofluid code. Comput. Phys. Com. **179**, 617 (2008)

Chapter 7
Conformal Decomposition

Abstract This is a technical chapter to prepare the following ones. We motivate and perform a conformal decomposition of the 3-metric on each hypersurface of a 3+1 slicing. To avoid dealing with tensor densities, we introduce a background flat 3-metric. The link between the connections of the physical 3-metric and the conformal one is exhibited, leading to the computation of the Ricci tensor of the conformal 3-metric. Two associated decompositions of the extrinsic curvature are presented, with two different conformal rescalings of the traceless part. The 3+1 Einstein equations are then rewritten in terms of the conformal quantities. Finally, we discuss the Isenberg-Wilson-Mathews approximation to general relativity, which amounts to assuming that the conformal 3-metric is flat and that the slicing is maximal.

7.1 Introduction

Historically, conformal decompositions in 3+1 general relativity have been introduced in two contexts. First of all, Lichnerowicz has introduced in 1944 [1][1] a decomposition of the induced metric $\boldsymbol{\gamma}$ of the hypersurfaces Σ_t of the type

$$\boldsymbol{\gamma} = \Psi^4 \tilde{\boldsymbol{\gamma}}, \tag{7.1}$$

where Ψ is some strictly positive scalar field and $\tilde{\boldsymbol{\gamma}}$ an auxiliary metric on Σ_t. Note that $\boldsymbol{\gamma}$ being a Riemannian metric (cf. Sect. 2.3.2), $\tilde{\boldsymbol{\gamma}}$ is necessarily of the same type. The relation (7.1) is called a ***conformal transformation*** and $\tilde{\boldsymbol{\gamma}}$ will be called hereafter the ***conformal metric*** . Lichnerowicz has shown that the conformal decomposition of $\boldsymbol{\gamma}$, along with some specific conformal decomposition of the extrinsic curvature provides a fruitful tool for the resolution of the constraint equations, in order to get valid initial data for the Cauchy problem. This will be discussed in Chap. 9.

[1] See also Ref. [2] which is freely accessible on the web.

É. Gourgoulhon, *3+1 Formalism in General Relativity*, Lecture Notes in Physics 846,
DOI: 10.1007/978-3-642-24525-1_7, © Springer-Verlag Berlin Heidelberg 2012

Then, in 1971–72, York [3, 4] has shown that conformal decompositions are also important for the time evolution problem, by demonstrating that the two degrees of freedom of the gravitational field are carried by the conformal equivalence classes of 3-metrics. A ***conformal equivalence class*** is defined as the set of all metrics that can be related to a given metric $\boldsymbol{\gamma}$ by a transformation like (7.1). The argument of York is based on the ***Cotton tensor*** [5], which is a rank-3 covariant tensor defined from the covariant derivative of the Ricci tensor $\boldsymbol{R}$ of $\boldsymbol{\gamma}$ by

$$\mathscr{C}_{ijk} := D_k\left(R_{ij} - \frac{1}{4}R\gamma_{ij}\right) - D_j\left(R_{ik} - \frac{1}{4}R\gamma_{ik}\right). \tag{7.2}$$

The Cotton tensor is conformally invariant and shows the same property with respect to 3-dimensional pseudo-Riemannian manifolds than the Weyl tensor [cf. Eq. (2.83)] for pseudo-Riemannian manifolds of dimension strictly greater than three, namely its vanishing is a necessary and sufficient condition for the metric to be ***conformally flat***, i.e. to be expressible as $\boldsymbol{\gamma} = \Psi^4 \boldsymbol{f}$, where Ψ is some strictly positive scalar field and $\boldsymbol{f}$ a flat metric. Let us recall that in dimension 3, the Weyl tensor vanishes identically (cf. Sect. 2.4.4). More precisely, York [3] constructed from the Cotton tensor the following rank-2 tensor

$$C^{ij} := -\frac{1}{2}\varepsilon^{ikl}\mathscr{C}_{mkl}\gamma^{mj} = \varepsilon^{ikl}D_k\left(R^j{}_l - \frac{1}{4}R\delta^j{}_l\right), \tag{7.3}$$

where $\boldsymbol{\varepsilon}$ is the Levi–Civita alternating tensor associated with the metric $\boldsymbol{\gamma}$ [cf. Eq. (6.93)]. The tensor $\boldsymbol{C}$ is called the ***Cotton–York tensor*** and exhibits the following properties:

- symmetry: $C^{ji} = C^{ij}$
- traceless: $\gamma_{ij}C^{ij} = 0$
- divergence-free (one says also *transverse*): $D_j C^{ij} = 0$

Moreover, if one consider, instead of $\boldsymbol{C}$, the following tensor density of weight $5/3$,

$$C_*^{ij} := \gamma^{5/6}C^{ij}, \tag{7.4}$$

where $\gamma := \det(\gamma_{ij})$, then one gets a conformally invariant quantity. Indeed, under a conformal transformation of the type (7.1), $\varepsilon^{ikl} = \Psi^{-6}\tilde{\varepsilon}^{ikl}$, $\mathscr{C}_{mkl} = \tilde{\mathscr{C}}_{mkl}$ (conformal invariance of the Cotton tensor), $\gamma^{ml} = \Psi^{-4}\tilde{\gamma}^{ml}$ and $\gamma^{5/6} = \Psi^{10}\tilde{\gamma}^{5/6}$, so that $C_*^{ij} = \tilde{C}_*^{ij}$. The traceless and transverse (TT) properties being characteristic of the pure spin 2 representations of the gravitational field (see e.g. [6]), the conformal invariance of C_*^{ij} shows that the true degrees of freedom of the gravitational field are carried by the conformal equivalence class.

Remark 7.1 The remarkable feature of the Cotton–York tensor is to be a TT object constructed from the physical metric $\boldsymbol{\gamma}$ alone, without the need of some extra-structure on the manifold Σ_t. Usually, TT objects are defined with respect to some extra-structure, such as privileged Cartesian coordinates or a flat background metric, as in the post-Newtonian approach to general relativity (see e.g. [7, 8]).

Remark 7.2 The Cotton and Cotton–York tensors involve third derivatives of the metric tensor.

7.2 Conformal Decomposition of the 3-Metric

7.2.1 Unit-Determinant Conformal "Metric"

A somewhat natural representative of a conformal equivalence class is the unit-determinant conformal "metric"

$$\hat{\boldsymbol{\gamma}} := \gamma^{-1/3}\boldsymbol{\gamma}, \tag{7.5}$$

where $\gamma := \det(\gamma_{ij})$. This would correspond to the choice $\Psi = \gamma^{1/12}$ in Eq. (7.1). All the metrics $\boldsymbol{\gamma}$ in the same conformal equivalence class lead to the same value of $\hat{\boldsymbol{\gamma}}$. However, since the determinant γ depends upon the choice of coordinates to express the components γ_{ij}, $\Psi = \gamma^{1/12}$ would not be a scalar field. Actually, the quantity $\hat{\boldsymbol{\gamma}}$ is not a tensor field, but a tensor density, of weight $-2/3$.

Let us recall that a ***tensor density of weight*** $n \in \mathbb{Q}$ is a quantity $\boldsymbol{\tau}$ such that

$$\boldsymbol{\tau} = \gamma^{n/2}\boldsymbol{T}, \tag{7.6}$$

where $\boldsymbol{T}$ is a tensor field.

Remark 7.3 The conformal "metric" (7.5) has been used notably in the BSSN formulation [9, 10] for the time evolution of the 3+1 Einstein system, to be discussed in Chap. 10. An "associated" connection $\hat{D}$ has been introduced, such that $\hat{D}\hat{\boldsymbol{\gamma}} = 0$. However, since $\hat{\boldsymbol{\gamma}}$ is a tensor density and not a tensor field, there is not a unique connection associated with it (the Levi–Civita connection, cf. Sect. 2.4.2). In particular one has $\boldsymbol{D}\hat{\boldsymbol{\gamma}} = 0$, so that the connection $\boldsymbol{D}$ associated with the metric $\boldsymbol{\gamma}$ is "associated" with $\hat{\boldsymbol{\gamma}}$, in addition to $\hat{\boldsymbol{D}}$. As a consequence, some of the formulæ presented in the original references [9, 10] for the BSSN formalism have a meaning only for Cartesian coordinates.

7.2.2 Background Metric

To clarify the meaning of $\hat{\boldsymbol{D}}$ (i.e. to avoid working with tensor densities) and to allow for the use of spherical coordinates, we introduce an extra structure on the hypersurfaces Σ_t, namely a ***background metric*** $\boldsymbol{f}$ [11]. It is asked that the signature of $\boldsymbol{f}$ is $(+,+,+)$, i.e. that $\boldsymbol{f}$ is a Riemannian metric, as $\boldsymbol{\gamma}$ (cf. Sect. 2.3.2). Moreover, we tight $\boldsymbol{f}$ to the coordinates (x^i) on Σ_t by demanding that the components f_{ij} of $\boldsymbol{f}$ with respect to (x^i) obey to

$$\frac{\partial f_{ij}}{\partial t} = 0. \tag{7.7}$$

An equivalent writing of this is

$$\mathcal{L}_{\boldsymbol{\partial}_t} \boldsymbol{f} = 0, \tag{7.8}$$

i.e. the metric $\boldsymbol{f}$ is Lie-dragged along the coordinate time evolution vector $\boldsymbol{\partial}_t$.

If the topology of Σ_t enables it, it is quite natural to choose $\boldsymbol{f}$ to be flat, i.e. such that its Riemann tensor vanishes. However, in this chapter, we shall not make such hypothesis, except in Sect. 7.6.

The inverse metric is denoted by f^{ij}:

$$f^{ik} f_{kj} = \delta^i_{\ j}. \tag{7.9}$$

In particular note that, except for the very special case $\gamma_{ij} = f_{ij}$, one has

$$f^{ij} \neq \gamma^{ik} \gamma^{jl} f_{kl}. \tag{7.10}$$

We denote by $\bar{\boldsymbol{D}}$ the Levi–Civita connection associated with $\boldsymbol{f}$:

$$\bar{D}_k f_{ij} = 0, \tag{7.11}$$

and define

$$\bar{D}^i = f^{ij} \bar{D}_j. \tag{7.12}$$

The Christoffel symbols of the connection $\bar{\boldsymbol{D}}$ with respect to the coordinates (x^i) are denoted by $\bar{\Gamma}^k{}_{ij}$; they are given by Eq. (2.62):

$$\bar{\Gamma}^k_{\ ij} = \frac{1}{2} f^{kl} \left(\frac{\partial f_{lj}}{\partial x^i} + \frac{\partial f_{il}}{\partial x^j} - \frac{\partial f_{ij}}{\partial x^l} \right). \tag{7.13}$$

7.2.3 Conformal Metric

Thanks to $\boldsymbol{f}$, we define

$$\boxed{\tilde{\boldsymbol{\gamma}} := \Psi^{-4} \boldsymbol{\gamma}}, \tag{7.14}$$

where

$$\boxed{\Psi := \left(\frac{\gamma}{f}\right)^{1/12}}, \quad \gamma := \det(\gamma_{ij}), \quad f := \det(f_{ij}). \tag{7.15}$$

The key point is that, contrary to γ, Ψ is a tensor field on Σ_t. Indeed a change of coordinates $(x^i) \mapsto (x^{i'})$ induces the following changes in the determinants:

$$\gamma' = (\det J)^2 \gamma \tag{7.16a}$$

$$f' = (\det J)^2 f, \tag{7.16b}$$

where J denotes the Jacobian matrix

$$J^i{}_{i'} := \frac{\partial x^i}{\partial x^{i'}}. \tag{7.17}$$

From Eqs. (7.16) it is obvious that $\gamma'/f' = \gamma/f$, which shows that γ/f, and hence Ψ, is a scalar field. Of course, this scalar field depends upon the choice of the background metric $\boldsymbol{f}$. Ψ being a scalar field, the quantity $\tilde{\boldsymbol{\gamma}}$ defined by (7.14) is a tensor field on Σ_t. Moreover, it is a Riemannian metric on Σ_t. We shall call it the ***conformal metric***. By construction, it satisfies

$$\boxed{\det(\tilde{\gamma}_{ij}) = f}. \tag{7.18}$$

This is the "unit-determinant" condition fulfilled by $\tilde{\boldsymbol{\gamma}}$. Indeed, if one uses for (x^i) Cartesian-type coordinates, then $f = 1$. But the condition (7.18) is more flexible and allows for the use of e.g. spherical type coordinates $(x^i) = (r, \theta, \varphi)$, for which $f = r^4 \sin^2\theta$.

We define the ***inverse conformal metric*** $\tilde{\gamma}^{ij}$ by the requirement

$$\tilde{\gamma}_{ik}\tilde{\gamma}^{kj} = \delta_i{}^j, \tag{7.19}$$

which is equivalent to

$$\tilde{\gamma}^{ij} = \Psi^4 \gamma^{ij}. \tag{7.20}$$

Hence, combining with Eq. (7.14),

$$\boxed{\gamma_{ij} = \Psi^4 \tilde{\gamma}_{ij}} \quad \text{and} \quad \boxed{\gamma^{ij} = \Psi^{-4}\tilde{\gamma}^{ij}}. \tag{7.21}$$

Note also that although we are using the same notation $\tilde{\gamma}$ for both $\tilde{\gamma}_{ij}$ and $\tilde{\gamma}^{ij}$, one has

$$\tilde{\gamma}^{ij} \neq \gamma^{ik}\gamma^{jl}\tilde{\gamma}_{kl}, \tag{7.22}$$

except in the special case $\Psi = 1$.

Example 7.1 A simple example of a conformal decomposition is provided by the Schwarzschild spacetime described with ***isotropic coordinates*** $(x^\alpha) = (t, r, \theta, \varphi)$; the latter are related to the standard ***Schwarzschild coordinates*** (t, R, θ, φ) by $R = r\left(1+\frac{m}{2r}\right)^2$. The components of the spacetime metric tensor in the isotropic coordinates are given by (see e.g. [12])

$$g_{\mu\nu}\mathrm{d}x^\mu \mathrm{d}x^\nu = -\left(\frac{1-\frac{m}{2r}}{1+\frac{m}{2r}}\right)^2 \mathrm{d}t^2 + \left(1+\frac{m}{2r}\right)^4 \left[\mathrm{d}r^2 + r^2(\mathrm{d}\theta^2 + \sin^2\theta \mathrm{d}\varphi^2)\right], \tag{7.23}$$

where the constant m is the mass of the Schwarzschild solution. If we define the background metric to be $f_{ij} = \mathrm{diag}(1, r^2, r^2\sin^2\theta)$, we read on this line element that $\boldsymbol{\gamma} = \Psi^4 \tilde{\boldsymbol{\gamma}}$ with

$$\Psi = 1 + \frac{m}{2r} \tag{7.24}$$

and $\tilde{\boldsymbol{\gamma}} = \boldsymbol{f}$. Notice that in this example, the background metric $\boldsymbol{f}$ is flat and that the conformal metric coincides with the background metric.

Example 7.2 Another example is provided by the weak field metric introduced in Sect. 6.2.3 to take Newtonian limits. We read on the line element (6.12) that the conformal metric is $\tilde{\boldsymbol{\gamma}} = \boldsymbol{f}$ and that the conformal factor is

$$\Psi = (1 - 2\Phi)^{1/4} \simeq 1 - \frac{1}{2}\Phi, \tag{7.25}$$

where $|\Phi| \ll 1$ and Φ reduces to the gravitational potential at the Newtonian limit. As a side remark, notice that if we identify expressions (7.24) and (7.25), we recover the standard expression $\Phi = -m/r$ (remember $G = 1$!) for the Newtonian gravitational potential outside a spherical distribution of mass.

7.2.4 Conformal Connection

$\tilde{\boldsymbol{\gamma}}$ being a well defined metric on Σ_t, let $\tilde{\boldsymbol{D}}$ be the Levi–Civita connection associated with it (cf. Sect. 2.4.2):

$$\tilde{\boldsymbol{D}}\tilde{\boldsymbol{\gamma}} = 0. \tag{7.26}$$

Let us denote by $\tilde{\Gamma}^k{}_{ij}$ the Christoffel symbols of $\tilde{\boldsymbol{D}}$ with respect to the coordinates (x^i) [cf. Eq. (2.62)]:

$$\tilde{\Gamma}^k_{ij} = \frac{1}{2}\tilde{\gamma}^{kl}\left(\frac{\partial\tilde{\gamma}_{lj}}{\partial x^i} + \frac{\partial\tilde{\gamma}_{il}}{\partial x^j} - \frac{\partial\tilde{\gamma}_{ij}}{\partial x^l}\right). \tag{7.27}$$

Given a tensor field $\boldsymbol{T}$ of type (p, q) on Σ_t, the covariant derivatives $\tilde{\boldsymbol{D}}\boldsymbol{T}$ and $\boldsymbol{DT}$ are related by the formula

$$D_k T^{i_1\ldots i_p}{}_{j_1\ldots j_q} = \tilde{D}_k T^{i_1\ldots i_p}{}_{j_1\ldots j_q} + \sum_{r=1}^{p} C^{i_r}{}_{kl} T^{i_1\ldots l\ldots i_p}{}_{j_1\ldots j_q}$$
$$- \sum_{r=1}^{q} C^l{}_{kj_r} T^{i_1\ldots i_p}{}_{j_1\ldots l\ldots j_q}, \tag{7.28}$$

where[2]

$$C^k{}_{ij} := \Gamma^k{}_{ij} - \tilde{\Gamma}^k_{ij}, \tag{7.29}$$

$\Gamma^k{}_{ij}$ being the Christoffel symbols of the connection $\boldsymbol{D}$. The formula (7.28) follows immediately from the expressions of $\boldsymbol{DT}$ and $\tilde{\boldsymbol{D}}\boldsymbol{T}$ in terms of respectively the Christoffel symbols $\Gamma^k{}_{ij}$ and $\tilde{\Gamma}^k_{ij}$ [cf. Eq. (2.54)]. Since $D_k T^{i_1\ldots i_p}{}_{j_1\ldots j_q} - \tilde{D}_k T^{i_1\ldots i_p}{}_{j_1\ldots j_q}$ are the components of a tensor field, namely $\boldsymbol{DT} - \tilde{\boldsymbol{D}}\boldsymbol{T}$, it follows from Eq. (7.28) that the $C^k{}_{ij}$ are also the components of a tensor field. Hence we recover a well known property: although the Christoffel symbols are not the components of any tensor field, the difference between two sets of them represents the components of a tensor field. We may express the tensor $C^k{}_{ij}$ in terms of the $\tilde{\boldsymbol{D}}$-derivatives of the metric $\boldsymbol{\gamma}$, by the same formula than the one for the Christoffel symbols $\Gamma^k{}_{ij}$, except that the partial derivatives are replaced by $\tilde{\boldsymbol{D}}$-derivatives:

$$C^k{}_{ij} = \frac{1}{2}\gamma^{kl}\left(\tilde{D}_i\gamma_{lj} + \tilde{D}_j\gamma_{il} - \tilde{D}_l\gamma_{ij}\right). \tag{7.30}$$

It is easy to establish this relation by evaluating the right-hand side, expressing the $\tilde{\boldsymbol{D}}$-derivatives of $\boldsymbol{\gamma}$ in terms of the Christoffel symbols $\tilde{\Gamma}^k_{ij}$:

$$\begin{aligned}
\frac{1}{2}\gamma^{kl}\left(\tilde{D}_i\gamma_{lj} + \tilde{D}_j\gamma_{il} - \tilde{D}_l\gamma_{ij}\right) &= \frac{1}{2}\gamma^{kl}\Bigg(\frac{\partial\gamma_{lj}}{\partial x^i} - \tilde{\Gamma}^m{}_{il}\gamma_{mj} - \tilde{\Gamma}^m{}_{ij}\gamma_{lm} + \frac{\partial\gamma_{il}}{\partial x^j} - \tilde{\Gamma}^m{}_{ji}\gamma_{ml} \\
&\quad - \tilde{\Gamma}^m{}_{jl}\gamma_{im} - \frac{\partial\gamma_{ij}}{\partial x^l} + \tilde{\Gamma}^m{}_{li}\gamma_{mj} + \tilde{\Gamma}^m{}_{lj}\gamma_{im}\Bigg) \\
&= \Gamma^k{}_{ij} + \frac{1}{2}\gamma^{kl}(-2)\tilde{\Gamma}^m{}_{ij}\gamma_{lm} \\
&= \Gamma^k{}_{ij} - \delta^k{}_m\tilde{\Gamma}^m_{ij} \\
&= C^k{}_{ij},
\end{aligned}$$

[2] The C^k_{ij} are not to be confused with the components of the Cotton tensor discussed in Sect. 7.1 Since we shall no longer make use of the latter, no confusion may arise.

where we have used the symmetry with respect to (i, j) of the Christoffel symbols $\tilde{\Gamma}^k{}_{ij}$ to get the second line.

Let us replace γ_{ij} and γ^{ij} in Eq. (7.30) by their expressions (7.21) in terms of $\tilde{\gamma}_{ij}$, $\tilde{\gamma}^{ij}$ and Ψ:

$$\begin{aligned} C^k{}_{ij} &= \frac{1}{2}\Psi^{-4}\tilde{\gamma}^{kl}\left[\tilde{D}_i(\Psi^4\tilde{\gamma}_{lj}) + \tilde{D}_j(\Psi^4\gamma_{il}) - \tilde{D}_l(\Psi^4\tilde{\gamma}_{ij})\right] \\ &= \frac{1}{2}\Psi^{-4}\tilde{\gamma}^{kl}\left(\tilde{\gamma}_{lj}\tilde{D}_i\Psi^4 + \tilde{\gamma}_{il}\tilde{D}_j\Psi^4 - \tilde{\gamma}_{ij}\tilde{D}_l\Psi^4\right) \\ &= \frac{1}{2}\Psi^{-4}\left(\delta^k{}_j\tilde{D}_i\Psi^4 + \delta^k{}_i\tilde{D}_j\Psi^4 - \tilde{\gamma}_{ij}\tilde{D}^k\Psi^4\right) \end{aligned}$$

Hence

$$\boxed{C^k{}_{ij} = 2\left(\delta^k{}_i\tilde{D}_j\ln\Psi + \delta^k{}_j\tilde{D}_i\ln\Psi - \tilde{D}^k\ln\Psi\,\tilde{\gamma}_{ij}\right)}. \quad (7.31)$$

A useful application of this formula is to derive the relation between the two covariant derivatives $\boldsymbol{Dv}$ and $\tilde{\boldsymbol{D}}\boldsymbol{v}$ of a vector field $\boldsymbol{v} \in \mathscr{T}(\Sigma_t)$. From Eq. (7.28), we have

$$D_j v^i = \tilde{D}_j v^i + C^i{}_{jk} v^k,$$

so that expression (7.31) yields

$$D_j v^i = \tilde{D}_j v^i + 2\left(v^k\tilde{D}_k\ln\Psi\,\delta^i{}_j + v^i\tilde{D}_j\ln\Psi - \tilde{D}^i\ln\Psi\,\tilde{\gamma}_{jk}v^k\right). \quad (7.32)$$

Taking the trace, we get a relation between the two divergences:

$$D_i v^i = \tilde{D}_i v^i + 6 v^i \tilde{D}_i \ln\Psi, \quad (7.33)$$

or equivalently,

$$\boxed{D_i v^i = \Psi^{-6}\tilde{D}_i\left(\Psi^6 v^i\right)}. \quad (7.34)$$

Remark 7.4 The above formula could have been obtained directly from the standard expression (2.65) of the divergence of a vector field in terms of partial derivatives and the determinant γ of $\boldsymbol{\gamma}$, both with respect to some coordinate system (x^i):

$$D_i v^i = \frac{1}{\sqrt{\gamma}}\frac{\partial}{\partial x^i}\left(\sqrt{\gamma}v^i\right). \quad (7.35)$$

Noticing that $\gamma_{ij} = \Psi^4\tilde{\gamma}_{ij}$ implies $\sqrt{\gamma} = \Psi^6\sqrt{\tilde{\gamma}}$, we get immediately Eq. (7.34).

7.3 Expression of the Ricci Tensor

In this section, we express the Ricci tensor $\boldsymbol{R}$, which appears in the 3+1 Einstein system (5.68–5.71), in terms of the Ricci tensor $\tilde{\boldsymbol{R}}$ associated with the metric $\tilde{\boldsymbol{\gamma}}$ and derivatives of the conformal factor Ψ.

7.3.1 General Formula Relating the Two Ricci Tensors

The starting point of the calculation is the Ricci identity (3.15) applied to a generic vector field $\boldsymbol{v} \in \mathscr{T}(\Sigma_t)$:

$$(D_i D_j - D_j D_i) v^k = R^k{}_{lij} v^l. \tag{7.36}$$

Contracting this relation on the indices i and k (and relabelling $i \leftrightarrow j$) makes the Ricci tensor appear:

$$R_{ij} v^j = D_j D_i v^j - D_i D_j v^j. \tag{7.37}$$

Expressing the $\boldsymbol{D}$-derivatives in term of the $\tilde{\boldsymbol{D}}$-derivatives via formula (7.28), we get

$$\begin{aligned}
R_{ij} v^j &= \tilde{D}_j (D_i v^j) - C^k{}_{ji} D_k v^j + C^j{}_{jk} D_i v^k - \tilde{D}_i (D_j v^j) \\
&= \tilde{D}_j (\tilde{D}_i v^j + C^j{}_{ik} v^k) - C^k{}_{ji} (\tilde{D}_k v^j + C^j{}_{kl} v^l) + C^j{}_{jk} (\tilde{D}_i v^k + C^k{}_{il} v^l) \\
&\quad - \tilde{D}_i (\tilde{D}_j v^j + C^j{}_{jk} v^k) \\
&= \tilde{D}_j \tilde{D}_i v^j + \tilde{D}_j C^j{}_{ik} v^k + C^j{}_{ik} \tilde{D}_j v^k - C^k{}_{ji} \tilde{D}_k v^j - C^k{}_{ji} C^j{}_{kl} v^l + C^j{}_{jk} \tilde{D}_i v^k \\
&\quad + C^j{}_{jk} C^k{}_{il} v^l - \tilde{D}_i \tilde{D}_j v^j - \tilde{D}_i C^j{}_{jk} v^k - C^j{}_{jk} \tilde{D}_i v^k \\
&= \tilde{D}_j \tilde{D}_i v^j - \tilde{D}_i \tilde{D}_j v^j + \tilde{D}_j C^j{}_{ik} v^k - C^k{}_{ji} C^j{}_{kl} v^l + C^j{}_{jk} C^k{}_{il} v^l - \tilde{D}_i C^j{}_{jk} v^k.
\end{aligned} \tag{7.38}$$

We can replace the first two terms in the right-hand side via the contracted Ricci identity similar to Eq. (7.37) but regarding the connection $\tilde{\boldsymbol{D}}$:

$$\tilde{D}_j \tilde{D}_i v^j - \tilde{D}_i \tilde{D}_j v^j = \tilde{R}_{ij} v^j \tag{7.39}$$

Then, after some relabelling $j \leftrightarrow k$ or $j \leftrightarrow l$ of dumb indices, Eq. (7.38) becomes

$$R_{ij} v^j = \tilde{R}_{ij} v^j + \tilde{D}_k C^k{}_{ij} v^j - \tilde{D}_i C^k{}_{jk} v^j + C^l{}_{lk} C^k{}_{ij} v^j - C^k{}_{li} C^l{}_{kj} v^j.$$

This relation being valid for any vector field $\boldsymbol{v}$, we conclude that

$$\boxed{R_{ij} = \tilde{R}_{ij} + \tilde{D}_k C^k{}_{ij} - \tilde{D}_i C^k{}_{kj} + C^k{}_{ij} C^l{}_{lk} - C^k{}_{il} C^l{}_{kj}}, \tag{7.40}$$

where we have used the symmetry of $C^k{}_{ij}$ in its two last indices.

Remark 7.5 Equation (7.40) is the general formula relating the Ricci tensors of two connections, with the $C^k{}_{ij}$'s being the differences of their Christoffel symbols [Eq. (7.29)]. This formula does not rely on the fact that the metrics $\boldsymbol{\gamma}$ and $\tilde{\boldsymbol{\gamma}}$ associated with the two connections are conformally related.

7.3.2 Expression in Terms of the Conformal Factor

Let now replace $C^k{}_{ij}$ in Eq. (7.40) by its expression in terms of the derivatives of Ψ, i.e. Eq. (7.31). First of all, by contracting Eq. (7.31) on the indices j and k, we have

$$C^k{}_{ki} = 2\left(\tilde{D}_i \ln\Psi + 3\tilde{D}_i \ln\Psi - \tilde{D}_i \ln\Psi\right),$$

i.e.

$$C^k{}_{ki} = 6\tilde{D}_i \ln\Psi, \tag{7.41}$$

whence $\tilde{D}_i C^k{}_{kj} = 6\tilde{D}_i\tilde{D}_j \ln\Psi$. Besides,

$$\begin{aligned}\tilde{D}_k C^k{}_{ij} &= 2\left(\tilde{D}_i\tilde{D}_j \ln\Psi + \tilde{D}_j\tilde{D}_i \ln\Psi - \tilde{D}_k\tilde{D}^k \ln\Psi\,\tilde{\gamma}_{ij}\right)\\ &= 4\tilde{D}_i\tilde{D}_j \ln\Psi - 2\tilde{D}_k\tilde{D}^k \ln\Psi\,\tilde{\gamma}_{ij}.\end{aligned}$$

Consequently, Eq. (7.40) becomes

$$\begin{aligned}R_{ij} = \tilde{R}_{ij} &+ 4\tilde{D}_i\tilde{D}_j \ln\Psi - 2\tilde{D}_k\tilde{D}^k \ln\Psi\,\tilde{\gamma}_{ij} - 6\tilde{D}_i\tilde{D}_j \ln\Psi\\ &+ 2\left(\delta^k{}_i\tilde{D}_j \ln\Psi + \delta^k{}_j\tilde{D}_i \ln\Psi - \tilde{D}^k \ln\Psi\,\tilde{\gamma}_{ij}\right) \times 6\tilde{D}_k \ln\Psi\\ &- 4\left(\delta^k{}_i\tilde{D}_l \ln\Psi + \delta^k{}_l\tilde{D}_i \ln\Psi - \tilde{D}^k \ln\Psi\,\tilde{\gamma}_{il}\right)\\ &\times\left(\delta^l{}_k\tilde{D}_j \ln\Psi + \delta^l{}_j\tilde{D}_k \ln\Psi - \tilde{D}^l \ln\Psi\,\tilde{\gamma}_{kj}\right).\end{aligned}$$

Expanding and simplifying, we get

$$\boxed{R_{ij} = \tilde{R}_{ij} - 2\tilde{D}_i\tilde{D}_j \ln\Psi - 2\tilde{D}_k\tilde{D}^k \ln\Psi\,\tilde{\gamma}_{ij} + 4\tilde{D}_i \ln\Psi\,\tilde{D}_j \ln\Psi - 4\tilde{D}_k \ln\Psi\,\tilde{D}^k \ln\Psi\,\tilde{\gamma}_{ij}}\,. \tag{7.42}$$

7.3.3 Formula for the Scalar Curvature

The relation between the scalar curvatures is obtained by taking the trace of Eq. (7.42) with respect to $\boldsymbol{\gamma}$:

$$
\begin{aligned}
R &= \gamma^{ij} R_{ij} = \Psi^{-4} \tilde{\gamma}^{ij} R_{ij} \\
&= \Psi^{-4} \big(\tilde{\gamma}^{ij} \tilde{R}_{ij} - 2\tilde{D}_i \tilde{D}^i \ln \Psi - 2\tilde{D}_k \tilde{D}^k \ln \Psi \times 3 + 4\tilde{D}_i \ln \Psi \tilde{D}^i \ln \Psi \\
&\quad - 4\tilde{D}_k \ln \Psi \tilde{D}^k \ln \Psi \times 3 \big) \\
R &= \Psi^{-4} \left[\tilde{R} - 8 \left(\tilde{D}_i \tilde{D}^i \ln \Psi + \tilde{D}_i \ln \Psi \tilde{D}^i \ln \Psi \right) \right],
\end{aligned}
\tag{7.43}
$$

where

$$
\boxed{\tilde{R} := \tilde{\gamma}^{ij} \tilde{R}_{ij}} \tag{7.44}
$$

is the scalar curvature associated with the conformal metric. Noticing that

$$
\tilde{D}_i \tilde{D}^i \ln \Psi = \Psi^{-1} \tilde{D}_i \tilde{D}^i \Psi - \tilde{D}_i \ln \Psi \tilde{D}^i \ln \Psi,
$$

we can rewrite the above formula as

$$
\boxed{R = \Psi^{-4} \tilde{R} - 8\Psi^{-5} \tilde{D}_i \tilde{D}^i \Psi}\,. \tag{7.45}
$$

7.4 Conformal Decomposition of the Extrinsic Curvature

7.4.1 Traceless Decomposition

The first step is to decompose the extrinsic curvature $\boldsymbol{K}$ of the hypersurface Σ_t into a trace part and a traceless one, the trace being taken with the metric $\boldsymbol{\gamma}$, i.e. we define

$$
\boldsymbol{A} := \boldsymbol{K} - \frac{1}{3} K \boldsymbol{\gamma}, \tag{7.46}
$$

where $K := \mathrm{tr}_{\boldsymbol{\gamma}} \boldsymbol{K} = K^i_{\ i} = \gamma^{ij} K_{ij}$ is the trace of $\boldsymbol{K}$ with respect to $\boldsymbol{\gamma}$, i.e. (minus three times) the mean curvature of Σ_t embedded in $(\mathscr{M}, \boldsymbol{g})$ (cf. Sect. 3.3.4). The bilinear form $\boldsymbol{A}$ is by construction traceless:

$$
\mathrm{tr}_{\boldsymbol{\gamma}} \boldsymbol{A} = \gamma^{ij} A_{ij} = 0. \tag{7.47}
$$

In what follows, we shall work occasionally with the twice contravariant version of $\boldsymbol{K}$, i.e. the tensor $\overset{\twoheadrightarrow}{\boldsymbol{K}}$, the components of which are [cf. Eq. (2.40)]

$$
K^{ij} = \gamma^{ik} \gamma^{jl} K_{kl}. \tag{7.48}
$$

Similarly, we define $\overset{\twoheadrightarrow}{\boldsymbol{A}}$ as the twice contravariant tensor, the components of which are

$$A^{ij} = \gamma^{ik}\gamma^{jl}A_{kl}. \tag{7.49}$$

Hence the traceless decomposition of $\boldsymbol{K}$ and $\vec{\vec{\boldsymbol{K}}}$:

$$\boxed{K_{ij} = A_{ij} + \frac{1}{3}K\gamma_{ij}} \quad \text{and} \quad \boxed{K^{ij} = A^{ij} + \frac{1}{3}K\gamma^{ij}}. \tag{7.50}$$

7.4.2 Conformal Decomposition of the Traceless Part

Let us now perform the conformal decomposition of the traceless part of $\boldsymbol{K}$, namely, let us write

$$A^{ij} = \Psi^{\alpha}\tilde{A}^{ij} \tag{7.51}$$

for some power α to be determined. Actually there are two natural choices: $\alpha = -4$ and $\alpha = -10$, as we discuss hereafter:

7.4.2.1 "Time-Evolution" Scaling: $\alpha = -4$

Let us consider Eq. (4.30) which express the time evolution of $\boldsymbol{\gamma}$ in terms of $\boldsymbol{K}$:

$$\boxed{\mathcal{L}_{\boldsymbol{m}}\gamma_{ij} = -2NK_{ij}}. \tag{7.52}$$

By means of Eqs. (7.21) and (7.50), this equation becomes

$$\mathcal{L}_{\boldsymbol{m}}\left(\Psi^4\tilde{\gamma}_{ij}\right) = -2NA_{ij} - \frac{2}{3}NK\gamma_{ij},$$

i.e.

$$\mathcal{L}_{\boldsymbol{m}}\tilde{\gamma}_{ij} = -2N\Psi^{-4}A_{ij} - \frac{2}{3}\left(NK + 6\mathcal{L}_{\boldsymbol{m}}\ln\Psi\right)\tilde{\gamma}_{ij}. \tag{7.53}$$

The trace of this relation with respect to $\tilde{\boldsymbol{\gamma}}$ is, since A_{ij} is traceless,

$$\tilde{\gamma}^{ij}\mathcal{L}_{\boldsymbol{m}}\tilde{\gamma}_{ij} = -2(NK + 6\mathcal{L}_{\boldsymbol{m}}\ln\Psi). \tag{7.54}$$

Now

$$\tilde{\gamma}^{ij}\mathcal{L}_{\boldsymbol{m}}\tilde{\gamma}_{ij} = \mathcal{L}_{\boldsymbol{m}}\ln\det(\tilde{\gamma}_{ij}). \tag{7.55}$$

This follows from the general law (2.64) for the variation of the determinant of an invertible matrix: applying Eq. (2.64) to $A = (\tilde{\gamma}_{ij})$ and $\delta = \mathcal{L}_{\boldsymbol{m}}$ gives Eq. (7.55). By construction, $\det(\tilde{\gamma}_{ij}) = f$ [Eq. (7.18)], so that, replacing $\boldsymbol{m}$ by $\boldsymbol{\partial}_t - \boldsymbol{\beta}$, we get

$$\mathcal{L}_{\boldsymbol{m}} \ln \det(\tilde{\gamma}_{ij}) = \left(\frac{\partial}{\partial t} - \mathcal{L}_{\boldsymbol{\beta}} \right) \ln f$$

But, as a consequence of Eq. (7.7), $\partial f / \partial t = 0$, so that

$$\mathcal{L}_{\boldsymbol{m}} \ln \det(\tilde{\gamma}_{ij}) = -\mathcal{L}_{\boldsymbol{\beta}} \ln f = -\mathcal{L}_{\boldsymbol{\beta}} \ln \det(\tilde{\gamma}_{ij}).$$

Applying again formula (2.64) to $A = (\tilde{\gamma}_{ij})$ and $\delta = \mathcal{L}_{\boldsymbol{\beta}}$, we get

$$\begin{aligned}
\mathcal{L}_{\boldsymbol{m}} \ln \det(\tilde{\gamma}_{ij}) &= -\tilde{\gamma}^{ij} \mathcal{L}_{\boldsymbol{\beta}} \tilde{\gamma}_{ij} \\
&= -\tilde{\gamma}^{ij} \Big(\beta^k \underbrace{\tilde{D}_k \tilde{\gamma}_{ij}}_{0} + \tilde{\gamma}_{kj} \tilde{D}_i \beta^k + \tilde{\gamma}_{ik} \tilde{D}_j \beta^k \Big) \\
&= -\delta^i{}_k \tilde{D}_i \beta^k - \delta^j{}_k \tilde{D}_j \beta^k \\
&= -2 \tilde{D}_i \beta^i .
\end{aligned}$$

Hence Eq. (7.55) becomes

$$\tilde{\gamma}^{ij} \mathcal{L}_{\boldsymbol{m}} \tilde{\gamma}_{ij} = -2 \tilde{D}_i \beta^i , \tag{7.56}$$

so that, after substitution into Eq. (7.54), we get

$$NK + 6 \mathcal{L}_{\boldsymbol{m}} \ln \Psi = \tilde{D}_i \beta^i , \tag{7.57}$$

i.e. the following evolution equation for the conformal factor:

$$\boxed{\left(\frac{\partial}{\partial t} - \mathcal{L}_{\boldsymbol{\beta}} \right) \ln \Psi = \frac{1}{6} \left(\tilde{D}_i \beta^i - NK \right)} . \tag{7.58}$$

Finally, substituting Eqs. (7.57) into (7.53) yields an evolution equation for the conformal metric:

$$\left(\frac{\partial}{\partial t} - \mathcal{L}_{\boldsymbol{\beta}} \right) \tilde{\gamma}_{ij} = -2N \Psi^{-4} A_{ij} - \frac{2}{3} \tilde{D}_k \beta^k \tilde{\gamma}_{ij} .$$

This suggests to introduce the quantity

$$\boxed{\tilde{A}_{ij} := \Psi^{-4} A_{ij}} \tag{7.59}$$

to write

$$\boxed{\left(\frac{\partial}{\partial t} - \mathcal{L}_{\boldsymbol{\beta}} \right) \tilde{\gamma}_{ij} = -2N \tilde{A}_{ij} - \frac{2}{3} \tilde{D}_k \beta^k \tilde{\gamma}_{ij}} . \tag{7.60}$$

Notice that, as an immediate consequence of Eq. (7.47), $\tilde{A}_{ij}$ is traceless:

$$\boxed{\tilde{\gamma}^{ij}\tilde{A}_{ij} = 0}. \tag{7.61}$$

Let us rise the indices of $\tilde{A}_{ij}$ with the conformal metric, defining

$$\tilde{A}^{ij} := \tilde{\gamma}^{ik}\tilde{\gamma}^{jl}\tilde{A}_{kl}. \tag{7.62}$$

Since $\tilde{\gamma}^{ij} = \Psi^4\gamma^{ij}$, we get

$$\boxed{\tilde{A}^{ij} = \Psi^4 A^{ij}}. \tag{7.63}$$

This corresponds to the scaling factor $\alpha = -4$ in Eq. (7.51). This choice of scaling has been first considered by Nakamura in 1994 [13].

We can deduce from Eq. (7.60) an evolution equation for the inverse conformal metric $\tilde{\gamma}^{ij}$. Indeed, raising the indices of Eq. (7.60) with $\tilde{\boldsymbol{\gamma}}$, we get

$$\begin{aligned}
\tilde{\gamma}^{ik}\tilde{\gamma}^{jl}\mathcal{L}_{\boldsymbol{m}}\tilde{\gamma}_{kl} &= -2N\tilde{A}^{ij} - \frac{2}{3}\tilde{D}_k\beta^k\tilde{\gamma}^{ij} \\
\tilde{\gamma}^{ik}\big[\mathcal{L}_{\boldsymbol{m}}(\underbrace{\tilde{\gamma}^{jl}\tilde{\gamma}_{kl}}_{\delta^j{}_k}) - \tilde{\gamma}_{kl}\mathcal{L}_{\boldsymbol{m}}\tilde{\gamma}^{jl}\big] &= -2N\tilde{A}^{ij} - \frac{2}{3}\tilde{D}_k\beta^k\tilde{\gamma}^{ij} \\
-\underbrace{\tilde{\gamma}^{ik}\tilde{\gamma}_{kl}}_{\delta^i{}_l}\mathcal{L}_{\boldsymbol{m}}\tilde{\gamma}^{jl} &= -2N\tilde{A}^{ij} - \frac{2}{3}\tilde{D}_k\beta^k\tilde{\gamma}^{ij},
\end{aligned}$$

hence

$$\boxed{\left(\frac{\partial}{\partial t} - \mathcal{L}_{\boldsymbol{\beta}}\right)\tilde{\gamma}^{ij} = 2N\tilde{A}^{ij} + \frac{2}{3}\tilde{D}_k\beta^k\tilde{\gamma}^{ij}}. \tag{7.64}$$

7.4.2.2 "Momentum-Constraint" Scaling: $\alpha = -10$

Whereas the scaling $\alpha = -4$ was suggested by the evolution equation (7.52) (or equivalently Eq. (5.68) of the 3+1 Einstein system), another scaling arises when contemplating the momentum constraint equation (5.71). In this equation appears the divergence of the extrinsic curvature, that we can write using the twice contravariant version of $\boldsymbol{K}$ and Eq. (7.50):

$$D_j K^{ij} = D_j A^{ij} + \frac{1}{3}D^i K. \tag{7.65}$$

Now, from Eqs. (7.28), (7.31) and (7.41),

$$\begin{aligned}
D_j A^{ij} &= \tilde{D}_j A^{ij} + C^i{}_{jk}A^{kj} + C^j{}_{jk}A^{ik} \\
&= \tilde{D}_j A^{ij} + 2\left(\delta^i{}_j\tilde{D}_k \ln\Psi + \delta^i{}_k\tilde{D}_j\ln\Psi - \tilde{D}^i\ln\Psi\,\tilde{\gamma}_{jk}\right)A^{kj} + 6\tilde{D}_k\ln\Psi\,A^{ik} \\
&= \tilde{D}_j A^{ij} + 10A^{ij}\tilde{D}_j\ln\Psi - 2\tilde{D}^i\ln\Psi\,\tilde{\gamma}_{jk}A^{jk}.
\end{aligned}$$

Since $\boldsymbol{A}$ is traceless, $\tilde{\gamma}_{jk}A^{jk} = \Psi^{-4}\gamma_{jk}A^{jk} = 0$. Then the above equation reduces to $D_jA^{ij} = \tilde{D}_jA^{ij} + 10A^{ij}\tilde{D}_j \ln \Psi$, which can be rewritten as

$$\boxed{D_jA^{ij} = \Psi^{-10}\tilde{D}_j\left(\Psi^{10}A^{ij}\right)}. \tag{7.66}$$

Notice that this identity is valid only because A^{ij} is symmetric and traceless.

Equation (7.66) suggests to introduce the quantity[3]

$$\boxed{\hat{A}^{ij} := \Psi^{10}A^{ij}}. \tag{7.67}$$

This corresponds to the scaling factor $\alpha = -10$ in Eq. (7.51). It has been first introduced by Lichnerowicz in 1944 [1]. Thanks to it and Eq. (7.65), the momentum constraint equation (5.71) can be rewritten as

$$\boxed{\tilde{D}_j\hat{A}^{ij} - \frac{2}{3}\Psi^6\tilde{D}^iK = 8\pi\Psi^{10}p^i}. \tag{7.68}$$

As for $\tilde{A}_{ij}$, we define $\hat{A}_{ij}$ as the tensor field deduced from $\hat{A}^{ij}$ by lowering the indices with the conformal metric:

$$\hat{A}_{ij} := \tilde{\gamma}_{ik}\tilde{\gamma}_{jl}\hat{A}^{kl} \tag{7.69}$$

Taking into account Eq. (7.67) and $\tilde{\gamma}_{ij} = \Psi^{-4}\gamma_{ij}$, we get

$$\boxed{\hat{A}_{ij} = \Psi^2 A_{ij}}. \tag{7.70}$$

7.5 Conformal Form of the 3+1 Einstein System

Having performed a conformal decomposition of $\boldsymbol{\gamma}$ and of the traceless part of $\boldsymbol{K}$, we are now in position to rewrite the 3+1 Einstein system (5.68)–(5.71) in terms of conformal quantities.

7.5.1 Dynamical Part of Einstein Equation

Let us consider Eq. (5.69), i.e. the so-called dynamical equation in the 3+1 Einstein system:

---

[3] Notice that we are using a hat, instead of a tilde, to distinguish this quantity from that defined by (7.63).

$$\mathcal{L}_{\boldsymbol{m}} K_{ij} = -D_i D_j N + N \left\{ R_{ij} + K K_{ij} - 2K_{ik} K^k_{\ j} + 4\pi \left[(S - E)\gamma_{ij} - 2S_{ij} \right] \right\}. \tag{7.71}$$

Let us substitute $A_{ij} + (K/3)\gamma_{ij}$ for K_{ij} [Eq. (7.50)]. The left-hand side of the above equation becomes

$$\mathcal{L}_{\boldsymbol{m}} K_{ij} = \mathcal{L}_{\boldsymbol{m}} A_{ij} + \frac{1}{3} \mathcal{L}_{\boldsymbol{m}} K \gamma_{ij} + \frac{1}{3} K \underbrace{\mathcal{L}_{\boldsymbol{m}} \gamma_{ij}}_{-2NK_{ij}} . \tag{7.72}$$

In this equation appears $\mathcal{L}_{\boldsymbol{m}} K$. We may express it by taking the trace of Eq. (7.71) and making use of Eq. (4.47):

$$\mathcal{L}_{\boldsymbol{m}} K = \gamma^{ij} \mathcal{L}_{\boldsymbol{m}} K_{ij} + 2N K_{ij} K^{ij},$$

hence

$$\mathcal{L}_{\boldsymbol{m}} K = -D_i D^i N + N \left[R + K^2 + 4\pi (S - 3E) \right]. \tag{7.73}$$

Let use the Hamiltonian constraint (5.70) to replace $R + K^2$ by $16\pi E + K_{ij} K^{ij}$. Then, writing $\mathcal{L}_{\boldsymbol{m}} K = (\frac{\partial}{\partial t} - \mathcal{L}_{\boldsymbol{\beta}}) K$,

$$\boxed{\left(\frac{\partial}{\partial t} - \mathcal{L}_{\boldsymbol{\beta}} \right) K = -D_i D^i N + N \left[4\pi (E + S) + K_{ij} K^{ij} \right]}. \tag{7.74}$$

Remark 7.6 At the Newtonian limit, as defined by Eqs. (6.13), (6.22) and (6.56), Eq. (7.74) reduces to the Poisson equation for the gravitational potential Φ:

$$\bar{D}_i \bar{D}^i \Phi = 4\pi \rho_0. \tag{7.75}$$

Substituting Eq. (7.73) for $\mathcal{L}_{\boldsymbol{m}} K$ and Eq. (7.71) for $\mathcal{L}_{\boldsymbol{m}} K_{ij}$ into Eq. (7.72) yields

$$\begin{aligned} \mathcal{L}_{\boldsymbol{m}} A_{ij} = & -D_i D_j N + N \left[R_{ij} + \frac{5}{3} K K_{ij} - 2K_{ik} K^k_{\ j} - 8\pi \left(S_{ij} - \frac{1}{3} S \gamma_{ij} \right) \right] \\ & + \frac{1}{3} \left[D_k D^k N - N (R + K^2) \right] \gamma_{ij}. \end{aligned} \tag{7.76}$$

Let us replace K_{ij} by its expression in terms of A_{ij} and K [Eq. (7.50)]: the terms in the right-hand side of the above equation that involve $\boldsymbol{K}$ are then written

$$\begin{aligned} \frac{5K}{3} K_{ij} - 2K_{ik} K^k_j - \frac{K^2}{3} \gamma_{ij} &= \frac{5K}{3} \left(A_{ij} + \frac{K}{3} \gamma_{ij} \right) - 2 \left(A_{ik} + \frac{K}{3} \gamma_{ik} \right) \left(A^k_j + \frac{K}{3} \delta^k_j \right) \\ & \quad - \frac{K^2}{3} \gamma_{ij} \\ &= \frac{1}{3} K A_{ij} - 2A_{ik} A^k_{\ j}. \end{aligned}$$

Accordingly Eq. (7.76) becomes

$$\mathcal{L}_{\boldsymbol{m}} A_{ij} = -D_i D_j N + N \left[R_{ij} + \frac{1}{3} K A_{ij} - 2 A_{ik} A^k{}_j - 8\pi \left(S_{ij} - \frac{1}{3} S \gamma_{ij} \right) \right] + \frac{1}{3} \left(D_k D^k N - N R \right) \gamma_{ij}. \tag{7.77}$$

Remark 7.7 Regarding the matter terms, this equation involves only the stress tensor $\boldsymbol{S}$ (more precisely its traceless part) and not the energy density E, contrary to the evolution equation (7.71) for K_{ij}, which involves both of them.

At this stage, we may say that we have split the dynamical Einstein equation (7.71) in two parts: a trace part: Eq. (7.74) and a traceless part: Eq. (7.77). Let us now perform the conformal decomposition of these relations, by introducing $\tilde{A}_{ij}$. We consider $\tilde{A}_{ij}$ and not $\hat{A}_{ij}$, i.e. the scaling $\alpha = -4$ and not $\alpha = -10$, since we are discussing time evolution equations.

Let us first transform Eq. (7.74). We can express the Laplacian of the lapse by applying the divergence relation (7.34) to the vector $v^i = D^i N = \gamma^{ij} D_j N = \Psi^{-4} \tilde{\gamma}^{ij} \tilde{D}_j N = \Psi^{-4} \tilde{D}^i N$

$$D_i D^i N = \Psi^{-6} \tilde{D}_i \left(\Psi^6 D^i N \right) = \Psi^{-6} \tilde{D}_i \left(\Psi^2 \tilde{D}^i N \right) = \Psi^{-4} \left(\tilde{D}_i \tilde{D}^i N + 2 \tilde{D}_i \ln \Psi \, \tilde{D}^i N \right). \tag{7.78}$$

Besides, from Eqs. (7.50), (7.59) and (7.63),

$$K_{ij} K^{ij} = \left(A_{ij} + \frac{K}{3} \gamma_{ij} \right) \left(A^{ij} + \frac{K}{3} \gamma^{ij} \right) = A_{ij} A^{ij} + \frac{K^2}{3} = \tilde{A}_{ij} \tilde{A}^{ij} + \frac{K^2}{3}. \tag{7.79}$$

In view of Eqs. (7.78), (7.79) and (7.74) becomes

$$\boxed{\left(\frac{\partial}{\partial t} - \mathcal{L}_{\boldsymbol{\beta}} \right) K = -\Psi^{-4} \left(\tilde{D}_i \tilde{D}^i N + 2 \tilde{D}_i \ln \Psi \, \tilde{D}^i N \right) + N \left[4\pi (E + S) + \tilde{A}_{ij} \tilde{A}^{ij} + \frac{K^2}{3} \right]}. \tag{7.80}$$

Let us now consider the traceless part, Eq. (7.77). We have, writing $A_{ij} = \Psi^4 \tilde{A}_{ij}$ and using Eq. (7.58),

$$\mathcal{L}_{\boldsymbol{m}} A_{ij} = \Psi^4 \mathcal{L}_{\boldsymbol{m}} \tilde{A}_{ij} + 4 \Psi^3 \mathcal{L}_{\boldsymbol{m}} \Psi \, \tilde{A}_{ij} = \Psi^4 \left[\mathcal{L}_{\boldsymbol{m}} \tilde{A}_{ij} + \frac{2}{3} \left(\tilde{D}_k \beta^k - N K \right) \tilde{A}_{ij} \right]. \tag{7.81}$$

Besides, from formulæ (7.28) and (7.31),

$$\begin{aligned} D_i D_j N &= D_i \tilde{D}_j N = \tilde{D}_i \tilde{D}_j N - C^k{}_{ij} \tilde{D}_k N \\ &= \tilde{D}_i \tilde{D}_j N - 2 \left(\delta^k{}_i \tilde{D}_j \ln \Psi + \delta^k{}_j \tilde{D}_i \ln \Psi - \tilde{D}^k \ln \Psi \, \tilde{\gamma}_{ij} \right) \tilde{D}_k N \\ &= \tilde{D}_i \tilde{D}_j N - 2 \left(\tilde{D}_i \ln \Psi \, \tilde{D}_j N + \tilde{D}_j \ln \Psi \, \tilde{D}_i N - \tilde{D}^k \ln \Psi \, \tilde{D}_k N \tilde{\gamma}_{ij} \right). \end{aligned} \tag{7.82}$$

In Eq. (7.77), we can now substitute expression (7.81) for $\mathcal{L}_m A_{ij}$, (7.82) for $D_i D_j N$, (7.42) for R_{ij}, (7.78) for $D_k D^k N$ and (7.43) for R. After some slight rearrangements, we get

$$\boxed{\begin{aligned}\left(\frac{\partial}{\partial t} - \mathcal{L}_{\boldsymbol{\beta}}\right)\tilde{A}_{ij} = &-\frac{2}{3}\tilde{D}_k\beta^k\tilde{A}_{ij} + N\left[K\tilde{A}_{ij} - 2\tilde{\gamma}^{kl}\tilde{A}_{ik}\tilde{A}_{jl} - 8\pi\left(\Psi^{-4}S_{ij} - \frac{S}{3}\tilde{\gamma}_{ij}\right)\right] \\ &+\Psi^{-4}\Bigg\{ - \tilde{D}_i\tilde{D}_jN + 2\tilde{D}_i\ln\Psi\,\tilde{D}_jN + 2\tilde{D}_j\ln\Psi\,\tilde{D}_iN \\ &+\frac{1}{3}\left(\tilde{D}_k\tilde{D}^kN - 4\tilde{D}_k\ln\Psi\,\tilde{D}^kN\right)\tilde{\gamma}_{ij} \\ &+N\Bigg[\tilde{R}_{ij} - \frac{1}{3}\tilde{R}\tilde{\gamma}_{ij} - 2\tilde{D}_i\tilde{D}_j\ln\Psi + 4\tilde{D}_i\ln\Psi\,\tilde{D}_j\ln\Psi \\ &+\frac{2}{3}\left(\tilde{D}_k\tilde{D}^k\ln\Psi - 2\tilde{D}_k\ln\Psi\,\tilde{D}^k\ln\Psi\right)\tilde{\gamma}_{ij}\Bigg]\Bigg\}.\end{aligned}} \tag{7.83}$$

7.5.2 Hamiltonian Constraint

Substituting Eq. (7.45) for R and Eq. (7.79) into the Hamiltonian constraint equation (5.70) yields

$$\boxed{\tilde{D}_i\tilde{D}^i\Psi - \frac{1}{8}\tilde{R}\Psi + \left(\frac{1}{8}\tilde{A}_{ij}\tilde{A}^{ij} - \frac{1}{12}K^2 + 2\pi E\right)\Psi^5 = 0}. \tag{7.84}$$

Let us consider the alternative scaling $\alpha = -10$ to re-express the term $\tilde{A}_{ij}\tilde{A}^{ij}$. By combining Eqs. (7.63), (7.59), (7.67) and (7.70), we get the following relations

$$\boxed{\hat{A}^{ij} = \Psi^6\tilde{A}^{ij}} \quad \text{and} \quad \boxed{\hat{A}_{ij} = \Psi^6\tilde{A}_{ij}}. \tag{7.85}$$

Hence $\tilde{A}_{ij}\tilde{A}^{ij} = \Psi^{-12}\hat{A}_{ij}\hat{A}^{ij}$ and Eq. (7.84) becomes

$$\boxed{\tilde{D}_i\tilde{D}^i\Psi - \frac{1}{8}\tilde{R}\Psi + \frac{1}{8}\hat{A}_{ij}\hat{A}^{ij}\Psi^{-7} + \left(2\pi E - \frac{1}{12}K^2\right)\Psi^5 = 0}. \tag{7.86}$$

This is the ***Lichnerowicz equation***. It has been obtained by Lichnerowicz in 1944 [1] in the special case $E = 0$ (vacuum) and $K = 0$ (maximal hypersurface[4]) (cf. also Eq. (11.7) in Ref. [2]).

Remark 7.8 If one regards Eqs. (7.84) and (7.86) as non-linear elliptic equations for Ψ, the negative power (–7) of Ψ in the $\hat{A}_{ij}\hat{A}^{ij}$ term in Eq. (7.86), as compared to the positive power (+5) in Eq. (7.84), makes a big difference about the mathematical properties of these two equations. This will be discussed in detail in Chap. 9.

---

[4] To be discussed in Sect. 10.2.2.

7.5.3 Momentum Constraint

The momentum constraint has been already written in terms of $\hat{A}^{ij}$: it is Eq. (7.68). Taking into account relation (7.85), we can easily rewrite it in terms of $\tilde{A}^{ij}$:

$$\boxed{\tilde{D}_j\tilde{A}^{ij} + 6\tilde{A}^{ij}\tilde{D}_j \ln \Psi - \frac{2}{3}\tilde{D}^i K = 8\pi\Psi^4 p^i}\,. \tag{7.87}$$

7.5.4 Summary: Conformal 3+1 Einstein System

Let us gather Eqs. (7.58), (7.60), (7.80), (7.83), (7.84) and (7.87):

$$\boxed{\left(\frac{\partial}{\partial t} - \mathscr{L}_{\boldsymbol{\beta}}\right)\Psi = \frac{\Psi}{6}\left(\tilde{D}_i\beta^i - NK\right)} \tag{7.88}$$

$$\boxed{\left(\frac{\partial}{\partial t} - \mathscr{L}_{\boldsymbol{\beta}}\right)\tilde{\gamma}_{ij} = -2N\tilde{A}_{ij} - \frac{2}{3}\tilde{D}_k\beta^k\tilde{\gamma}_{ij}} \tag{7.89}$$

$$\boxed{\left(\frac{\partial}{\partial t} - \mathscr{L}_{\boldsymbol{\beta}}\right)K = -\Psi^{-4}\left(\tilde{D}_i\tilde{D}^iN + 2\tilde{D}_i\ln\Psi\,\tilde{D}^iN\right) + N\left[4\pi(E+S) + \tilde{A}_{ij}\tilde{A}^{ij} + \frac{K^2}{3}\right]} \tag{7.90}$$

$$\boxed{\begin{aligned}\left(\frac{\partial}{\partial t} - \mathscr{L}_{\boldsymbol{\beta}}\right)\tilde{A}_{ij} = &-\frac{2}{3}\tilde{D}_k\beta^k\tilde{A}_{ij} + N\left[K\tilde{A}_{ij} - 2\tilde{\gamma}^{kl}\tilde{A}_{ik}\tilde{A}_{jl} - 8\pi\left(\Psi^{-4}S_{ij} - \frac{S}{3}\tilde{\gamma}_{ij}\right)\right]\\ &+\Psi^{-4}\Bigg\{-\tilde{D}_i\tilde{D}_jN + 2\tilde{D}_i\ln\Psi\,\tilde{D}_jN + 2\tilde{D}_j\ln\Psi\,\tilde{D}_iN\\ &\quad+\frac{1}{3}\left(\tilde{D}_k\tilde{D}^kN - 4\tilde{D}_k\ln\Psi\,\tilde{D}^kN\right)\tilde{\gamma}_{ij}\\ &\quad+N\Bigg[\tilde{R}_{ij} - \frac{1}{3}\tilde{R}\tilde{\gamma}_{ij} - 2\tilde{D}_i\tilde{D}_j\ln\Psi + 4\tilde{D}_i\ln\Psi\,\tilde{D}_j\ln\Psi\\ &\quad\quad+\frac{2}{3}\left(\tilde{D}_k\tilde{D}^k\ln\Psi - 2\tilde{D}_k\ln\Psi\,\tilde{D}^k\ln\Psi\right)\tilde{\gamma}_{ij}\Bigg]\Bigg\}.\end{aligned}} \tag{7.91}$$

$$\boxed{\tilde{D}_i\tilde{D}^i\Psi - \frac{1}{8}\tilde{R}\Psi + \left(\frac{1}{8}\tilde{A}_{ij}\tilde{A}^{ij} - \frac{1}{12}K^2 + 2\pi E\right)\Psi^5 = 0} \tag{7.92}$$

$$\boxed{\tilde{D}_j\tilde{A}^{ij} + 6\tilde{A}^{ij}\tilde{D}_j \ln \Psi - \frac{2}{3}\tilde{D}^i K = 8\pi\Psi^4 p^i}\,. \tag{7.93}$$

For the last two equations, which are the constraints, we have the alternative forms (7.86) and (7.68) in terms of $\hat{A}^{ij}$ (instead of $\tilde{A}^{ij}$):

$$\boxed{\tilde{D}_i\tilde{D}^i\Psi - \frac{1}{8}\tilde{R}\Psi + \frac{1}{8}\hat{A}_{ij}\hat{A}^{ij}\Psi^{-7} + \left(2\pi E - \frac{1}{12}K^2\right)\Psi^5 = 0}, \quad (7.94)$$

$$\boxed{\tilde{D}_j\hat{A}^{ij} - \frac{2}{3}\Psi^6\tilde{D}^iK = 8\pi\Psi^{10}p^i}. \quad (7.95)$$

Equations (7.88)–(7.93) constitute the conformal 3+1 Einstein system. An alternative form is constituted by Eqs. (7.88)–(7.91) and (7.94)–(7.95). In terms of the original 3+1 Einstein system (5.68)–(5.71), Eq. (7.88) corresponds to the trace of the kinematic equation (5.68) and Eq. (7.89) to its traceless part, Eq. (7.90) corresponds to the trace of the dynamical Einstein equation (5.69) and Eq. (7.91) to its traceless part, Eq. (7.92) or (7.94) is the Hamiltonian constraint (5.70), whereas Eq. (7.93) or (7.95) is the momentum constraint.

If the system (7.88)–(7.93) is solved in terms of $\tilde{\gamma}_{ij}$, $\tilde{A}_{ij}$ (or $\hat{A}_{ij}$), Ψ and $\boldsymbol{K}$, then the physical metric $\boldsymbol{\gamma}$ and the extrinsic curvature K are recovered by

$$\gamma_{ij} = \Psi^4\tilde{\gamma}_{ij} \quad (7.96)$$

$$K_{ij} = \Psi^4\left(\tilde{A}_{ij} + \frac{1}{3}K\tilde{\gamma}_{ij}\right) = \Psi^{-2}\hat{A}_{ij} + \frac{1}{3}K\Psi^4\tilde{\gamma}_{ij}. \quad (7.97)$$

7.6 Isenberg–Wilson–Mathews Approximation to General Relativity

In 1978, Isenberg [14] was looking for some approximation to general relativity beyond the Newtonian theory but without any gravitational wave. The simplest of the approximations that he found amounts to imposing that the 3-metric $\boldsymbol{\gamma}$ is conformally flat. In the framework of the discussion of Sect. 7.1, this is very natural since this means that $\boldsymbol{\gamma}$ belongs to the conformal equivalence class of a flat metric and there are no gravitational waves in a flat spacetime. This approximation has been reintroduced by Wilson and Mathews in 1989 [15], who were not aware of Isenberg's work [14] (unpublished at that time, except for the proceeding [16]). It is now designed as the ***Isenberg–Wilson–Mathews approximation*** (***IWM***) to General Relativity [17], or sometimes the ***conformal flatness condition*** (***CFC***).

In our notations, the IWM approximation amounts to setting

$$\tilde{\boldsymbol{\gamma}} = \boldsymbol{f} \tag{7.98}$$

and demanding that the background metric $\boldsymbol{f}$ is flat. Moreover the foliation $(\Sigma_t)_{t\in\mathbb{R}}$ must be chosen so that

$$K = 0, \tag{7.99}$$

i.e. the hypersurfaces Σ_t have a vanishing mean curvature. Equivalently Σ_t is a hypersurface of maximal volume, as it will be explained in Chap. 10. For this reason, foliations with $K=0$ are called ***maximal slicings***.

Notice that while the condition (7.99) can always be satisfied by selecting a maximal slicing for the foliation $(\Sigma_t)_{t\in\mathbb{R}}$, the requirement (7.98) is possible only if the Cotton tensor of $(\Sigma_t, \boldsymbol{\gamma})$ vanishes identically, as we have seen in Sect. 7.1. Otherwise, one deviates from general relativity.

Immediate consequences of (7.98) are that the connection $\tilde{\boldsymbol{D}}$ is simply $\bar{\boldsymbol{D}}$ and that the Ricci tensor $\tilde{\boldsymbol{R}}$ vanishes identically, since $\boldsymbol{f}$ is flat. The conformal 3+1 Einstein system (7.88)–(7.93) then reduces to

$$\left(\frac{\partial}{\partial t} - \mathcal{L}_{\boldsymbol{\beta}}\right)\Psi = \frac{\Psi}{6}\bar{D}_i\beta^i \tag{7.100}$$

$$\left(\frac{\partial}{\partial t} - \mathcal{L}_{\boldsymbol{\beta}}\right)f_{ij} = -2N\tilde{A}_{ij} - \frac{2}{3}\bar{D}_k\beta^k f_{ij} \tag{7.101}$$

$$0 = -\Psi^{-4}\left(\bar{D}_i\bar{D}^iN + 2\bar{D}_i\ln\Psi\,\bar{D}^iN\right) + N\left[4\pi(E+S) + \tilde{A}_{ij}\tilde{A}^{ij}\right] \tag{7.102}$$

$$\begin{aligned}\left(\frac{\partial}{\partial t} - \mathcal{L}_{\boldsymbol{\beta}}\right)\tilde{A}_{ij} = {}& -\frac{2}{3}\bar{D}_k\beta^k\tilde{A}_{ij} + N\left[-2f^{kl}\tilde{A}_{ik}\tilde{A}_{jl} - 8\pi\left(\Psi^{-4}S_{ij} - \frac{S}{3}f_{ij}\right)\right]\\ & + \Psi^{-4}\Bigg\{ - \bar{D}_i\bar{D}_jN + 2\bar{D}_i\ln\Psi\,\bar{D}_jN + 2\bar{D}_j\ln\Psi\,\bar{D}_iN\\ & \quad + \frac{1}{3}\left(\bar{D}_k\bar{D}^kN - 4\bar{D}_k\ln\Psi\,\bar{D}^kN\right)f_{ij}\\ & \quad + N\Bigg[- 2\bar{D}_i\bar{D}_j\ln\Psi + 4\bar{D}_i\ln\Psi\,\bar{D}_j\ln\Psi\\ & \qquad + \frac{2}{3}\left(\bar{D}_k\bar{D}^k\ln\Psi - 2\bar{D}_k\ln\Psi\,\bar{D}^k\ln\Psi\right)f_{ij}\Bigg]\Bigg\}\end{aligned} \tag{7.103}$$

$$\bar{D}_i\bar{D}^i\Psi + \left(\frac{1}{8}\tilde{A}_{ij}\tilde{A}^{ij} + 2\pi E\right)\Psi^5 = 0 \tag{7.104}$$

$$\bar{D}_j\tilde{A}^{ij} + 6\tilde{A}^{ij}\bar{D}_j\ln\Psi = 8\pi\Psi^4p^i. \tag{7.105}$$

Let us consider Eq. (7.101). By hypothesis $\partial f_{ij}/\partial t = 0$ [Eq. (7.7)]. Moreover,

$$\mathcal{L}_{\beta} f_{ij} = \beta^k \underbrace{\bar{D}_k f_{ij}}_{0} + f_{kj}\bar{D}_i\beta^k + f_{ik}\bar{D}_j\beta^k = f_{kj}\bar{D}_i\beta^k + f_{ik}\bar{D}_j\beta^k,$$

so that Eq. (7.101) becomes

$$2N\tilde{A}_{ij} = f_{kj}\bar{D}_i\beta^k + f_{ik}\bar{D}_j\beta^k - \frac{2}{3}\bar{D}_k\beta^k f_{ij}.$$

Using $\tilde{A}^{ij} = f^{ik}f^{jl}\tilde{A}_{kl}$, we may rewrite this equation as

$$\tilde{A}^{ij} = \frac{1}{2N}(L\beta)^{ij}, \tag{7.106}$$

where

$$(L\beta)^{ij} := \bar{D}^i\beta^j + \bar{D}^j\beta^i - \frac{2}{3}\bar{D}_k\beta^k f^{ij} \tag{7.107}$$

is the ***conformal Killing operator*** associated with the metric $\boldsymbol{f}$ (cf. Appendix A). Consequently, the term $\bar{D}_j\tilde{A}^{ij}$ which appears in Eq. (7.105) is expressible in terms of $\boldsymbol{\beta}$ as

$$\begin{aligned}\bar{D}_j\tilde{A}^{ij} &= \bar{D}_j\left[\frac{1}{2N}(L\beta)^{ij}\right] = \frac{1}{2N}\bar{D}_j\left(\bar{D}^i\beta^j + \bar{D}^j\beta^i - \frac{2}{3}\bar{D}_k\beta^k f^{ij}\right) - \frac{1}{2N^2}(L\beta)^{ij}\bar{D}_jN \\ &= \frac{1}{2N}\left(\bar{D}_j\bar{D}^j\beta^i + \frac{1}{3}\bar{D}^i\bar{D}_j\beta^j - 2\tilde{A}^{ij}\bar{D}_jN\right),\end{aligned} \tag{7.108}$$

where we have used $\bar{D}_j\bar{D}^i\beta^j = \bar{D}^i\bar{D}_j\beta^j$ since $\boldsymbol{f}$ is flat. Inserting Eq. (7.108) into Eq. (7.105) yields

$$\bar{D}_j\bar{D}^j\beta^i + \frac{1}{3}\bar{D}^i\bar{D}_j\beta^j + 2\tilde{A}^{ij}\left(6N\bar{D}_j\ln\Psi - \bar{D}_jN\right) = 16\pi N\Psi^4 p^i. \tag{7.109}$$

The IWM system is formed by Eqs. (7.102), (7.104) and (7.109), which we rewrite as

$$\boxed{\Delta N + 2\bar{D}_i\ln\Psi\,\bar{D}^iN = N\Psi^4\left[4\pi(E+S) + \tilde{A}_{ij}\tilde{A}^{ij}\right]} \tag{7.110}$$

$$\boxed{\Delta\Psi + \left(\frac{1}{8}\tilde{A}_{ij}\tilde{A}^{ij} + 2\pi E\right)\Psi^5 = 0} \tag{7.111}$$

$$\boxed{\Delta\beta^i + \frac{1}{3}\bar{D}^i\bar{D}_j\beta^j + 2\tilde{A}^{ij}\left(6N\bar{D}_j\ln\Psi - \bar{D}_jN\right) = 16\pi N\Psi^4 p^i}, \tag{7.112}$$

where

$$\Delta := \bar{D}_i \bar{D}^i \tag{7.113}$$

is the flat-space Laplacian. In the above equations, $\tilde{A}^{ij}$ is to be understood, not as an independent variable, but as the function of N and β^i defined by Eq. (7.106).

The IWM system (7.110)–(7.112) is a system of three elliptic equations (two scalar equations and one vector equation) for the three unknowns N, Ψ and β^i. The physical 3-metric is fully determined by Ψ

$$\gamma_{ij} = \Psi^4 f_{ij}, \tag{7.114}$$

so that, once the IWM system is solved, the full spacetime metric $\boldsymbol{g}$ can be reconstructed via Eq. (5.49).

Remark 7.9 In the original article [14], Isenberg has derived the system (7.110)–(7.112) from a variational principle based on the Hilbert action (5.100), by restricting γ_{ij} to take the form (7.114) and requiring that the momentum conjugate to Ψ vanishes.

That the IWM scheme constitutes some approximation to general relativity is clear because the solutions (N, Ψ, β^i) to the IWM system (7.110)–(7.112) do not in general satisfy the remaining equations of the full conformal 3+1 Einstein system, i.e. Eqs. (7.100) and (7.103). However, the IWM approximation

- is exact for spherically symmetric spacetimes, because (i) any such spacetime symmetric spacetime admits locally a maximal slicing ($K = 0$) [18] and (ii) the Cotton tensor vanishes for any spherically symmetric 3-space $(\Sigma_t, \boldsymbol{\gamma})$; a concrete example is Schwarzschild spacetime (cf. Example 7.1);
- is very accurate for axisymmetric rotating neutron stars [19];
- is correct at the 1-PN order in the post-Newtonian expansion of general relativity.

The IWM approximation has been widely used in relativistic astrophysics, to compute binary neutron star mergers [20–22] gravitational collapses of stellar cores [23–27], as well as quasi-equilibrium configurations of binary neutron stars or binary black holes (cf. Sect. 9.4).

It turned out that in some highly relativistic situations, the IWM system (7.110–7.112) may suffer from some non-uniqueness issue that prevents numerical codes to converge [28]. This non-uniqueness is related to that of the XCTS system that will be discussed in Sect. 9.3.4. A solution has been found [26, 28, 29] and consists in solving an extra vector elliptic equation, in addition to the shift equation (7.112). The resulting scheme is then called *XCFC* (for *extended conformal flatness condition*) [28, 30]. A recent implementation of XCFC can be found in [31].

References

1. Lichnerowicz, A.: L'intégration des équations de la gravitation relativiste et le problème des n corps, J. Math. Pures Appl. **23**, 37 (1944). Reprinted in A. Lichnerowicz : Choix d'œuvres mathématiques, Hermann, Paris (1982), p. 4
2. Lichnerowicz, A.: Sur les équations relativistes de la gravitation, Bulletin de la S.M.F. **80**, 237 (1952). Available at http://www.numdam.org/item?id=BSMF_1952__80__237_0
3. York, J.W.: Gravitational degrees of freedom and the initial-value problem. Phys. Rev. Lett. **26**, 1656 (1971)
4. York, J.W.: Role of conformal three-geometry in the dynamics of gravitation. Phys. Rev. Lett. **28**, 1082 (1972)
5. Cotton, E.: Sur les variétés à trois dimensions, Annales de la faculté des sciences de Toulouse Sér. 2, **1**, 385 (1899). Available at http://www.numdam.org/item?id=AFST_1899_2_1_4_385_0
6. Damour, T.: Advanced General Relativity, lectures at Institut Henri Poincaré, Paris (2006). Available at http://www.luth.obspm.fr/IHP06/
7. Blanchet, L.: Gravitational Radiation from Post-Newtonian Sources and Inspiralling Compact Binaries, Living Rev. Relativity **9**, 4 (2006). http://www.livingreviews.org/lrr-2006-4
8. Blanchet, L.: Theory of Gravitational Wave Emission, lectures at Institut Henri Poincaré, Paris (2006). Available at http://www.luth.obspm.fr/IHP06/
9. Shibata, M., Nakamura, T.: Evolution of three-dimensional gravitational waves: harmonic slicing case. Phys. Rev. D **52**, 5428 (1995)
10. Baumgarte, T.W., Shapiro, S.L.: Numerical integration of Einstein's field equations. Phys. Rev. D **59**, 024007 (1999)
11. Bonazzola, S., Gourgoulhon, E., Grandclément, P., Novak, J.: Constrained scheme for the Einstein equations based on the Dirac gauge and spherical coordinates. Phys. Rev. D **70**, 104007 (2004)
12. Misner, C.W., Thorne, K.S., Wheeler, J.A.: Gravitation. Freeman, New York (1973)
13. Nakamura, T.: 3D Numerical Relativity. In: Sasaki M. (eds) Relativistic Cosmology, Proceedings of the 8th Nishinomiya-Yukawa Memorial Symposium. Universal Academy Press, Tokyo, (1994) pp. 155
14. Isenberg, J.A.: Waveless Approximation Theories of Gravity, preprint University of Maryland (1978). Published in Int. J. Mod. Phys. D **17**, 265 (2008). Available as http://arxiv.org/abs/gr-qc/0702113 an abridged version can be found in Ref. [16].
15. Wilson, J.R., Mathews, G.J.: Relativistic hydrodynamics. In: Evans, C.R., Finn, L.S., Hobill, D.W. (eds) Frontiers in numerical relativity., pp. 306. Cambridge Univ. Press, Cambridge (1989)
16. Isenberg, J., Nester, J.: Canonical Gravity. In: Held, A. (eds) General Relativity and Gravitation, one hundred Years after the Birth of Albert Einstein, pp. 23. Plenum Press, New York (1980)
17. Friedman, J.L., Uryu, K. and Shibata, M.: Thermodynamics of binary black holes and neutron stars, Phys. Rev. D **65**, 064035 (2002), Erratum in Phys. Rev. D **70**, 129904(E) (2004)
18. Cordero-Carrión, I., Ibáñez, J.M., Morales-Lladosa, J.A.: Maximal slicings in spherical symmetry: local existence and construction. J. Math. Phys. **52**, 112501 (2011)
19. Cook, G.B., Shapiro, S.L., Teukolsky, S.A.: Testing a simplified version of Einstein's equations for numerical relativity. Phys. Rev. D **53**, 5533 (1996)
20. Mathews, G.J., Wilson, J.R.: Revised relativistic hydrodynamical model for neutron-star binaries. Phys. Rev. D **61**, 127304 (2000)
21. Faber, J.A., Grandclément, P., Rasio, F.A.: Mergers of irrotational neutron star binaries in conformally flat gravity. Phys. Rev. D **69**, 124036 (2004)
22. Oechslin, R., Uryu, K., Poghosyan, G., Thielemann, F.K.: The Influence of Quark Matter at High Densities on Binary Neutron Star Mergers. Mon. Not. Roy. Astron. Soc. **349**, 1469 (2004)
23. Dimmelmeier, H., Font, J.A., Müller, E.: Relativistic simulations of rotational core collapse I. Methods, initial models, and code tests. Astron. Astrophys **388**, 917 (2002)

24. Dimmelmeier, H., Font, J.A., Müller, E.: Relativistic simulations of rotational core collapse II. Collapse dynamics and gravitational radiation. Astron. Astrophys **393**, 523 (2002)
25. Dimmelmeier, H., Novak, J., Font, J.A., Ibáñez, J.M., Müller, E.: Combining spectral and shock-capturing methods: A new numerical approach for 3D relativistic core collapse simulations. Phys. Rev. D **71**, 064023 (2005)
26. Saijo, M.: The collapse of differentially rotating supermassive stars: conformally flat simulations. Astrophys. J. **615**, 866 (2004)
27. Saijo, M.: Dynamical bar instability in a relativistic rotational core collapse. Phys. Rev. D **71**, 104038 (2005)
28. Cordero-Carrión, I., Cerdá-Durán, P., Dimmelmeier, H., Jaramillo, J.L., Novak, J., Gourgoulhon, E.: Improved constrained scheme for the Einstein equations: an approach to the uniqueness issue. Phys. Rev. D **79**, 024017 (2009)
29. Shibata, M., Uryu, K.: Merger of black hole-neutron star binaries: nonspinning black hole case. Phys. Rev. D **74**, 121503(R) (2006)
30. Gourgoulhon, E.: Constrained schemes for evolving the 3+1 Einstein equations, presentation at the CoCoNuT Meeting 2009 (Valencia, Spain, 4–6 November 2009). Available at http://www.mpa-garching.mpg.de/hydro/COCONUT/valencia2009/intro.php
31. Bucciantini, N., Del Zanna, L.: General relativistic magnetohydrodynamics in axisymmetric dynamical spacetimes: the X-ECHO code. Astron. Astrophys. **528**, A101 (2011)

Chapter 8
Asymptotic Flatness and Global Quantities

Abstract After providing a definition of asymptotic flatness, we introduce the global quantities that one may associate to the spacetime or to each slice of the 3 + 1 foliation: the ADM mass, the ADM linear momentum, the total angular momentum, the Komar mass and the Komar angular momentum. For each of these quantities, we derive expressions in terms of the 3+1 objects and provide some concrete examples.

8.1 Introduction

Global quantities are important characterizations of a given spacetime or a given hypersurface of a 3+1 slicing. Such quantities encompass various notions of mass, linear momentum and angular momentum. There are many such concepts in general relativity and we refer the reader to Refs. [1, 2] for reviews in this vast topic. Here we limit ourselves to quantities which are directly connected to the 3+1 formalism. In particular, we do not discuss quantities associated with null infinity, like the Bondi mass. In the absence of any symmetry, the global quantities are defined only for asymptotically flat spacetimes. So we shall start by defining this notion.

8.2 Asymptotic Flatness

The concept of asymptotic flatness applies to stellar type objects, modeled as if they were alone in an otherwise empty universe (the so-called ***isolated bodies***). Of course, most cosmological spacetimes are not asymptotically flat.

É. Gourgoulhon, *3+1 Formalism in General Relativity*, Lecture Notes in Physics 846,
DOI: 10.1007/978-3-642-24525-1_8,

8.2.1 Definition

We consider a globally hyperbolic[1] spacetime $(\mathscr{M}, \boldsymbol{g})$ foliated by a family $(\Sigma_t)_{t\in\mathbb{R}}$ of spacelike hypersurfaces. Let $\boldsymbol{\gamma}$ and $\boldsymbol{K}$ be respectively the induced metric and extrinsic curvature of the hypersurfaces Σ_t. One says that the spacetime is ***asymptotically flat*** iff there exists, on each slice Σ_t, a Riemannian "background" metric $\boldsymbol{f}$ such that [3, 4, 5]

- $\boldsymbol{f}$ is flat ($\mathbf{Riem}(\boldsymbol{f}) = 0$), except possibly on a compact domain $\mathscr{B}$ of Σ_t (the "strong field region");
- there exists a coordinate system $(x^i) = (x, y, z)$ on Σ_t such that outside $\mathscr{B}$, the components of $\boldsymbol{f}$ are $f_{ij} = \mathrm{diag}(1, 1, 1)$ ("Cartesian-type coordinates") and the variable $r := \sqrt{x^2 + y^2 + z^2}$ can take arbitrarily large values on Σ_t;
- when $r \to +\infty$, the components of $\boldsymbol{\gamma}$ with respect to the coordinates (x^i) satisfy

$$\gamma_{ij} = f_{ij} + O(r^{-1}), \tag{8.1a}$$

$$\frac{\partial \gamma_{ij}}{\partial x^k} = O(r^{-2}); \tag{8.1b}$$

- when $r \to +\infty$, the components of $\boldsymbol{K}$ with respect to the coordinates (x^i) satisfy

$$K_{ij} = O(r^{-2}), \tag{8.2a}$$

$$\frac{\partial K_{ij}}{\partial x^k} = O(r^{-3}). \tag{8.2b}$$

The "region" $r \to +\infty$ is called ***spatial infinity*** and is denoted i^0.

Remark 8.1 There exist other definitions of *asymptotic flatness* which are not based on any coordinate system nor background flat metric (see e.g. Ref. [6] or Chap. 11 in Wald's textbook [7]). In particular, the spatial infinity i^0 can be rigorously defined as a single point in some "extended" spacetime $(\hat{\mathscr{M}}, \hat{\boldsymbol{g}})$ in which $(\mathscr{M}, \boldsymbol{g})$ can be embedded with $\boldsymbol{g}$ conformal to $\hat{\boldsymbol{g}}$. However the present definition is perfectly adequate for our purposes.

Remark 8.2 The requirement (8.1b) excludes the presence of gravitational waves at spatial infinity. Indeed for gravitational waves propagating in the radial direction:

$$\gamma_{ij} = f_{ij} + \frac{F_{ij}(t-r)}{r} + O(r^{-2}),$$

where F_{ij} is an oscillating function (for instance $F_{ij}(t) = \cos(\omega t)$), which satisfies $F_{ij}(t-r) = O(1)$ and $F'_{ij}(t-r) = O(1)$. This fulfills condition (8.1a) but

[1] See Sect. 4.2.1.

$$\frac{\partial \gamma_{ij}}{\partial x^k} = -\frac{F'_{ij}(t-r)}{r}\frac{x^k}{r} - \frac{F_{ij}(t-r)}{r^2}\frac{x^k}{r} + O(r^{-2})$$

is $O(r^{-1})$ since $F'_{ij}(t-r) = O(1)$. This violates condition (8.1b). Notice that the absence of gravitational waves at spatial infinity is not a serious physical restriction, since one may consider that any isolated system has started to emit gravitational waves at a finite time in the past and that these waves have not reached the spatial infinity yet.

8.2.2 Asymptotic Coordinate Freedom

Obviously the above definition of asymptotic flatness depends both on the foliation $(\Sigma_t)_{t\in\mathbb{R}}$ and on the coordinates (x^i) chosen on each leaf Σ_t. It is of course important to assess whether this dependence is strong or not. In other words, we would like to determine the class of coordinate changes $(x^\alpha) = (t, x^i) \to (x'^\alpha) = (t', x'^i)$ which preserve the asymptotic properties (8.1)–(8.2). The answer is that the coordinates (x'^α) must be related to the coordinates (x^α) by [8]

$$x'^\alpha = \Lambda^\alpha{}_\mu x^\mu + c^\alpha(\theta, \varphi) + O(r^{-1}) \tag{8.3}$$

where $\Lambda^\alpha{}_\beta$ is a Lorentz matrix and the c^α's are four functions of the angles (θ, φ) related to the coordinates $(x^i) = (x, y, z)$ by the standard formulæ:

$$x = r\sin\theta\cos\varphi, \quad y = r\sin\theta\sin\varphi, \quad z = r\cos\theta. \tag{8.4}$$

The group of transformations generated by (8.3) is related to the ***Spi group*** (for *Spatial infinity*) introduced by Ashtekar and Hansen [6, 9]. However the precise relation is not clear because the definition of asymptotic flatness used by these authors is not expressed as decay conditions for γ_{ij} and K_{ij}, as in Eqs. (8.1)–(8.2).

Notice that ***Poincaré transformations*** are contained in the transformation group defined by (8.3): they simply correspond to the case $c^\alpha(\theta, \varphi) = \text{const}$. The transformations with $c^\alpha(\theta, \varphi) \neq \text{const}$ and $\Lambda^\alpha{}_\beta = \delta^\alpha{}_\beta$ constitute "angle-dependent translations" and are called ***supertranslations***.

Note that if the Lorentz matrix $\Lambda^\alpha{}_\beta$ involves a boost, the transformation (8.3) implies a change of the 3+1 foliation $(\Sigma_t)_{t\in\mathbb{R}}$, whereas if $\Lambda^\alpha{}_\beta$ corresponds only to some spatial rotation and the c^α's are constant, the transformation (8.3) describes some change of Cartesian-type coordinates (x^i) (rotation + translation) within the same hypersurface Σ_t.

8.3 ADM Mass

8.3.1 *Definition from the Hamiltonian Formulation of GR*

In the short introduction to the Hamiltonian formulation of general relativity given in Sect. 5.5, we have for simplicity discarded any boundary term in the action. However, because the gravitational Lagrangian density (the scalar curvature ${}^4R$) contains second order derivatives of the metric tensor (and not only first order ones, which is a particularity of general relativity with respect to other field theories), the precise action should be [7, 8, 10, 11]

$$S = \int_{\mathscr{V}} {}^4R\sqrt{-g}\,\mathrm{d}^4x + 2\oint_{\partial\mathscr{V}} (Y - Y_0)\sqrt{h}\,\mathrm{d}^3y, \tag{8.5}$$

where $\partial\mathscr{V}$ is the boundary of the domain $\mathscr{V}$ ($\partial\mathscr{V}$ is assumed to be a timelike hypersurface), Y the trace of the extrinsic curvature [i.e. minus three times the mean curvature, cf. Eq. (3.21)] of $\partial\mathscr{V}$ embedded in $(\mathscr{M}, \boldsymbol{g})$ and Y_0 the trace of the extrinsic curvature of $\partial\mathscr{V}$ embedded in $(\mathscr{M}, \boldsymbol{\eta})$, where $\boldsymbol{\eta}$ is a Lorentzian metric on $\mathscr{M}$ that is *flat* around $\partial\mathscr{V}$. Finally $\sqrt{h}\,\mathrm{d}^3y$ is the volume element induced by $\boldsymbol{g}$ on the hypersurface $\partial\mathscr{V}$, h being the determinant of the components of the induced metric $\boldsymbol{h}$ on $\partial\mathscr{V}$ with respect to the coordinates (y^i) on $\partial\mathscr{V}$. The boundary term in (8.5) guarantees that the variation of S with the values of $\boldsymbol{g}$ (and not its derivatives) held fixed at $\partial\mathscr{V}$ leads to the Einstein equation. Otherwise, from the volume term alone (Hilbert action), one has to held fixed $\boldsymbol{g}$ *and* all its derivatives at $\partial\mathscr{V}$.

Let

$$\mathscr{S}_t := \partial\mathscr{V} \cap \Sigma_t. \tag{8.6}$$

We assume that $\mathscr{S}_t$ has the topology of a sphere. The gravitational Hamiltonian derived from the action (8.5) (see [11] for details) contains an additional boundary term with respect to the Hamiltonian (5.111) obtained in Sect. 5.5:

$$H = -\int_{\Sigma_t^{\mathrm{int}}} \left(NC_0 - 2\beta^i C_i\right)\sqrt{\gamma}\,\mathrm{d}^3x - 2\oint_{\mathscr{S}_t} \left[N(\kappa - \kappa_0) + \beta_i(K_{ij} - K\gamma_{ij})s^j\right]\sqrt{q}\,\mathrm{d}^2y, \tag{8.7}$$

where Σ_t^{int} is the part of Σ_t bounded by $\mathscr{S}_t$, κ is the trace of the extrinsic curvature of $\mathscr{S}_t$ embedded in $(\Sigma_t, \boldsymbol{\gamma})$, and κ_0 the trace of the extrinsic curvature of $\mathscr{S}_t$ embedded in $(\Sigma_t, \boldsymbol{f})$ ($\boldsymbol{f}$ being the metric introduced in Sect. 8.2), $\boldsymbol{s}$ is the unit normal to $\mathscr{S}_t$ in Σ_t, oriented towards the asymptotic region, and $\sqrt{q}\,\mathrm{d}^2y$ denotes the surface element induced by the spacetime metric on $\mathscr{S}_t$, $\boldsymbol{q}$ being the induced metric, $y^a = (y^1, y^2)$ some coordinates on $\mathscr{S}_t$ [for instance $y^a = (\theta, \varphi)$] and $q := \det(q_{ab})$.

For solutions of Einstein equation, the constraints are satisfied: $C_0 = 0$ and $C_i = 0$, so that the value of the Hamiltonian reduces to

$$H_{\mathrm{solution}} = -2\oint_{\mathscr{S}_t} \left[N(\kappa - \kappa_0) + \beta^i(K_{ij} - K\gamma_{ij})s^j\right]\sqrt{q}\,\mathrm{d}^2y. \tag{8.8}$$

The ***total energy*** contained in Σ_t is then defined as the numerical value of the Hamiltonian for solutions, taken on a surface $\mathcal{S}_t$ at spatial infinity (i.e. for $r \to +\infty$) and for coordinates (t, x^i) that could be associated with some asymptotically inertial observer, i.e. such that $N = 1$ and $\boldsymbol{\beta} = 0$. From Eq. (8.8), we get (after restoration of a $(16\pi)^{-1}$ factor)

$$\boxed{M_{\mathrm{ADM}} := -\frac{1}{8\pi} \lim_{\mathcal{S}_t \to \infty} \oint_{\mathcal{S}_t} (\kappa - \kappa_0) \sqrt{q}\, \mathrm{d}^2 y}\,. \tag{8.9}$$

This total energy is called the ***ADM mass*** of the slice Σ_t. By evaluating the extrinsic curvature traces κ and κ_0, it can be shown that Eq. (8.9) can be written

$$\boxed{M_{\mathrm{ADM}} = \frac{1}{16\pi} \lim_{\mathcal{S}_t \to \infty} \oint_{\mathcal{S}_t} \left[\bar{D}^j \gamma_{ij} - \bar{D}_i (f^{kl} \gamma_{kl}) \right] s^i \sqrt{q}\, \mathrm{d}^2 y}\,, \tag{8.10}$$

where $\bar{\boldsymbol{D}}$ stands for the connection associated with the metric $\boldsymbol{f}$ and, as above, s^i stands for the components of unit normal to $\mathcal{S}_t$ within Σ_t and oriented towards the exterior of $\mathcal{S}_t$. In particular, if one uses the Cartesian-type coordinates (x^i) involved in the definition of asymptotic flatness (Sect. 8.2), then $\bar{D}_i = \partial/\partial x^i$ and $f^{kl} = \delta^{kl}$ and the above formula becomes

$$M_{\mathrm{ADM}} = \frac{1}{16\pi} \lim_{\mathcal{S}_t \to \infty} \oint_{\mathcal{S}_t} \left(\frac{\partial \gamma_{ij}}{\partial x^j} - \frac{\partial \gamma_{jj}}{\partial x^i} \right) s^i \sqrt{q}\, \mathrm{d}^2 y. \tag{8.11}$$

Notice that thanks to the asymptotic flatness requirement (8.1b), this integral takes a finite value: the $O(r^2)$ part of $\sqrt{q}\, \mathrm{d}^2 y$ is compensated by the $O(r^{-2})$ parts of $\partial \gamma_{ij}/\partial x^j$ and $\partial \gamma_{jj}/\partial x^i$.

Example 8.1 Let us consider Schwarzschild spacetime and use the standard ***Schwarzschild coordinates*** $(x^\alpha) = (t, r, \theta, \phi)$:

$$g_{\mu\nu} \mathrm{d}x^\mu \mathrm{d}x^\nu = -\left(1 - \frac{2m}{r}\right) \mathrm{d}t^2 + \left(1 - \frac{2m}{r}\right)^{-1} \mathrm{d}r^2 + r^2 (\mathrm{d}\theta^2 + \sin^2\theta\, \mathrm{d}\varphi^2). \tag{8.12}$$

Let us take for Σ_t the hypersurface of constant Schwarzschild coordinate time t. Then we read on (8.12) the components of the induced metric in the coordinates $(x^i) = (r, \theta, \varphi)$:

$$\gamma_{ij} = \mathrm{diag}\left[\left(1 - \frac{2m}{r}\right)^{-1}, r^2, r^2 \sin^2\theta \right]. \tag{8.13}$$

On the other side, the components of the flat metric in the same coordinates are

$$f_{ij} = \mathrm{diag}\left(1, r^2, r^2 \sin^2\theta\right) \qquad \text{and} \qquad f^{ij} = \mathrm{diag}\left(1, r^{-2}, r^{-2} \sin^{-2}\theta\right). \tag{8.14}$$

Let us now evaluate $M_{\rm ADM}$ by means of the integral (8.10) (we cannot use formula (8.11) because the coordinates (x^i) are not Cartesian-like). It is quite natural to take for $\mathscr{S}_t$ the sphere $r = \text{const}$ in the hypersurface Σ_t. Then $y^a = (\theta, \varphi)$, $\sqrt{q} = r^2 \sin\theta$ and, at spatial infinity, $s^i \sqrt{q}\, \mathrm{d}^2 y = r^2 \sin\theta \, \mathrm{d}\theta \, \mathrm{d}\varphi (\boldsymbol{\partial}_r)^i$, where $\boldsymbol{\partial}_r$ is the natural basis vector associated with the coordinate r: $(\boldsymbol{\partial}_r)^i = (1, 0, 0)$. Consequently, Eq. (8.10) becomes

$$M_{\rm ADM} = \frac{1}{16\pi} \lim_{r\to\infty} \oint_{r=\text{const}} \left[\bar{D}^j \gamma_{rj} - \bar{D}_r (f^{kl} \gamma_{kl})\right] r^2 \sin\theta \, \mathrm{d}\theta \, \mathrm{d}\varphi, \tag{8.15}$$

with

$$f^{kl}\gamma_{kl} = \gamma_{rr} + \frac{1}{r^2}\gamma_{\theta\theta} + \frac{1}{r^2\sin^2\theta}\gamma_{\varphi\varphi} = \left(1 - \frac{2m}{r}\right)^{-1} + 2,$$

and since $f^{kl}\gamma_{kl}$ is a scalar field,

$$\bar{D}_r (f^{kl}\gamma_{kl}) = \frac{\partial}{\partial r}(f^{kl}\gamma_{kl}) = -\left(1 - \frac{2m}{r}\right)^{-2} \frac{2m}{r^2}. \tag{8.16}$$

There remains to evaluate $\bar{D}^j \gamma_{rj}$. One has

$$\bar{D}^j \gamma_{rj} = f^{jk} \bar{D}_k \gamma_{rj} = \bar{D}_r \gamma_{rr} + \frac{1}{r^2} \bar{D}_\theta \gamma_{r\theta} + \frac{1}{r^2 \sin^2\theta} \bar{D}_\varphi \gamma_{r\varphi},$$

with the covariant derivatives given by (taking into account the form (8.13) of γ_{ij})

$$\begin{aligned}
\bar{D}_r \gamma_{rr} &= \frac{\partial \gamma_{rr}}{\partial r} - 2\bar{\Gamma}^i{}_{rr}\gamma_{ir} = \frac{\partial \gamma_{rr}}{\partial r} - 2\bar{\Gamma}^r{}_{rr}\gamma_{rr} \\
\bar{D}_\theta \gamma_{r\theta} &= \frac{\partial \gamma_{r\theta}}{\partial \theta} - \bar{\Gamma}^i{}_{\theta r}\gamma_{i\theta} - \bar{\Gamma}^i{}_{\theta\theta}\gamma_{ri} = -\bar{\Gamma}^\theta{}_{\theta r}\gamma_{\theta\theta} - \bar{\Gamma}^r{}_{\theta\theta}\gamma_{rr} \\
\bar{D}_\varphi \gamma_{r\varphi} &= \frac{\partial \gamma_{r\varphi}}{\partial \varphi} - \bar{\Gamma}^i{}_{\varphi r}\gamma_{i\varphi} - \bar{\Gamma}^i{}_{\varphi\varphi}\gamma_{ri} = -\bar{\Gamma}^\varphi{}_{\varphi r}\gamma_{\varphi\varphi} - \bar{\Gamma}^r{}_{\varphi\varphi}\gamma_{rr},
\end{aligned}$$

where the $\bar{\Gamma}^k{}_{ij}$'s are the Christoffel symbols of the connection $\bar{\boldsymbol{D}}$ with respect to the coordinates (x^i). The non-vanishing ones are

$$\bar{\Gamma}^r{}_{\theta\theta} = -r \quad \text{and} \quad \bar{\Gamma}^r{}_{\varphi\varphi} = -r\sin^2\theta \tag{8.17a}$$

$$\bar{\Gamma}^\theta{}_{r\theta} = \bar{\Gamma}^\theta{}_{\theta r} = \frac{1}{r} \quad \text{and} \quad \bar{\Gamma}^\theta{}_{\varphi\varphi} = -\cos\theta\sin\theta \tag{8.17b}$$

$$\bar{\Gamma}^\varphi{}_{r\varphi} = \bar{\Gamma}^\varphi{}_{\varphi r} = \frac{1}{r} \quad \text{and} \quad \bar{\Gamma}^\varphi{}_{\theta\varphi} = \bar{\Gamma}^\varphi{}_{\varphi\theta} = \frac{1}{\tan\theta}. \tag{8.17c}$$

Hence

$$\bar{D}^j\gamma_{rj} = \frac{\partial}{\partial r}\left[\left(1-\frac{2m}{r}\right)^{-1}\right] + \frac{1}{r^2}\left[-\frac{1}{r}\times r^2 + r\times\left(1-\frac{2m}{r}\right)^{-1}\right]$$
$$+\frac{1}{r^2\sin^2\theta}\left[-\frac{1}{r}\times r^2\sin^2\theta + r\sin^2\theta\times\left(1-\frac{2m}{r}\right)^{-1}\right]$$

$$\bar{D}^j\gamma_{rj} = \frac{2m}{r^2}\left(1-\frac{2m}{r}\right)^{-2}\left(1-\frac{4m}{r}\right). \tag{8.18}$$

Combining Eqs. (8.16) and (8.18), we get

$$\bar{D}^j\gamma_{rj} - \bar{D}_r(f^{kl}\gamma_{kl}) = \frac{2m}{r^2}\left(1-\frac{2m}{r}\right)^{-2}\left(1-\frac{4m}{r}+1\right) = \frac{4m}{r^2}\left(1-\frac{2m}{r}\right)^{-1}$$
$$\sim \frac{4m}{r^2} \quad \text{when } r\to\infty,$$

so that the integral (8.15) results in

$$M_{\mathrm{ADM}} = m. \tag{8.19}$$

We conclude that the ADM mass of any hypersurface $t = \text{const}$ of Schwarzschild spacetime is nothing but the mass parameter m of the Schwarzschild solution.

Example 8.2 (counter-example) On Schwarzschild spacetime, the ***Painlevé–Gullstrand coordinates*** (t, r, θ, φ) are defined as follows (see e.g. Ref. [12]): r is nothing but the standard Schwarzschild radial coordinate, whereas the Painlevé–Gullstrand coordinate t is related to the Schwarzschild time coordinate t_S (i.e. the coordinate t of Example 8.1) by

$$t = t_S + 4m\left(\sqrt{\frac{r}{2m}} + \frac{1}{2}\ln\left|\frac{\sqrt{r/2m}-1}{\sqrt{r/2m}+1}\right|\right). \tag{8.20}$$

The metric components with respect to Painlevé–Gullstrand coordinates are remarkably simple, being given by

$$g_{\mu\nu}\mathrm{d}x^\mu\mathrm{d}x^\nu = -\mathrm{d}t^2 + \left(\mathrm{d}r + \sqrt{\frac{2m}{r}}\mathrm{d}t\right)^2 + r^2(\mathrm{d}\theta^2 + \sin^2\theta\,\mathrm{d}\varphi^2). \quad (8.21)$$

By comparing with the general line element (5.50), we read on the above expression that

$$N = 1 \quad (8.22a)$$

$$\beta^i = \left(\sqrt{\frac{2m}{r}}, 0, 0\right) \quad (8.22b)$$

$$\gamma_{ij} = \mathrm{diag}(1, \mathrm{r}^2, \mathrm{r}^2\sin^2\theta). \quad (8.22c)$$

We notice that the metric $\boldsymbol{\gamma}$ on the hypersurfaces Σ_t is flat : $\boldsymbol{\gamma} = \boldsymbol{f}$. Hence if we apply naively formula (8.10), we get $M_{\mathrm{ADM}} = 0$, whereas one would have expected $M_{\mathrm{ADM}} = m$ as in Example 8.1! This surprising result stems from the fact that the Painlevé–Gullstrand slicing $(\Sigma_t)_{t\in\mathbb{R}}$ is not asymptotically flat in the sense defined in Sect. 8.2.1: whereas the conditions (8.1) are obviously satisfied by the flat metric $\boldsymbol{\gamma}$, the conditions (8.2) on the extrinsic curvature are violated. Indeed, let us evaluate K_{ij} via formulas (5.68) and (5.75) with $\gamma_{ij} = f_{ij}$ and $N = 1$. We get

$$K_{ij} = \frac{1}{2}\left(\frac{\partial\beta_i}{\partial x^j} + \frac{\partial\beta_j}{\partial x^i} - 2\bar{\Gamma}^k_{\ ij}\beta_k\right).$$

Given the values (8.22b) for β^i and (8.17) for $\bar{\Gamma}^k_{\ ij}$, we conclude that the only non-vanishing components of K_{ij} are

$$K_{rr} = -\sqrt{\frac{m}{2r^3}}, \quad K_{\theta\theta} = \sqrt{2mr}, \quad K_{\varphi\varphi} = \sqrt{2mr}\sin^2\theta. \quad (8.23)$$

This implies that the Cartesian components of $\boldsymbol{K}$ have the asymptotic behavior

$$K_{ij} = O(r^{-3/2}).$$

This decay is too slow to comply with the asymptotic flatness condition (8.2a).

8.3.2 Expression in Terms of the Conformal Decomposition

Let us introduce the conformal metric $\tilde{\boldsymbol{\gamma}}$ and conformal factor Ψ associated to $\boldsymbol{\gamma}$ according to the prescription given in Sect. 7.2.3, taking for the background metric $\boldsymbol{f}$ the *same* metric as that involved in the definition of asymptotic flatness and ADM mass:

$$\boldsymbol{\gamma} = \Psi^4 \tilde{\boldsymbol{\gamma}}, \tag{8.24}$$

with, in the Cartesian-type coordinates $(x^i) = (x, y, z)$ introduced in Sect. 8.2:

$$\det(\tilde{\gamma}_{ij}) = 1. \tag{8.25}$$

This is the property (7.18) since $f = \det(f_{ij}) = 1$ ($f_{ij} = \mathrm{diag}(1, 1, 1)$). The asymptotic flatness conditions (8.1) impose

$$\Psi = 1 + O(r^{-1}) \quad \text{and} \quad \frac{\partial \Psi}{\partial x^k} = O(r^{-2}) \tag{8.26}$$

and

$$\tilde{\gamma}_{ij} = f_{ij} + O(r^{-1}) \quad \text{and} \quad \frac{\partial \tilde{\gamma}_{ij}}{\partial x^k} = O(r^{-2}). \tag{8.27}$$

Thanks to the decomposition (8.24), the integrand of the ADM mass formula (8.10) is

$$\begin{aligned}\bar{D}^j \gamma_{ij} - \bar{D}_i(f^{kl}\gamma_{kl}) = {} & 4 \underbrace{\Psi^3}_{\sim 1} \bar{D}^j \Psi \underbrace{\tilde{\gamma}_{ij}}_{\sim f_{ij}} + \underbrace{\Psi^4}_{\sim 1} \bar{D}^j \tilde{\gamma}_{ij} - 4 \underbrace{\Psi^3}_{\sim 1} \bar{D}_i \Psi \underbrace{f^{kl} \tilde{\gamma}_{kl}}_{\sim 3} \\ & - \underbrace{\Psi^4}_{\sim 1} \bar{D}_i(f^{kl}\tilde{\gamma}_{kl}),\end{aligned}$$

where the $\sim$'s denote values when $r \to \infty$, taking into account (8.26) and (8.27). Thus we have

$$\bar{D}^j \gamma_{ij} - \bar{D}_i(f^{kl}\gamma_{kl}) \sim -8\bar{D}_i \Psi + \bar{D}^j \tilde{\gamma}_{ij} - \bar{D}_i(f^{kl}\tilde{\gamma}_{kl}). \tag{8.28}$$

From (8.26) and (8.27), $\bar{D}_i \Psi = O(r^{-2})$ and $\bar{D}^j \tilde{\gamma}_{ij} = O(r^{-2})$. Let us show that the unit determinant condition (8.25) implies $\bar{D}_i(f^{kl}\tilde{\gamma}_{kl}) = O(r^{-3})$ so that this term actually does not contribute to the ADM mass integral. Let us write

$$\tilde{\gamma}_{ij} =: f_{ij} + \varepsilon_{ij}, \tag{8.29}$$

with according to Eq. (8.27), $\varepsilon_{ij} = O(r^{-1})$. Then

$$f^{kl}\tilde{\gamma}_{kl} = 3 + \varepsilon_{xx} + \varepsilon_{yy} + \varepsilon_{zz} \tag{8.30}$$

and

$$\bar{D}_i(f^{kl}\tilde{\gamma}_{kl}) = \frac{\partial}{\partial x^i}(f^{kl}\tilde{\gamma}_{kl}) = \frac{\partial}{\partial x^i}\left(\varepsilon_{xx} + \varepsilon_{yy} + \varepsilon_{zz}\right). \tag{8.31}$$

Now the determinant of $\tilde{\gamma}_{ij}$ is

$$\begin{aligned}\det(\tilde{\gamma}_{ij}) &= \det\begin{pmatrix} 1+\varepsilon_{xx} & \varepsilon_{xy} & \varepsilon_{xz} \\ \varepsilon_{xy} & 1+\varepsilon_{yy} & \varepsilon_{yz} \\ \varepsilon_{xz} & \varepsilon_{yz} & 1+\varepsilon_{zz} \end{pmatrix} \\ &= 1 + \varepsilon_{xx} + \varepsilon_{yy} + \varepsilon_{zz} + \varepsilon_{xx}\varepsilon_{yy} + \varepsilon_{xx}\varepsilon_{zz} + \varepsilon_{yy}\varepsilon_{zz} - \varepsilon_{xy}^2 - \varepsilon_{xz}^2 - \varepsilon_{yz}^2 \\ &\quad + \varepsilon_{xx}\varepsilon_{yy}\varepsilon_{zz} + 2\varepsilon_{xy}\varepsilon_{xz}\varepsilon_{yz} - \varepsilon_{xx}\varepsilon_{yz}^2 - \varepsilon_{yy}\varepsilon_{xz}^2 - \varepsilon_{zz}\varepsilon_{xy}^2.\end{aligned}$$

Requiring $\det(\tilde{\gamma}_{ij}) = 1$ implies then

$$\begin{aligned}\varepsilon_{xx} + \varepsilon_{yy} + \varepsilon_{zz} = &- \varepsilon_{xx}\varepsilon_{yy} - \varepsilon_{xx}\varepsilon_{zz} - \varepsilon_{yy}\varepsilon_{zz} + \varepsilon_{xy}^2 + \varepsilon_{xz}^2 + \varepsilon_{yz}^2 \\ &- \varepsilon_{xx}\varepsilon_{yy}\varepsilon_{zz} - 2\varepsilon_{xy}\varepsilon_{xz}\varepsilon_{yz} + \varepsilon_{xx}\varepsilon_{yz}^2 + \varepsilon_{yy}\varepsilon_{xz}^2 + \varepsilon_{zz}\varepsilon_{xy}^2.\end{aligned}$$

Since according to (8.27), $\varepsilon_{ij} = O(r^{-1})$ and $\partial\varepsilon_{ij}/\partial x^k = O(r^{-2})$, we conclude that

$$\frac{\partial}{\partial x^i}\left(\varepsilon_{xx} + \varepsilon_{yy} + \varepsilon_{zz}\right) = O(r^{-3}),$$

i.e. in view of (8.31),

$$\bar{D}_i(f^{kl}\tilde{\gamma}_{kl}) = O(r^{-3}). \tag{8.32}$$

Thus in Eq. (8.28), only the first two terms in the right-hand side contribute to the ADM mass integral, so that formula (8.10) becomes

$$\boxed{M_{\rm ADM} = -\frac{1}{2\pi}\lim_{\mathscr{S}_t\to\infty}\oint_{\mathscr{S}_t} s^i\left(\bar{D}_i\Psi - \frac{1}{8}\bar{D}^j\tilde{\gamma}_{ij}\right)\sqrt{q}\,\mathrm{d}^2y}\,. \tag{8.33}$$

Example 8.3 Let us return to Example 7.1 (Sect. 7.2.3), namely Schwarzschild spacetime in isotropic coordinates (t, r, θ, φ) (although we use the same symbol, the r used here is different from the Schwarzschild coordinate r of Example 8.1 above). The conformal factor was found to be $\Psi = 1 + m/(2r)$ [Eq. (7.24)] and the conformal metric to be $\tilde{\boldsymbol{\gamma}} = \boldsymbol{f}$. Then $\bar{D}^j\tilde{\gamma}_{ij} = 0$ and only the first term remains in the integral (8.33):

$$M_{\rm ADM} = -\frac{1}{2\pi} \lim_{r\to\infty} \oint_{r=\rm const} \frac{\partial\Psi}{\partial r} r^2 \sin\theta \, d\theta \, d\varphi,$$

with

$$\frac{\partial\Psi}{\partial r} = \frac{\partial}{\partial r}\left(1 + \frac{m}{2r}\right) = -\frac{m}{2r^2},$$

so that we get

$$M_{\rm ADM} = m,$$

i.e. we recover the result (8.19), which was obtained by means of different coordinates (Schwarzschild coordinates).

8.3.3 Newtonian Limit

To check that at the Newtonian limit, the ADM mass reduces to the usual definition of mass, let us consider the weak field metric given by Eq. (6.12). We have found in Sect. 7.2.3 that the corresponding conformal metric is $\tilde{\boldsymbol{\gamma}} = \boldsymbol{f}$ and the conformal factor $\Psi = 1 - \Phi/2$ [Eq. (7.25)], where Φ reduces to the gravitational potential at the Newtonian limit. Accordingly, $\bar{D}^j \tilde{\gamma}_{ij} = 0$ and $\bar{D}_i \Psi = -\frac{1}{2}\bar{D}_i \Phi$, so that Eq. (8.33) becomes

$$M_{\rm ADM} = \frac{1}{4\pi} \lim_{\mathscr{S}_t\to\infty} \oint_{\mathscr{S}_t} s^i \bar{D}_i \Phi \sqrt{q} \, d^2 y.$$

To take the Newtonian limit, we may assume that Σ_t has the topology of $\mathbb{R}^3$ and transform the above surface integral to a volume one by means of the Gauss–Ostrogradsky theorem:

$$M_{\rm ADM} = \frac{1}{4\pi} \int_{\Sigma_t} \bar{D}_i \bar{D}^i \Phi \sqrt{f} \, d^3 x. \tag{8.34}$$

Now, at the Newtonian limit, Φ is a solution of the Poisson equation

$$\bar{D}_i \bar{D}^i \Phi = 4\pi\rho, \tag{8.35}$$

where ρ is the mass density (remember we are using units in which Newton's gravitational constant G is unity). Hence Eq. (8.34) becomes

$$M_{\rm ADM} = \int_{\Sigma_t} \rho \sqrt{f} \, d^3 x, \tag{8.36}$$

which shows that at the Newtonian limit, the ADM mass is nothing but the total mass of the considered system.

8.3.4 Positive Energy Theorem

Since the ADM mass represents the total energy of a gravitational system, it is important to show that it is always positive, at least for "reasonable" models of matter (take $\rho < 0$ in Eq. (8.36) and you will get $M_{\rm ADM} < 0 \ldots$). If negative values of the energy would be possible, then a gravitational system could decay to lower and lower values and thereby emit an unbounded amount of energy via gravitational radiation.

The positivity of the ADM mass has been hard to establish. The complete proof was eventually given in 1981 by Schoen and Yau [13]. A simplified proof has been found shortly after by Witten [14]. Schoen, Yau and Witten have shown that if the matter content of spacetime obeys the dominant energy condition, then $M_{\rm ADM} \geq 0$. Furthermore, $M_{\rm ADM} = 0$ if and only if Σ_t is a hypersurface of Minkowski spacetime.

The ***dominant energy condition*** is the following requirement on the matter stress-energy tensor $\boldsymbol{T}$: for any timelike and future-directed vector $\boldsymbol{v}$, the vector $-\vec{\boldsymbol{T}}(\boldsymbol{v})$ defined by Eq. (2.39)[2] must be a future-directed timelike or null vector. If $\boldsymbol{v}$ is the 4-velocity of some observer, $-\vec{\boldsymbol{T}}(\boldsymbol{v})$ is the energy-momentum density 4-vector as measured by the observer and the dominant energy condition means that this vector must be causal. In particular, the dominant energy condition implies the ***weak energy condition***, namely that for any timelike and future-directed vector $\boldsymbol{v}$, $\boldsymbol{T}(\boldsymbol{v}, \boldsymbol{v}) \geq 0$. If again $\boldsymbol{v}$ is the 4-velocity of some observer, the quantity $\boldsymbol{T}(\boldsymbol{v}, \boldsymbol{v})$ is nothing but the energy density as measured by that observer [cf. Eq. (5.4)], and the weak energy condition simply stipulates that this energy density must be non-negative. In short, the dominant energy condition means that the matter energy must be positive and that it must not travel faster than light.

The dominant energy condition is easily expressible in terms of the matter energy density E and momentum density $\boldsymbol{p}$, both measured by the Eulerian observer and introduced in Sect. 5.2.1. Indeed, from the 3+1 split (5.14) of $\boldsymbol{T}$, the energy-momentum density 4-vector relative to the Eulerian observer is found to be

$$\boldsymbol{J} := -\vec{\boldsymbol{T}}(\boldsymbol{n}) = E\boldsymbol{n} + \vec{\boldsymbol{p}}. \tag{8.37}$$

Then, since $\boldsymbol{n} \cdot \vec{\boldsymbol{p}} = 0$, $\boldsymbol{J} \cdot \boldsymbol{J} = -E^2 + \vec{\boldsymbol{p}} \cdot \vec{\boldsymbol{p}}$. Requiring that $\boldsymbol{J}$ is timelike or null means $\boldsymbol{J} \cdot \boldsymbol{J} \leq 0$ and that it is future-oriented amounts to $E \geq 0$ (since $\boldsymbol{n}$ is itself future-oriented). Hence the dominant energy condition is equivalent to the two conditions $E^2 \geq \vec{\boldsymbol{p}} \cdot \vec{\boldsymbol{p}}$ and $E \geq 0$. Since $\vec{\boldsymbol{p}}$ is always a spacelike vector, these two conditions are actually equivalent to the single requirement

---

[2] In index notation, $-\vec{\boldsymbol{T}}(\boldsymbol{v})$ is the vector $-T^{\alpha}{}_{\mu} v^{\mu}$.

$$\boxed{E \geq \sqrt{\vec{p} \cdot \vec{p}}}. \tag{8.38}$$

This justifies the term *dominant* energy condition.

8.3.5 Constancy of the ADM Mass

Since the Hamiltonian H given by Eq. (8.7) depends on the configuration variables $(\gamma_{ij}, N, \beta^i)$ and their conjugate momenta $(\pi^{ij}, \pi^N = 0, \pi^{\boldsymbol{\beta}} = 0)$, but not explicitly on the time t, the associated energy is a constant of motion:

$$\boxed{\frac{\mathrm{d}}{\mathrm{d}t} M_{\mathrm{ADM}} = 0}. \tag{8.39}$$

Note that this property is not obvious when contemplating formula (8.10), which expresses M_{ADM} as an integral over $\mathscr{S}_t$.

8.4 ADM Momentum

8.4.1 Definition

As the ADM mass is associated with time translations at infinity [taking $N = 1$ and $\boldsymbol{\beta} = 0$ in Eq. (8.8)], the ADM momentum is defined as the conserved quantity associated with the invariance of the action with respect to spatial translations. With respect to the Cartesian-type coordinates (x^i) introduced in Sect. 8.2, three privileged directions for translations at spatial infinity are given by the three vectors $(\boldsymbol{\partial}_i)_{i\in\{1,2,3\}}$. The three conserved quantities are then obtained by setting $N = 0$ and $\beta^i = (\boldsymbol{\partial}_j)^i$ in Eq. (8.8) [8, 10]:

$$\boxed{P_i := \frac{1}{8\pi} \lim_{\mathscr{S}_t \to \infty} \oint_{\mathscr{S}_t} \left(K_{jk} - K\gamma_{jk}\right) (\boldsymbol{\partial}_i)^j s^k \sqrt{q}\, \mathrm{d}^2 y}, \qquad i \in \{1, 2, 3\}. \tag{8.40}$$

Notice that the asymptotic flatness condition (8.2a) ensures that P_i is a finite quantity. The three numbers (P_1, P_2, P_3) define the ***ADM momentum*** of the hypersurface Σ_t. The values P_i depend upon the choice of the coordinates (x^i) but the set (P_1, P_2, P_3) transforms as the components of a linear form under a change of Cartesian coordinates $(x^i) \to (x'^i)$ that asymptotically corresponds to a rotation and/or a translation. Therefore (P_1, P_2, P_3) can be regarded as a linear form which "lives" at the "edge" of Σ_t. It can be regarded as well as a vector since the duality vector/linear forms is trivial in the asymptotically Euclidean space.

Example 8.4 For foliations associated with the standard coordinates of Schwarzschild spacetime (e.g. Schwarzschild coordinates (8.12) or isotropic coordinates (7.23)), the extrinsic curvature vanishes identically: $\boldsymbol{K} = 0$, so that Eq. (8.40) yields

$$P_i = 0. \tag{8.41}$$

For a non trivial example based on a "boosted" Schwarzschild solution, see Ref. [4].

8.4.2 ADM 4-Momentum

Not only (P_1, P_2, P_3) behaves as the components of a linear form, but the set of four numbers

$$\boxed{P_\alpha^{\rm ADM} := (-M_{\rm ADM}, P_1, P_2, P_3)} \tag{8.42}$$

behaves as the components of a 4-dimensional linear form any under coordinate change $(x^\alpha) = (t, x^i) \to (x'^\alpha) = (t', x'^i)$ which preserves the asymptotic conditions (8.1)–(8.2), i.e. any coordinate change of the form (8.3). In particular, $P_\alpha^{\rm ADM}$ is transformed in the proper way under the Poincaré group:

$$P'^{\rm ADM}_\alpha = (\Lambda^{-1})^\mu{}_\alpha \, P^{\rm ADM}_\mu. \tag{8.43}$$

This last property has been shown first by Arnowitt, Deser and Misner [15]. For this reason, $P_\alpha^{\rm ADM}$ is considered as a linear form which "lives" at spatial infinity and is called the ***ADM 4-momentum***.

8.5 Angular Momentum

8.5.1 The Supertranslation Ambiguity

Generically, the angular momentum is the conserved quantity associated with the invariance of the action with respect to rotations, in the same manner as the linear momentum is associated with the invariance with respect to translations. Then one might naively define the total angular momentum of a given slice Σ_t by an integral of the type (8.40) but with $\boldsymbol{\partial}_i$ being replaced by a rotational Killing vector $\boldsymbol{\phi}$ of the flat metric $\boldsymbol{f}$. More precisely, in terms of the Cartesian coordinates $(x^i) = (x, y, z)$ introduced in Sect. 8.2, the three vectors $(\boldsymbol{\phi}_i)_{i\in\{1,2,3\}}$ defined by

$$\boldsymbol{\phi}_x = -z\boldsymbol{\partial}_y + y\boldsymbol{\partial}_z \tag{8.44a}$$

$$\boldsymbol{\phi}_y = -x\boldsymbol{\partial}_z + z\boldsymbol{\partial}_x \tag{8.44b}$$

$$\boldsymbol{\phi}_z = -y\boldsymbol{\partial}_x + x\boldsymbol{\partial}_y \tag{8.44c}$$

are three independent Killing vectors of $\boldsymbol{f}$, corresponding to a rotation about respectively the x-axis, y-axis and the z-axis. Then one may defined the three numbers

$$J_i := \frac{1}{8\pi} \lim_{\mathscr{S}_t \to \infty} \oint_{\mathscr{S}_t} \left(K_{jk} - K\gamma_{jk}\right) (\boldsymbol{\phi}_i)^j s^k \sqrt{q}\, \mathrm{d}^2 y, \qquad i \in \{1, 2, 3\}. \tag{8.45}$$

The problem is that the quantities J_i hence defined depend upon the choice of the coordinates and, contrary to P_α^{ADM}, do not transform as the components of a vector under any coordinate change $(x^\alpha) = (t, x^i) \to (x'^\alpha) = (t', x'^i)$ that preserves the asymptotic properties (8.1)–(8.2), i.e. a transformation of the type (8.3). As discussed by York [3, 4], the problem arises because of the existence of the supertranslations (cf. Sect. 8.2.2) in the permissible coordinate changes (8.3).

Remark 8.3 Independently of the above coordinate ambiguity, one may notice that the asymptotic flatness conditions (8.1)–(8.2) are not sufficient, *by themselves*, to guarantee that the integral (8.45) takes a finite value when $\mathscr{S}_t \to \infty$, i.e. when $r \to \infty$. Indeed, Eq. (8.44) shows that the Cartesian components of the rotational vectors behave like $(\boldsymbol{\phi}_i)^j \sim O(r)$, so that Eq. (8.2a) implies only $\left(K_{jk} - K\gamma_{jk}\right)(\boldsymbol{\phi}_i)^j = O(r^{-1})$. It is the contraction with the unit normal vector s^k which ensures

$$\left(K_{jk} - K\gamma_{jk}\right) (\boldsymbol{\phi}_i)^j s^k = O(r^{-2})$$

and hence that J_i is finite. This is clear for the $K\gamma_{jk}(\boldsymbol{\phi}_i)^j s^k$ part because the vectors $\boldsymbol{\phi}_i$ given by Eq. (8.44) are all orthogonal to $\boldsymbol{s} \sim x/r\boldsymbol{\partial}_x + y/r\boldsymbol{\partial}_y + z/r\boldsymbol{\partial}_z$. For the $K_{jk}(\boldsymbol{\phi}_i)^j s^k$ part, this turns out to be true in practice, as we shall see on the specific example of Kerr spacetime in Sect. 8.6.3.

8.5.2 The "Cure"

In view of the above coordinate dependence problem, one may define the angular momentum as a quantity which remains invariant only with respect to a subclass of the coordinate changes (8.3). This is made by imposing decay conditions stronger than (8.1)–(8.2). For instance, York [3] has proposed the following conditions[3] on the flat divergence of the conformal metric and the trace of the extrinsic curvature:

$$\frac{\partial \tilde{\gamma}_{ij}}{\partial x^j} = O(r^{-3}), \tag{8.46}$$

[3] Actually the first condition proposed by York, Eq. (90) of Ref. [3], is not exactly (8.46) but can be shown to be equivalent to it; see also Sec. V of Ref. [16].

$$K = O(r^{-3}). \tag{8.47}$$

Clearly these conditions are stronger than respectively (8.27) and (8.2a). Actually they are so severe that they exclude some well known coordinates that one would like to use to describe asymptotically flat spacetimes, for instance the standard Schwarzschild coordinates (8.12) for the Schwarzschild solution. For this reason, conditions (8.46) and (8.47) are considered as asymptotic *gauge conditions*, i.e. conditions restricting the choice of coordinates, rather than conditions on the nature of spacetime at spatial infinity. Condition (8.46) is called the ***quasi-isotropic gauge***. The isotropic coordinates (7.23) of the Schwarzschild solution trivially belong to this gauge (since $\tilde{\gamma}_{ij} = f_{ij}$ for them). Condition (8.47) is called the ***asymptotically maximal gauge***, since for maximal hypersurfaces K vanishes identically. York has shown that in the gauge (8.46)–(8.47), the angular momentum as defined by the integral (8.45) is carried by the $O(r^{-3})$ piece of $\boldsymbol{K}$ (the $O(r^{-2})$ piece carrying the linear momentum P_i) and is invariant (i.e. behaves as a vector) for any coordinate change within this gauge.

Alternative decay requirements have been proposed by other authors to fix the ambiguities in the angular momentum definition (see e.g. [17] and references therein). For instance, Regge and Teitelboim [10] impose a specific form and some parity conditions on the coefficient of the $O(r^{-1})$ term in Eq. (8.1a) and on the coefficient of the $O(r^{-2})$ term in Eq. (8.2a) (cf. also [8]).

As we shall see in Sect. 8.6.3, in the particular case of an axisymmetric spacetime, there exists a unique definition of the angular momentum, which is independent of any coordinate system.

Remark 8.4 In the literature, there is often mention of the "*ADM angular momentum*", on the same footing as the ADM mass and ADM linear momentum. But as discussed above, there is no such thing as the "ADM angular momentum". One has to specify a gauge first and define the angular momentum within that gauge. In particular, there is no mention whatsoever of angular momentum in the original ADM article [15].

8.5.3 ADM Mass in the Quasi-Isotropic Gauge

In the quasi-isotropic gauge, the ADM mass can be expressed entirely in terms of the flux at infinity of the gradient of the conformal factor Ψ. Indeed, thanks to (8.46), the term $\bar{D}^j \tilde{\gamma}_{ij}$ Eq. (8.33) does not contribute to the integral and we get

$$\boxed{M_{\mathrm{ADM}} = -\frac{1}{2\pi} \lim_{\mathscr{S}_t \to \infty} \oint_{\mathscr{S}_t} s^i \bar{D}_i \Psi \sqrt{q}\, \mathrm{d}^2 y} \quad \text{(quasi-isotropic gauge)}. \tag{8.48}$$

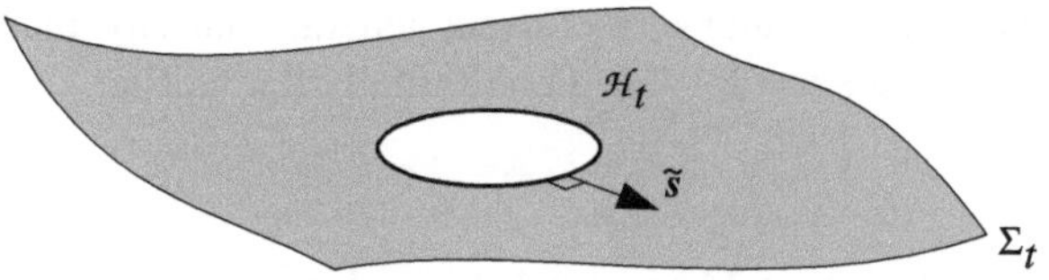

Fig. 8.1 Hypersurface Σ_t with a hole defining an inner boundary $\mathscr{H}_t$.

Thanks to the Gauss–Ostrogradsky theorem, we may transform this formula into a volume integral. More precisely, let us assume that Σ_t is diffeomorphic to either $\mathbb{R}^3$ or $\mathbb{R}^3$ minus a ball. In the latter case, Σ_t has an inner boundary, that we may call a ***hole*** and denote by $\mathscr{H}_t$ (cf. Fig. 8.1). We assume that $\mathscr{H}_t$ has the topology of a sphere. Actually this case is relevant for black hole spacetimes when black holes are treated via the so-called excision technique. The Gauss–Ostrogradsky formula enables to transform expression (8.48) into

$$M_{\text{ADM}} = -\frac{1}{2\pi} \int_{\Sigma_t} \tilde{D}_i \tilde{D}^i \Psi \sqrt{\tilde{\gamma}}\, \mathrm{d}^3x + M_{\mathscr{H}}, \tag{8.49}$$

where $M_{\mathscr{H}}$ is defined by

$$M_{\mathscr{H}} := -\frac{1}{2\pi} \oint_{\mathscr{H}_t} \tilde{s}^i \tilde{D}_i \Psi \sqrt{\tilde{q}}\, \mathrm{d}^2y. \tag{8.50}$$

In this last equation, $\tilde{q} := \det(\tilde{q}_{ab})$, $\tilde{\boldsymbol{q}}$ being the metric induced on $\mathscr{H}_t$ by $\tilde{\boldsymbol{\gamma}}$, and $\tilde{\boldsymbol{s}}$ is the unit vector with respect to $\tilde{\boldsymbol{\gamma}}$ ($\tilde{\boldsymbol{\gamma}}(\tilde{\boldsymbol{s}}, \tilde{\boldsymbol{s}}) = 1$) tangent to Σ_t, normal to $\mathscr{H}_t$ and oriented towards the exterior of the hole (cf. Fig. 8.1). If Σ_t is diffeomorphic to $\mathbb{R}^3$, we use formula (8.49) with $M_{\mathscr{H}} = 0$.

Let us now use the Lichnerowicz equation (7.86) to express $\tilde{D}_i \tilde{D}^i \Psi$ in Eq. (8.49). We get

$$\boxed{M_{\text{ADM}} = \int_{\Sigma_t} \left[\Psi^5 E + \frac{1}{16\pi} \left(\hat{A}_{ij} \hat{A}^{ij} \Psi^{-7} - \tilde{R}\Psi - \frac{2}{3} K^2 \Psi^5 \right) \right] \sqrt{\tilde{\gamma}}\, \mathrm{d}^3x + M_{\mathscr{H}}.} \tag{8.51}$$

For the computation of the ADM mass in a numerical code, this formula may result in a greater precision that the surface integral at infinity (8.48).

Remark 8.5 On the formula (8.51), we get immediately the Newtonian limit (8.36) by making $\Psi \to 1$, $E \to \rho$, $\hat{A}^{ij} \to 0$, $\tilde{R} \to 0$, $K \to 0$, $\tilde{\gamma} \to f$ and $M_{\mathscr{H}} = 0$.

For the IWM approximation of general relativity considered in Sect. 7.6, the coordinates belong to the quasi-isotropic gauge (since $\tilde{\boldsymbol{\gamma}} = \boldsymbol{f}$), so we may apply (8.51). Moreover, as a consequence of $\tilde{\boldsymbol{\gamma}} = \boldsymbol{f}$, $\tilde{R} = 0$ and in the IWM approximation, $K = 0$. Therefore Eq.(8.51) simplifies to

$$M_{\text{ADM}} = \int_{\Sigma_t} \left(\Psi^5 E + \frac{1}{16\pi} \hat{A}_{ij} \hat{A}^{ij} \Psi^{-7} \right) \sqrt{\tilde{\gamma}}\, \mathrm{d}^3x + M_{\mathscr{H}}. \tag{8.52}$$

Within the framework of exact general relativity, the above formula is valid for any maximal slice Σ_t with a conformally flat metric.

8.6 Komar Mass and Angular Momentum

When the spacetime $(\mathscr{M}, \boldsymbol{g})$ has some symmetries, one may define global quantities in a coordinate-independent way by means of a general technique introduced by Komar [18]. It consists in taking flux integrals of the derivative of the Killing vector associated with the symmetry over closed 2-surfaces surrounding the matter sources. The quantities thus obtained are conserved in the sense that they do not depend upon the choice of the integration 2-surface, as long as the latter stays outside the matter. We discuss here two important cases: the *Komar mass* resulting from time symmetry (stationarity) and the *Komar angular momentum* resulting from axisymmetry.

8.6.1 Komar Mass

Let us assume that the spacetime $(\mathscr{M}, \boldsymbol{g})$ is ***stationary***. This means that the metric tensor $\boldsymbol{g}$ is invariant by Lie transport along the field lines of a timelike vector field $\boldsymbol{k}$:

$$\mathscr{L}_{\boldsymbol{k}} \boldsymbol{g} = 0, \tag{8.53}$$

where $\mathscr{L}_{\boldsymbol{k}}$ is the Lie derivative along $\boldsymbol{k}$ (cf. Sect. 2.5). The latter is called a ***Killing vector***. Provided that it is normalized so that $\boldsymbol{k} \cdot \boldsymbol{k} = -1$ at spatial infinity, it is then unique. Given a 3+1 foliation $(\Sigma_t)_{t\in\mathbb{R}}$ of $\mathscr{M}$, and a closed 2-surface $\mathscr{S}_t$ in Σ_t, with the topology of a sphere, the ***Komar mass*** is defined by

$$\boxed{M_{\mathrm{K}} := -\frac{1}{8\pi} \oint_{\mathscr{S}_t} \nabla^\mu k^\nu \, \mathrm{d}S_{\mu\nu}}, \tag{8.54}$$

with the 2-surface element

$$\mathrm{d}S_{\mu\nu} = (s_\mu n_\nu - n_\mu s_\nu)\sqrt{q}\, \mathrm{d}^2 y, \tag{8.55}$$

where $\boldsymbol{n}$ is the unit timelike normal to Σ_t, $\boldsymbol{s}$ is the unit normal to $\mathscr{S}_t$ within Σ_t oriented towards the exterior of $\mathscr{S}_t$, $(y^a) = (y^1, y^2)$ are coordinates spanning $\mathscr{S}_t$, and $q := \det(q_{ab})$, the q_{ab}'s being the components with respect to (y^a) of the metric $\boldsymbol{q}$ induced by $\boldsymbol{\gamma}$ (or equivalently by $\boldsymbol{g}$) on $\mathscr{S}_t$. Actually the Komar mass can be defined over any closed 2-surface, but in the present context it is quite natural to consider only 2-surfaces lying in the hypersurfaces Σ_t of the 3+1 foliation.

A priori the quantity M_{K} as defined by (8.54) should depend on the choice of the 2-surface $\mathscr{S}_t$. However, thanks to the fact that $\boldsymbol{k}$ is a Killing vector, this is not the

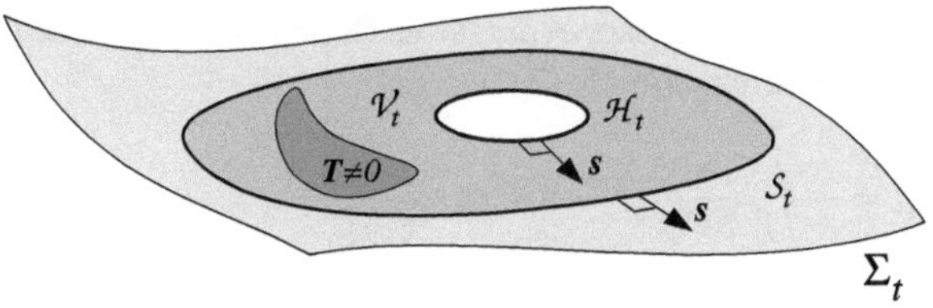

Fig. 8.2 Integration surface $\mathscr{S}_t$ for the computation of Komar mass. $\mathscr{S}_t$ is the external boundary of a part $\mathscr{V}_t$ of Σ_t which contains all the matter sources ($\boldsymbol{T} \neq 0$). $\mathscr{V}_t$ has possibly some inner boundary, in the form of one (or more) hole $\mathscr{H}_t$

case, as long as $\mathscr{S}_t$ is located outside any matter content of spacetime. In order to show this, let us transform the surface integral (8.54) into a volume integral. As in Sect. 8.5.4, we suppose that Σ_t is diffeomorphic to either $\mathbb{R}^3$ or $\mathbb{R}^3$ minus one hole, the results being easily generalized to an arbitrary number of holes (see Fig. 8.2). The hole, the surface of which is denoted by $\mathscr{H}_t$ as in Sect. 8.5.3, must be totally enclosed within the surface $\mathscr{S}_t$. Let us then denote by $\mathscr{V}_t$ the part of Σ_t delimited by $\mathscr{H}_t$ and $\mathscr{S}_t$.

The starting point is to notice that since $\boldsymbol{k}$ is a Killing vector the $\nabla^\mu k^\nu$'s in the integrand of Eq. (8.54) are the components of an antisymmetric tensor. Indeed, $\boldsymbol{k}$ obeys ***Killing equation***[4]:

$$\boxed{\nabla_\alpha k_\beta + \nabla_\beta k_\alpha = 0}. \tag{8.56}$$

Now for any antisymmetric tensor $\boldsymbol{A}$ of type $(2,0)$, the following identity holds:

$$2\int_{\mathscr{V}_t} \nabla_\nu A^{\mu\nu}\,\mathrm{d}V_\mu = \oint_{\mathscr{S}_t} A^{\mu\nu}\,\mathrm{d}S_{\mu\nu} + \oint_{\mathscr{H}_t} A^{\mu\nu}\,\mathrm{d}S^{\mathscr{H}}_{\mu\nu}, \tag{8.57}$$

with $\mathrm{d}V_\mu$ is the volume element on Σ_t:

$$\mathrm{d}V_\mu = -n_\mu\sqrt{\gamma}\,\mathrm{d}^3x \tag{8.58}$$

and $\mathrm{d}S^{\mathscr{H}}_{\mu\nu}$ is the surface element on $\mathscr{H}_t$ and is given by a formula similar to Eq. (8.55), using the same notation for the coordinates and the induced metric on $\mathscr{H}_t$:

$$\mathrm{d}S^{\mathscr{H}}_{\mu\nu} = (n_\mu s_\nu - s_\mu n_\nu)\sqrt{q}\,\mathrm{d}^2y. \tag{8.59}$$

The change of sign with respect to Eq. (8.55) arises because we choose the unit vector $\boldsymbol{s}$ normal to $\mathscr{H}_t$ to be oriented towards the interior of $\mathscr{V}_t$ (cf. Fig. 8.2). Let us establish (8.57). Expressing the divergence of the antisymmetric tensor $\boldsymbol{A}$ via

[4] The Killing equation follows immediately from Eq. (8.53) with the Lie derivative expressed via Eq. (2.92), along with $\nabla \boldsymbol{g} = 0$.

the identity (2.66), and using the expression (8.58) of dV_μ with the components $n_\mu = (-N, 0, 0, 0)$ given by Eq.(5.37), we get

$$\int_{\mathscr{V}_t} \nabla_\nu A^{\mu\nu}\, \mathrm{d}V_\mu = -\int_{\mathscr{V}_t} \frac{\partial}{\partial x^\nu}\left(\sqrt{-g}A^{\mu\nu}\right) n_\mu \frac{\sqrt{\gamma}}{\sqrt{-g}}\, \mathrm{d}^3x = \int_{\mathscr{V}_t} \frac{\partial}{\partial x^\nu}\left(\sqrt{\gamma} N A^{0\nu}\right) \mathrm{d}^3x,$$

where we have also invoked the relation (5.55) between the determinants of $\boldsymbol{g}$ and $\boldsymbol{\gamma}$: $\sqrt{-g} = N\sqrt{\gamma}$. Now, since $A^{\alpha\beta}$ is antisymmetric, $A^{00} = 0$ and we can write $\partial/\partial x^\nu \left(\sqrt{\gamma} N A^{0\nu}\right) = \partial/\partial x^i \left(\sqrt{\gamma} V^i\right)$ where $V^i = N A^{0i}$ are the components of the vector $\boldsymbol{V} \in \mathscr{T}(\Sigma_t)$ defined by $\boldsymbol{V} := -\vec{\boldsymbol{\gamma}}(\boldsymbol{n} \cdot \boldsymbol{A})$. The above integral then becomes

$$\int_{\mathscr{V}_t} \nabla_\nu A^{\mu\nu}\, \mathrm{d}V_\mu = \int_{\mathscr{V}_t} \frac{1}{\sqrt{\gamma}} \frac{\partial}{\partial x^i}\left(\sqrt{\gamma} V^i\right) \sqrt{\gamma}\, \mathrm{d}^3x = \int_{\mathscr{V}_t} D_i V^i \sqrt{\gamma}\, \mathrm{d}^3x.$$

We can now use the Gauss–Ostrogradsky theorem to get

$$\int_{\mathscr{V}_t} \nabla_\nu A^{\mu\nu}\, \mathrm{d}V_\mu = \oint_{\partial\mathscr{V}_t} V^i s_i \sqrt{q}\, \mathrm{d}^2y.$$

Noticing that $\partial\mathscr{V}_t = \mathscr{H}_t \cup \mathscr{S}_t$ (cf. Fig. 8.2) and (from the antisymmetry of $A^{\mu\nu}$)

$$V^i s_i = V^\nu s_\nu = -n_\mu A^{\mu\nu} s_\nu = \frac{1}{2} A^{\mu\nu}(s_\mu n_\nu - n_\mu s_\nu),$$

we get the identity (8.57).

Remark 8.6 Equation (8.57) can also be derived by applying Stokes' theorem to the 2-form ${}^4\varepsilon_{\alpha\beta\mu\nu} A^{\mu\nu}$, where ${}^4\varepsilon_{\alpha\beta\mu\nu}$ is the Levi–Civita tensor associated with the spacetime metric $\boldsymbol{g}$ (cf. Sect. 11.2.10) (see e.g. derivation of Eq. (11.2.10) in Wald's book [7]).

Applying formula (8.57) to $A^{\mu\nu} = \nabla^\mu k^\nu$ we get, in view of the definition (8.54),

$$M_{\mathrm{K}} = -\frac{1}{4\pi} \int_{\mathscr{V}_t} \nabla_\nu \nabla^\mu k^\nu\, \mathrm{d}V_\mu + M_{\mathrm{K}}^{\mathscr{H}}, \tag{8.60}$$

where

$$M_{\mathrm{K}}^{\mathscr{H}} := \frac{1}{8\pi} \oint_{\mathscr{H}_t} \nabla^\mu k^\nu\, \mathrm{d}S^{\mathscr{H}}_{\mu\nu} \tag{8.61}$$

will be called the ***Komar mass of the hole***. Now, from the contracted Ricci identity (2.68),

$$\nabla_\nu \nabla^\mu k^\nu - \nabla^\mu \underbrace{\nabla_\nu k^\nu}_{0} = {}^4R^\mu{}_\nu k^\nu,$$

where the "$= 0$" is a consequence of Killing's equation (8.56). Equation (8.60) becomes then

$$M_K = -\frac{1}{4\pi}\int_{\mathscr{V}_t} {}^4R^{\mu}{}_{\nu}k^{\nu}\,\mathrm{d}V_{\mu} + M_K^{\mathscr{H}} = \frac{1}{4\pi}\int_{\mathscr{V}_t} {}^4R_{\mu\nu}k^{\nu}n^{\mu}\sqrt{\gamma}\,\mathrm{d}^3x + M_K^{\mathscr{H}}.$$

At this point, we can use Einstein equation in the form (5.2) to express the Ricci tensor ${}^4\boldsymbol{R}$ in terms of the matter stress-energy tensor $\boldsymbol{T}$. We obtain

$$M_K = 2\int_{\mathscr{V}_t}\left(T_{\mu\nu} - \frac{1}{2}T g_{\mu\nu}\right)n^{\mu}k^{\nu}\sqrt{\gamma}\,\mathrm{d}^3x + M_K^{\mathscr{H}}. \tag{8.62}$$

The support of the integral over $\mathscr{V}_t$ is reduced to the location of matter, i.e. the domain where $\boldsymbol{T} \neq 0$. It is then clear on formula (8.62) that M_K is independent of the choice of the 2-surface $\mathscr{S}_t$, provided all the matter is contained in $\mathscr{S}_t$. In particular, we may extend the integration to all Σ_t and write formula (8.62) as

$$\boxed{M_K = 2\int_{\Sigma_t}\left[\boldsymbol{T}(\boldsymbol{n},\boldsymbol{k}) - \frac{1}{2}T\boldsymbol{n}\cdot\boldsymbol{k}\right]\sqrt{\gamma}\,\mathrm{d}^3x + M_K^{\mathscr{H}}}. \tag{8.63}$$

The Komar mass then appears as a global quantity defined for stationary spacetimes.

Remark 8.7 One may have $M_K^{\mathscr{H}} < 0$, with $M_K > 0$, provided that the matter integral in Eq. (8.63) compensates for the negative value of $M_K^{\mathscr{H}}$. Such spacetimes exist, as demonstrated by Ansorg and Petroff [19]: these authors have numerically constructed spacetimes containing a black hole with $M_K^{\mathscr{H}} < 0$ surrounded by a ring of matter (incompressible perfect fluid) such that the total Komar mass is positive.

8.6.2 3+1 Expression of the Komar Mass and Link with the ADM Mass

In stationary spacetimes, it is natural to use coordinates adapted to the symmetry, i.e. coordinates (t, x^i) such that

$$\boxed{\boldsymbol{\partial}_t = \boldsymbol{k}}. \tag{8.64}$$

Then Eq. (5.59) results in the following 3+1 decomposition of the Killing vector in terms of the lapse and shift:

$$\boldsymbol{k} = N\boldsymbol{n} + \boldsymbol{\beta}. \tag{8.65}$$

Let us insert this relation into the integrand in the definition (8.54):

$$\begin{aligned}\nabla^\mu k^\nu \, \mathrm{d}S_{\mu\nu} &= \nabla_\mu k_\nu (s^\mu n^\nu - n^\mu s^\nu)\sqrt{q}\, \mathrm{d}^2 y \\ &= 2\nabla_\mu k_\nu s^\mu n^\nu \sqrt{q}\, \mathrm{d}^2 y \\ &= 2\left(\nabla_\mu N n_\nu + N\nabla_\mu n_\nu + \nabla_\mu \beta_\nu\right) s^\mu n^\nu \sqrt{q}\, \mathrm{d}^2 y \\ &= 2\left(-s^\mu \nabla_\mu N + 0 - s^\mu \beta_\nu \nabla_\mu n^\nu\right)\sqrt{q}\, \mathrm{d}^2 y \\ &= -2\left(s^i D_i N - K_{ij} s^i \beta^j\right)\sqrt{q}\, \mathrm{d}^2 y,\end{aligned} \tag{8.66}$$

where we have used Killing's equation (8.56) to get the second line, the orthogonality of $\boldsymbol{n}$ and $\boldsymbol{\beta}$ to get the fourth one and expression (4.26) for $\nabla_\mu n^\nu$ to get the last line. Inserting Eq. (8.66) into Eq. (8.54) yields the 3+1 expression of the Komar mass:

$$\boxed{M_{\mathrm{K}} = \frac{1}{4\pi}\oint_{\mathscr{S}_t}\left(s^i D_i N - K_{ij} s^i \beta^j\right)\sqrt{q}\, \mathrm{d}^2 y}\,. \tag{8.67}$$

Example 8.5. A simple prototype of a stationary spacetime is of course the Schwarzschild spacetime. Let us compute its Komar mass by means of the above formula and the foliation $(\Sigma_t)_{t\in\mathbb{R}}$ defined by the standard Schwarzschild coordinates (8.12). For this foliation, $K_{ij} = 0$, which reduces Eq. (8.67) to the flux of the lapse's gradient across $\mathscr{S}_t$. Taking advantage of the spherical symmetry, we choose $\mathscr{S}_t$ to be a surface $r =$ const. Then $y^a = (\theta, \varphi)$. The unit normal $\boldsymbol{s}$ is read from the line element (8.12); its components with respect to the Schwarzschild coordinates (r, θ, φ) are

$$s^i = \left(\left(1 - \frac{2m}{r}\right)^{1/2}, 0, 0\right). \tag{8.68}$$

N and $\sqrt{q}$ are also read on the line element (8.12): $N = (1 - 2m/r)^{1/2}$ and $\sqrt{q} = r^2 \sin\theta$, so that Eq. (8.67) results in

$$M_{\mathrm{K}} = \frac{1}{4\pi}\oint_{r=\mathrm{const}}\left(1 - \frac{2m}{r}\right)^{1/2}\frac{\partial}{\partial r}\left[\left(1 - \frac{2m}{r}\right)^{1/2}\right] r^2 \sin\theta \, \mathrm{d}\theta \, \mathrm{d}\varphi.$$

All the terms containing r simplify and we get

$$M_{\mathrm{K}} = m. \tag{8.69}$$

On this particular example, we have verified that the value of M_{K} does not depend upon the choice of $\mathscr{S}_t$ (i.e. upon the value of r).

Let us now turn to the volume expression (8.63) of the Komar mass. By using the 3+1 decomposition (5.14) and (5.15) of respectively $\boldsymbol{T}$ and T, we get

$$T(\boldsymbol{n},\boldsymbol{k}) - \frac{1}{2}T\boldsymbol{n}\cdot\boldsymbol{k} = -\langle \boldsymbol{p},\boldsymbol{k}\rangle - E\langle \underline{\boldsymbol{n}},\boldsymbol{k}\rangle - \frac{1}{2}(S-E)\boldsymbol{n}\cdot\boldsymbol{k}$$
$$= -\langle \boldsymbol{p},\boldsymbol{\beta}\rangle + EN + \frac{1}{2}(S-E)N = \frac{1}{2}N(E+S) - \langle \boldsymbol{p},\boldsymbol{\beta}\rangle.$$

Hence formula (8.63) becomes

$$\boxed{M_{\mathrm{K}} = \int_{\Sigma_t}\left[N(E+S) - 2\langle \boldsymbol{p},\boldsymbol{\beta}\rangle\right]\sqrt{\gamma}\,\mathrm{d}^3x + M_{\mathrm{K}}^{\mathscr{H}}}, \tag{8.70}$$

with the Komar mass of the hole given by an expression identical to Eq. (8.67), except for $\mathscr{S}_t$ replaced by $\mathscr{H}_t$ [notice the double change of sign: first in Eq. (8.61) and secondly in Eq. (8.59), so that at the end we get an expression identical to Eq. (8.67)]:

$$\boxed{M_{\mathrm{K}}^{\mathscr{H}} = \frac{1}{4\pi}\oint_{\mathscr{H}_t}\left(s^i D_i N - K_{ij}s^i\beta^j\right)\sqrt{q}\,\mathrm{d}^2y}. \tag{8.71}$$

It is easy to take the Newtonian limit Eq. (8.70), by making $N \to 1$, $E \to \rho$, $S \ll E$ [Eq. (6.22)], $\boldsymbol{\beta} \to 0$, $\gamma \to f$ and $M_{\mathrm{K}}^{\mathscr{H}} = 0$. We get

$$M_{\mathrm{K}} = \int_{\Sigma_t}\rho\sqrt{f}\,\mathrm{d}^3x. \tag{8.72}$$

Hence at the Newtonian limit, the Komar mass reduces to the standard total mass. This, along with the result (8.69) for Schwarzschild spacetime, justifies the name Komar *mass*.

A natural question which arises then is how does the Komar mass relate to the ADM mass of Σ_t? The answer is not obvious if one compares the defining formulæ (8.9) and (8.54). It is even not obvious if one compares the 3+1 expressions (8.33) and (8.67): Eq. (8.33) involves the flux of the gradient of the conformal factor Ψ of the 3-metric, whereas Eq. (8.67) involves the flux of the gradient of the lapse function N. Moreover, in Eq. (8.33) the integral must be evaluated at spatial infinity, whereas in Eq. (8.67) it can be evaluated at any finite distance (outside the matter sources). Besides, in the quasi-isotropic gauge, we have obtained a volume expression of the ADM mass, Eq. (8.51), that we may compare to the volume expression (8.70) of the Komar mass. Even when there is no hole, the two expressions are pretty different. In particular, the Komar mass integral has a compact support (the matter domain), whereas the ADM mass integral has not.

The answer to the above question has been obtained in 1978 by Beig [20], as well as by Ashtekar and Magnon-Ashtekar the year after [21]: for any foliation $(\Sigma_t)_{t\in\mathbb{R}}$ whose unit normal vector $\boldsymbol{n}$ coincides with the timelike Killing vector $\boldsymbol{k}$ at spatial infinity [i.e. $N \to 1$ and $\boldsymbol{\beta} \to 0$ in Eq. (8.65)],

$$\boxed{M_{\mathrm{K}} = M_{\mathrm{ADM}}}. \tag{8.73}$$

Remark 8.8 In expression (8.67) of $M_{\rm K}$, there is no longer any mention of the timelike Killing vector $\boldsymbol{k}$. Therefore (8.67) is sometimes used to define the Komar mass in non-stationary spacetimes. The integral must then be performed at spatial infinity, otherwise it would depend on the choice of the 2-surface $\mathscr{S}_t$.

8.6.3 Komar Angular Momentum

If the spacetime $(\mathscr{M}, \boldsymbol{g})$ is axisymmetric, its ***Komar angular momentum*** is defined by a surface integral similar to that of the Komar mass, Eq. (8.54), but with the Killing vector $\boldsymbol{k}$ replaced by the Killing vector $\boldsymbol{\phi}$ associated with the axisymmetry:

$$\boxed{J_{\rm K} := \frac{1}{16\pi} \oint_{\mathscr{S}_t} \nabla^{\mu} \phi^{\nu} \, \mathrm{d}S_{\mu\nu}} \,. \tag{8.74}$$

Notice a factor -2 of difference with respect to formula (8.54) (the so-called *Komar's anomalous factor* [22]).

For the same reason as for $M_{\rm K}$, $J_{\rm K}$ is actually independent of the surface $\mathscr{S}_t$ as long as the latter is outside all the possible matter sources and $J_{\rm K}$ can be expressed by a volume integral over the matter by a formula similar to (8.63) (except for the factor -2):

$$J_{\rm K} = -\int_{\Sigma_t} \left[\boldsymbol{T}(\boldsymbol{n}, \boldsymbol{\phi}) - \frac{1}{2} T \boldsymbol{n} \cdot \boldsymbol{\phi} \right] \sqrt{\gamma} \, \mathrm{d}^3 x + J_{\rm K}^{\mathscr{H}} \,, \tag{8.75}$$

with

$$J_{\rm K}^{\mathscr{H}} := -\frac{1}{16\pi} \oint_{\mathscr{H}_t} \nabla^{\mu} \phi^{\nu} \, \mathrm{d}S_{\mu\nu}^{\mathscr{H}} \,. \tag{8.76}$$

Let us now establish the 3+1 expression of the Komar angular momentum. It is natural to choose a foliation adapted to the axisymmetry in the sense that the Killing vector $\boldsymbol{\phi}$ is tangent to the hypersurfaces Σ_t. Then $\boldsymbol{n} \cdot \boldsymbol{\phi} = 0$ and the integrand in the definition (8.74) is

$$\begin{aligned} \nabla^{\mu} \phi^{\nu} \, \mathrm{d}S_{\mu\nu} &= \nabla_{\mu} \phi_{\nu} (s^{\mu} n^{\nu} - n^{\mu} s^{\nu}) \sqrt{q} \, \mathrm{d}^2 y \;=\; 2 \nabla_{\mu} \phi_{\nu} s^{\mu} n^{\nu} \sqrt{q} \, \mathrm{d}^2 y \\ &= -2 s^{\mu} \phi_{\nu} \nabla_{\mu} n^{\nu} \sqrt{q} \, \mathrm{d}^2 y \;=\; 2 K_{ij} s^i \phi^j \sqrt{q} \, \mathrm{d}^2 y. \end{aligned}$$

Accordingly Eq. (8.74) becomes

$$\boxed{J_{\rm K} = \frac{1}{8\pi} \oint_{\mathscr{S}_t} K_{ij} s^i \phi^j \sqrt{q} \, \mathrm{d}^2 y} \,. \tag{8.77}$$

Remark 8.9 Contrary to the 3+1 expression of the Komar mass which turned out to be very different from the expression of the ADM mass, the 3+1 expression of the Komar angular momentum as given by Eq. (8.77) is very similar to the expression of the angular momentum deduced from the Hamiltonian formalism, i.e. Eq. (8.45). The only differences are that it is no longer necessary to take the limit $\mathscr{S}_t \to \infty$ and that there is no trace term $K\gamma_{ij}s^i\phi^j$ in Eq. (8.77). Moreover, if one evaluates the Hamiltonian expression in the asymptotically maximal gauge (8.47) then $K = O(r^{-3})$ and thanks to the asymptotic orthogonality of $\boldsymbol{s}$ and $\boldsymbol{\phi}$, $\gamma_{ij}s^i\phi^j = O(1)$, so that $K\gamma_{ij}s^i\phi^j$ does not contribute to the integral and expressions (8.77) and (8.45) are then identical.

Example 8.6. A trivial example is provided by Schwarzschild spacetime, which among other things is axisymmetric. For the 3+1 foliation associated with the Schwarzschild coordinates (8.12), the extrinsic curvature tensor $\boldsymbol{K}$ vanishes identically, so that Eq. (8.77) yields immediately $J_{\mathrm{K}} = 0$. For other foliations, like that associated with Eddington–Finkelstein coordinates, $\boldsymbol{K}$ is no longer zero but is such that $K_{ij}s^i\phi^j = 0$, yielding again $J_{\mathrm{K}} = 0$ (as it should be since the Komar angular momentum is independent of the foliation). Explicitly for Eddington–Finkelstein coordinates,

$$K_{ij}s^i = \left(-\frac{2m}{r^2}\frac{1+\frac{m}{r}}{1+\frac{2m}{r}}, 0, 0 \right), \tag{8.78}$$

(see e.g. Eq. (D.25) in Ref. [23]) and $\phi^j = (0, 0, 1)$, so that obviously $K_{ij}s^i\phi^j = 0$.

Example 8.7 The most natural non trivial example is certainly that of Kerr spacetime. Let us use the 3+1 foliation associated with the standard Boyer–Lindquist coordinates (t, r, θ, φ) and evaluate the integral (8.77) by choosing for $\mathscr{S}_t$ a sphere $r =$ const. Then $y^a = (\theta, \varphi)$. The Boyer–Lindquist components of $\boldsymbol{\phi}$ are $\phi^i = (0, 0, 1)$ and those of $\boldsymbol{s}$ are $s^i = (s^r, 0, 0)$ since γ_{ij} is diagonal is these coordinates. The formula (8.77) then reduces to

$$J_{\mathrm{K}} = \frac{1}{8\pi}\oint_{r=\mathrm{const}} K_{r\varphi}s^r\sqrt{q}\,\mathrm{d}\theta\,\mathrm{d}\varphi.$$

The extrinsic curvature component $K_{r\varphi}$ can be evaluated via formula (5.68), which reduces to $2NK_{ij} = \mathscr{L}_{\boldsymbol{\beta}}\gamma_{ij}$ since $\partial\gamma_{ij}/\partial t = 0$. From the Boyer–Lindquist line element (see e.g. Eq. (5.29) in Ref. [24]), we read the components of the shift:

$$(\beta^r, \beta^\theta, \beta^\varphi) = \left(0, 0, -\frac{2amr}{(r^2+a^2)(r^2+a^2\cos^2\theta)+2a^2mr\sin^2\theta}\right), \tag{8.79}$$

where m and a are the two parameters of the Kerr solution. Then, using Eq. (2.90),

$$K_{r\varphi} = \frac{1}{2N}\mathcal{L}_{\boldsymbol{\beta}}\gamma_{r\varphi} = \frac{1}{2N}\Bigg(\beta^\varphi \underbrace{\frac{\partial\gamma_{r\varphi}}{\partial\varphi}}_{0} + \gamma_{\varphi\varphi}\frac{\partial\beta^\varphi}{\partial r} + \gamma_{r\varphi}\underbrace{\frac{\partial\beta^\varphi}{\partial\varphi}}_{0}\Bigg) = \frac{1}{2N}\gamma_{\varphi\varphi}\frac{\partial\beta^\varphi}{\partial r}.$$

Hence

$$J_{\mathrm{K}} = \frac{1}{16\pi}\oint_{r=\mathrm{const}} \frac{s^r}{N}\gamma_{\varphi\varphi}\frac{\partial\beta^\varphi}{\partial r}\sqrt{q}\,\mathrm{d}\theta\,\mathrm{d}\varphi.$$

The values of s^r, N, $\gamma_{\varphi\varphi}$ and $\sqrt{q}$ can all be read on the Boyer–Lindquist line element. However this is a bit tedious. To simplify things, let us evaluate J_{K} only in the limit $r \to \infty$. Then $s^r \sim 1$, $N \sim 1$, $\gamma_{\varphi\varphi} \sim r^2\sin^2\theta$, $\sqrt{q} \sim r^2\sin\theta$ and, from Eq. (8.79), $\beta^\varphi \sim -2am/r^3$, so that

$$J_{\mathrm{K}} = \frac{1}{16\pi}\oint_{r=\mathrm{const}} r^2\sin^2\theta\frac{6am}{r^4}r^2\sin\theta\,\mathrm{d}\theta\,\mathrm{d}\varphi = \frac{3am}{8\pi}\times 2\pi\times\int_0^\pi \sin^3\theta\,\mathrm{d}\theta.$$

Hence, as expected,

$$J_{\mathrm{K}} = am. \tag{8.80}$$

Let us now find the 3+1 expression of the volume version (8.75) of the Komar angular momentum. We have $\boldsymbol{n}\cdot\boldsymbol{\phi} = 0$ and, from the 3+1 decomposition (5.14) of $\boldsymbol{T}$:

$$\boldsymbol{T}(\boldsymbol{n}, \boldsymbol{\phi}) = -\langle \boldsymbol{p}, \boldsymbol{\phi}\rangle.$$

Hence formula (8.75) becomes

$$\boxed{J_{\mathrm{K}} = \int_{\Sigma_t} \langle \boldsymbol{p}, \boldsymbol{\phi}\rangle\sqrt{\gamma}\,\mathrm{d}^3x + J_{\mathrm{K}}^{\mathscr{H}}}, \tag{8.81}$$

with

$$\boxed{J_{\mathrm{K}}^{\mathscr{H}} = \frac{1}{8\pi}\oint_{\mathscr{H}_t} K_{ij}s^i\phi^j\sqrt{q}\,\mathrm{d}^2y}. \tag{8.82}$$

Example 8.8 Let us consider a perfect fluid. Then $\boldsymbol{p} = (E+P)\underline{U}$ [Eq. (6.57)], so that

$$J_{\rm K} = \int_{\Sigma_t} (E + P)\boldsymbol{U} \cdot \boldsymbol{\phi}\sqrt{\gamma}\, \mathrm{d}^3x + J_{\rm K}^{\mathscr{H}}. \tag{8.83}$$

Taking $\boldsymbol{\phi} = -y\boldsymbol{\partial}_x + x\boldsymbol{\partial}y$ (symmetry axis = z-axis), the Newtonian limit of this expression is then

$$J_{\rm K} = \int_{\Sigma_t} \rho(-yU^x + xU^y)\, \mathrm{d}x\, \mathrm{d}y\, \mathrm{d}z, \tag{8.84}$$

i.e. we recognize the standard expression for the angular momentum around the z-axis.

References

1. Jaramillo, J.L., Gourgoulhon, E.: Mass and angular momentum in general relativity. In: Blanchet, L., Spallicci, A., Whiting, B. (eds.) Mass and motion in general relativity. Fundamental Theories of Physics, vol. 162, p. 87. Springer, Dordrecht (2011)
2. Szabados, L.B.: Quasi-local energy-momentum and angular momentum in general relativity. Living Rev. Relativity **12**, 4. http://www.livingreviews.org/lrr-2009-4 (2009)
3. York, J.W.: Kinematics and dynamics of general relativity. In: Smarr, L.L. (ed.) Sources of Gravitational Radiation, p. 83. Cambridge University Press, Cambridge (1979)
4. York, J.W.: Energy and momentum of the gravitational field. In: Tipler, F.J. (ed.) Essays in General Relativity, a Festschrift for Abraham Taub, p. 39. Academic Press, New York (1980)
5. Straumann, N.: General Relavity, with Applications to Astrophysics. Springer, Berlin (2004)
6. Ashtekar, A.: Asymptotic structure of the gravitational field at spatial infinity. In: Held, A. (ed.) General Relativity and Gravitation, One Hundred Years After the Birth of Albert Einstein, vol. 2, p. 37. Plenum Press, New York (1980)
7. Wald, R.M.: General Relativity. University of Chicago Press, Chicago (1984)
8. Henneaux, M.: Hamiltonian Formalism of General Relativity. Lectures at Institut Henri Poincaré, Paris, http://www.luth.obspm.fr/IHP06/ (2006)
9. Ashtekar, A., Hansen, R.O.: A unified treatment of null and spatial infinity in general relativity. I. Universal structure, asymptotic symmetries, and conserved quantities at spatial infinity. J. Math. Phys. **19**, 1542 (1978)
10. Regge, T., Teitelboim, C.: Role of surface integrals in the Hamiltonian formulation of general relativity. Ann. Phys. (N.Y.) **88**, 286 (1974)
11. Poisson, E.: A Relativist's Toolkit, The Mathematics of Black-Hole Mechanics. Cambridge University Press, Cambridge, http://www.physics.uoguelph.ca/poisson/toolkit/ (2004)
12. Martel, K., Poisson, E.: Regular coordinate systems for Schwarzschild and other spherical spacetimes. Am. J. Phys. **69**, 476 (2001)
13. Schoen, R., Yau, S.-T.: Proof of the positive mass theorem. II. Commun. Math. Phys. **79**, 231 (1981)
14. Witten, E.: A new proof of the positive energy theorem. Commun. Math. Phys. **80**, 381 (1981)

15. Arnowitt, R., Deser, S., Misner, C.W.: The dynamics of general relativity. In: Witten, L. (ed.) Gravitation: An Introduction to Current Research, p. 227. Wiley, New York, available at http://arxiv.org/abs/gr-qc/0405109 (1962)
16. Smarr L., York J.W.: Radiation gauge in general relativity. Phys. Rev. **D 17**, 1945 (1978)
17. Chruściel P.T.: On angular momentum at spatial infinity. Class. Quantum Grav. **4**, L205 (1987)
18. Komar A.: Covariant conservation laws in general relativity. Phys. Rev. **113**, 934 (1959)
19. Ansorg, M., Petroff, D.: Negative Komar mass of single objects in regular, asymptotically flat spacetimes. Class. Quantum Grav. **23**, L81 (2006)
20. Beig, R.: Arnowitt-Deser-Misner energy and g_{00}. Phys. Lett. **69A**, 153 (1978)
21. Ashtekar, A., Magnon-Ashteka, A.: On conserved quantities in general relativity. J. Math. Phys. **20**, 793 (1979)
22. Katz, J.: A note on Komar's anomalous factor. Class. Quantum Grav. **2**, 423 (1985)
23. Gourgoulhon, E., Jaramillo, J.L.: A 3+1 perspective on null hypersurfaces and isolated horizons. Phys. Rep. **423**, 159 (2006)
24. Hawking, S.W., Ellis, G.F.R.: The large scale structure of space-time. Cambridge University Press, Cambridge (1973)

Chapter 9
The Initial Data Problem

Abstract The problem of solving the constraint equations to get valid initial data for the time evolution is discussed. We focus on two methods based on the conformal decomposition introduced in Chap. 7: the conformal transverse-traceless method and the conformal thin sandwich method. Both methods are illustrated by initial data in Schwarzschild spacetime. Finally, we give a survey of the construction of initial for binary compact objects, which are of major interest in numerical relativity.

9.1 Introduction

9.1.1 The Initial Data Problem

We have seen in Chap. 5 that thanks to the 3+1 decomposition, the resolution of Einstein equation amounts to solving a Cauchy problem, namely to evolve "forward in time" some initial data. This is however a Cauchy problem with constraints. This makes the set up of initial data a non trivial task, because these data must obey the constraints. Actually one may distinguish two problems:

- *The mathematical problem*: given some hypersurface Σ_0, find a Riemannian metric $\boldsymbol{\gamma}$, a symmetric bilinear form $\boldsymbol{K}$ and some matter distribution $(E, \boldsymbol{p})$ on Σ_0 such that the Hamiltonian constraint (5.70) and the momentum constraint (5.71) are satisfied:

$$\boxed{R + K^2 - K_{ij}K^{ij} = 16\pi E} \tag{9.1}$$

$$\boxed{D_j K^i{}_j - D_i K = 8\pi p_i}\,. \tag{9.2}$$

  In addition, the matter distribution $(E, \boldsymbol{p})$ may have some constraints from its own. We shall not discuss them here.

É. Gourgoulhon, *3+1 Formalism in General Relativity*, Lecture Notes in Physics 846,
DOI: 10.1007/978-3-642-24525-1_9,

- *The astrophysical problem*: make sure that the solution to the constraint equations has something to do with the physical system that one wish to study.

Notice that Eqs. (9.1)–(9.2) involve a single hypersurface Σ_0, not a foliation $(\Sigma_t)_{t\in\mathbb{R}}$. In particular, neither the lapse function nor the shift vector appear in these equations. Facing them, a naive way to proceed would be to choose freely the metric $\boldsymbol{\gamma}$, thereby fixing the connection $\boldsymbol{D}$ and the scalar curvature R, and to solve Eqs. (9.1)–(9.2) for $\boldsymbol{K}$. Indeed, for fixed $\boldsymbol{\gamma}$, E, and $\boldsymbol{p}$, Eqs. (9.1)–(9.2) form a quasi-linear system of first order for the components K_{ij}. However, as discussed by Choquet-Bruhat [1], this approach is not satisfactory because we have only four equations for six unknowns K_{ij} and there is no natural prescription for choosing arbitrarily two among the six components K_{ij}.

Lichnerowicz [2] has shown that a much more satisfactory split of the initial data $(\boldsymbol{\gamma}, \boldsymbol{K})$ between freely chosable parts and parts obtained by solving Eqs. (9.1)–(9.2) is provided by the conformal decomposition introduced in Chap. 7. Lichnerowicz method has been extended by Choquet-Bruhat [1, 3], by York and Ó Murchadha [4–7] and by York and Pfeiffer [8, 9]. Actually, conformal decompositions are by far the most employed techniques to get initial data for the 3+1 Cauchy problem. Alternative methods exist, such as the quasi-spherical ansatz introduced by Bartnik in 1993 [10] or a procedure developed by Corvino [11] and Isenberg et al. [12] for gluing together known solutions of the constraints, thereby producing new ones (see also [13]). Here we shall limit ourselves to the conformal methods. Standard reviews on this subject are the articles by York [7] and Choquet-Bruhat and York [14]. More recent reviews are the articles by Cook [15], Pfeiffer [16] and Bartnik and Isenberg [17], Gourgoulhon [18] and Chruściel et al. [13], as well as the devoted chapters of the textbooks [19–21].

9.1.2 *Conformal Decomposition of the Constraints*

The conformal form of the constraint equations has been derived in Chap. 7. We have introduced there the conformal metric $\tilde{\boldsymbol{\gamma}}$ and the conformal factor Ψ such that the metric $\boldsymbol{\gamma}$ induced by the spacetime metric on some hypersurface Σ_0 is [cf. Eq. (7.2)]

$$\gamma_{ij} = \Psi^4 \tilde{\gamma}_{ij}, \tag{9.3}$$

and have decomposed the traceless part A^{ij} of the extrinsic curvature K^{ij} according to [cf. Eq. (7.67)]

$$A^{ij} = \Psi^{-10} \hat{A}^{ij}. \tag{9.4}$$

We consider here the decomposition involving $\hat{A}^{ij}$ [$\alpha = -10$ in Eq. (7.51)] and not the alternative one, which uses $\tilde{A}^{ij}$ ($\alpha = -4$), because we have seen in Sect. 7.4.2

that the former is well adapted to the momentum constraint. Using the decompositions (9.3) and (9.4), we have rewritten the Hamiltonian constraint (9.1) and the momentum constraint (9.2) as respectively the Lichnerowicz equation [Eq. (7.94)] and an equation involving the divergence of $\hat{A}^{ij}$ with respect to the conformal metric [Eq. (7.95)] :

$$\boxed{\tilde{D}_i\tilde{D}^i\Psi - \frac{1}{8}\tilde{R}\Psi + \frac{1}{8}\hat{A}_{ij}\hat{A}^{ij}\Psi^{-7} + 2\pi\tilde{E}\Psi^{-3} - \frac{1}{12}K^2\Psi^5 = 0}, \quad (9.5)$$

$$\boxed{\tilde{D}_j\hat{A}^{ij} - \frac{2}{3}\Psi^6\tilde{D}^iK = 8\pi\tilde{p}^i}, \quad (9.6)$$

where the following rescaled matter quantities have been introduced:

$$\tilde{E} := \Psi^8 E \quad (9.7)$$

and

$$\tilde{p}^i := \Psi^{10} p^i. \quad (9.8)$$

The definition of $\tilde{p}^i$ is clearly motivated by Eq. (7.95). On the contrary the power 8 in the definition of $\tilde{E}$ is not the only possible choice. As we shall see in Sect. 9.2.4, it is chosen (i) to guarantee a negative power of Ψ in the $\tilde{E}$ term in Eq. (9.5), resulting in some uniqueness property of the solution and (ii) to allow for an easy implementation of the dominant energy condition.

9.2 Conformal Transverse-Traceless Method

9.2.1 Longitudinal / Transverse Decomposition of $\hat{A}^{ij}$

In order to solve the system (9.5)–(9.6), York [5, 7, 22] has decomposed $\hat{A}^{ij}$ into a longitudinal part and a transverse one, by setting

$$\boxed{\hat{A}^{ij} = (\tilde{L}X)^{ij} + \hat{A}^{ij}_{\rm TT}}, \quad (9.9)$$

where $\hat{A}^{ij}_{\rm TT}$ is both traceless and transverse (i.e. divergence-free) with respect to the metric $\tilde{\boldsymbol{\gamma}}$:

$$\tilde{\gamma}_{ij}\hat{A}^{ij}_{\rm TT} = 0 \quad \text{and} \quad \tilde{D}_j\hat{A}^{ij}_{\rm TT} = 0, \quad (9.10)$$

and $(\tilde{L}X)^{ij}$ is the ***conformal Killing operator*** associated with the metric $\tilde{\boldsymbol{\gamma}}$ and acting on the vector field $\boldsymbol{X}$:

$$\boxed{(\tilde{L}X)^{ij} := \tilde{D}^i X^j + \tilde{D}^j X^i - \frac{2}{3}\tilde{D}_k X^k \tilde{\gamma}^{ij}}. \tag{9.11}$$

The properties of this linear differential operator are detailed in Appendix A. Let us retain here that $(\tilde{L}X)^{ij}$ is by construction traceless:

$$\tilde{\gamma}_{ij}(\tilde{L}X)^{ij} = 0 \tag{9.12}$$

(it must be so because in Eq. (9.9) both $\hat{A}^{ij}$ and $\hat{A}^{ij}_{\mathrm{TT}}$ are traceless) and the kernel of $\tilde{\boldsymbol{L}}$ is made of the ***conformal Killing vectors*** of the metric $\tilde{\boldsymbol{\gamma}}$, i.e. the generators of the conformal isometries (cf. Sect. A.1.3). The symmetric tensor $(\tilde{L}X)^{ij}$ is called the ***longitudinal part*** of $\hat{A}^{ij}$, whereas $\hat{A}^{ij}_{\mathrm{TT}}$ is called the ***transverse part***.

Given $\hat{A}^{ij}$, the vector $\boldsymbol{X}$ is determined by taking the divergence of Eq. (9.9): taking into account property (9.10), we get

$$\tilde{D}_j(\tilde{L}X)^{ij} = \tilde{D}_j\hat{A}^{ij}. \tag{9.13}$$

The second order operator $\tilde{D}_j(\tilde{L}X)^{ij}$ acting on the vector $\boldsymbol{X}$ is the ***conformal vector Laplacian*** $\tilde{\boldsymbol{\Delta}}_L$:

$$\boxed{\tilde{\Delta}_L X^i := \tilde{D}_j(\tilde{L}X)^{ij} = \tilde{D}_j\tilde{D}^j X^i + \frac{1}{3}\tilde{D}^i\tilde{D}_j X^j + \tilde{R}^i_{\ j} X^j}, \tag{9.14}$$

where the second equality follows from Eq. (A.7). The basic properties of $\tilde{\boldsymbol{\Delta}}_L$ are investigated in Appendix A, where it is shown that this operator is elliptic and that its kernel is, in practice, reduced to the conformal Killing vectors of $\tilde{\boldsymbol{\gamma}}$, if any. We rewrite Eq. (9.13) as

$$\tilde{\Delta}_L X^i = \tilde{D}_j\hat{A}^{ij}. \tag{9.15}$$

The existence and uniqueness of the longitudinal/transverse decomposition (9.9) depend on the existence and uniqueness of solutions $\boldsymbol{X}$ to Eq. (9.15). We shall consider two cases:

- Σ_0 is a ***closed manifold***, i.e. is compact without boundary;
- $(\Sigma_0, \boldsymbol{\gamma})$ is an ***asymptotically flat manifold***, in the sense made precise in Sect. 8.2.

In the first case, it is shown in Appendix A that solutions to Eq. (9.15) exist provided that the source $\tilde{D}_j\hat{A}^{ij}$ is orthogonal to all conformal Killing vectors of $\tilde{\boldsymbol{\gamma}}$, in the sense that [cf. Eq. (A.20)]:

$$\forall \boldsymbol{C} \in \ker\tilde{\boldsymbol{L}}, \quad \int_{\Sigma_0} \tilde{\gamma}_{ij} C^i \tilde{D}_k \hat{A}^{jk} \sqrt{\tilde{\gamma}}\, \mathrm{d}^3x = 0. \tag{9.16}$$

But this is easy to check: using the fact that the source is a pure divergence and that Σ_0 is closed, we may integrate by parts and get, for any vector field $\boldsymbol{C}$,

$$\int_{\Sigma_0} \tilde{\gamma}_{ij} C^i \tilde{D}_k \hat{A}^{jk} \sqrt{\tilde{\gamma}}\, \mathrm{d}^3x = -\frac{1}{2} \int_{\Sigma_0} \tilde{\gamma}_{ij} \tilde{\gamma}_{kl} (\tilde{L}C)^{ik} \hat{A}^{jl} \sqrt{\tilde{\gamma}}\, \mathrm{d}^3x.$$

Then, obviously, when $\boldsymbol{C}$ is a conformal Killing vector, the right-hand side of the above equation vanishes. So the condition (9.16) is fulfilled and there exists a solution to Eq. (9.15); this solution is unique up to the addition of a conformal Killing vector. However, given a solution $\boldsymbol{X}$, for any conformal Killing vector $\boldsymbol{C}$, the solution $\boldsymbol{X}+\boldsymbol{C}$ yields the same value of $\tilde{\boldsymbol{L}}\boldsymbol{X}$, since $\boldsymbol{C}$ is by definition in the kernel of $\tilde{\boldsymbol{L}}$. Therefore we conclude that the decomposition (9.9) of $\hat{A}^{ij}$ is unique, although the vector $\boldsymbol{X}$ may not be if $(\Sigma_0, \tilde{\boldsymbol{\gamma}})$ admits some conformal isometries.

In the case of an asymptotically flat manifold, the existence and uniqueness is guaranteed by the Cantor theorem mentioned in Sect. A.2.4. We shall then require the decay condition

$$\frac{\partial^2 \tilde{\gamma}_{ij}}{\partial x^k \partial x^l} = O(r^{-3}) \tag{9.17}$$

in addition to the asymptotic flatness conditions (8.27) introduced in Chap. 8. This guarantees that [cf. Eq. (A.23)]

$$\tilde{R}_{ij} = O(r^{-3}). \tag{9.18}$$

In addition, we notice that $\hat{A}^{ij}$ obeys the decay condition $\hat{A}^{ij} = O(r^{-2})$ which is inherited from the asymptotic flatness condition (8.2a). Then $\tilde{D}_j \hat{A}^{ij} = O(r^{-3})$ so that condition (A.21) is satisfied. Then all conditions are fulfilled to conclude that Eq. (9.15) admits a unique solution $\boldsymbol{X}$ which vanishes at infinity.

To summarize, for all considered cases (asymptotic flatness with the additional condition (9.17) and closed manifold), any symmetric and traceless tensor $\hat{A}^{ij}$ (decaying as $O(r^{-2})$ in the asymptotically flat case) admits a unique longitudinal/transverse decomposition of the form (9.9).

9.2.2 Conformal Transverse-Traceless Form of the Constraints

Inserting the longitudinal/transverse decomposition (9.9) into the constraint equations (9.5) and (9.6) and making use of Eq. (9.15) leads to the system

$$\boxed{\begin{aligned}\tilde{D}_i \tilde{D}^i \Psi - \frac{1}{8} \tilde{R} \Psi + \frac{1}{8} \left[(\tilde{L}X)_{ij} + \hat{A}^{\mathrm{TT}}_{ij} \right] \left[(\tilde{L}X)^{ij} + \hat{A}^{ij}_{\mathrm{TT}} \right] \Psi^{-7} \\ + 2\pi \tilde{E} \Psi^{-3} - \frac{1}{12} K^2 \Psi^5 = 0,\end{aligned}} \tag{9.19}$$

$$\boxed{\tilde{\Delta}_L X^i - \frac{2}{3} \Psi^6 \tilde{D}^i K = 8\pi \tilde{p}^i}\,, \tag{9.20}$$

where

$$(\tilde{L}X)_{ij} := \tilde{\gamma}_{ik}\tilde{\gamma}_{jl}(\tilde{L}X)^{kl} \tag{9.21}$$

$$\hat{A}^{\mathrm{TT}}_{ij} := \tilde{\gamma}_{ik}\tilde{\gamma}_{jl}\hat{A}^{kl}_{\mathrm{TT}}. \tag{9.22}$$

With the constraint equations written as (9.19) (the Lichnerowicz equation) and (9.20), we see clearly which part of the initial data on Σ_0 can be freely chosen and which part is "constrained":

- free data:
  - conformal metric $\tilde{\boldsymbol{\gamma}}$;
  - symmetric traceless and transverse tensor $\hat{A}^{ij}_{\mathrm{TT}}$ (traceless and transverse are meant with respect to $\tilde{\boldsymbol{\gamma}}$: $\tilde{\gamma}_{ij}\hat{A}^{ij}_{\mathrm{TT}} = 0$ and $\tilde{D}_j\hat{A}^{ij}_{\mathrm{TT}} = 0$);
  - scalar field K;
  - conformal matter variables: $(\tilde{E}, \tilde{p}^i)$;
- constrained data (or "determined data"):
  - conformal factor Ψ, obeying the non-linear elliptic equation (9.19) (Lichnerowicz equation)
  - vector $\boldsymbol{X}$, obeying the linear elliptic equation (9.20).

Accordingly the general strategy to get valid initial data for the Cauchy problem is to choose $(\tilde{\gamma}_{ij}, \hat{A}^{ij}_{\mathrm{TT}}, K, \tilde{E}, \tilde{p}^i)$ on Σ_0 and solve the system (9.19)–(9.20) to get Ψ and X^i. Then one constructs

$$\gamma_{ij} = \Psi^4\tilde{\gamma}_{ij} \tag{9.23}$$

$$K^{ij} = \Psi^{-10}\left((\tilde{L}X)^{ij} + \hat{A}^{ij}_{\mathrm{TT}}\right) + \frac{1}{3}\Psi^{-4}K\tilde{\gamma}^{ij} \tag{9.24}$$

$$E = \Psi^{-8}\tilde{E} \tag{9.25}$$

$$p^i = \Psi^{-10}\tilde{p}^i \tag{9.26}$$

and obtains a set $(\boldsymbol{\gamma}, \boldsymbol{K}, E, \boldsymbol{p})$ which satisfies the constraint equations (9.1)–(9.2). This method has been proposed by York [7] and is naturally called the ***conformal transverse traceless*** (***CTT***) method.

9.2.3 Decoupling on Hypersurfaces of Constant Mean Curvature

Equations (9.19) and (9.20) are coupled, but we notice that if, among the free data, we choose K to be a constant field on Σ_0,

$$K = \mathrm{const}, \tag{9.27}$$

then they decouple partially : condition (9.27) implies $\tilde{D}^i K = 0$, so that the momentum constraint (9.20) becomes independent of Ψ:

$$\tilde{\Delta}_L X^i = 8\pi \tilde{p}^i \qquad (K = \text{const}). \tag{9.28}$$

The condition (9.27) on the extrinsic curvature of Σ_0 defines what is called a ***constant mean curvature* (*CMC*)** hypersurface. Indeed let us recall that K is nothing but minus three times the mean curvature of $(\Sigma_0, \boldsymbol{\gamma})$ embedded in $(\mathscr{M}, \boldsymbol{g})$ [cf. Eq. (3.21)].

> *Example 9.1* A maximal hypersurface, having $K = 0$, is of course a special case of a CMC hypersurface. Another example is provided by the hyperbolic slice of Minkowski spacetime considered in Examples 3.4 and 4.1, for which $K = -3/b$, with b constant [Eq. (3.48)].

On a CMC hypersurface, the task of obtaining initial data is greatly simplified: one has first to solve the linear elliptic equation (9.28) to get $\boldsymbol{X}$ and plug the solution in Eq. (9.19) to form an equation for Ψ. Equation (9.28) is the conformal vector Poisson equation studied in Appendix A. It is shown in Sect. A.2.4 that it always solvable for the two cases of interest mentioned in Sect. 9.2.1: closed or asymptotically flat manifold. Moreover, the solutions $\boldsymbol{X}$ are such that the value of $\tilde{\boldsymbol{L}}\boldsymbol{X}$ is unique.

9.2.4 Existence and Uniqueness of Solutions to Lichnerowicz Equation

Taking into account the CMC decoupling, the difficult problem is to solve Lichnerowicz equation (9.19) for Ψ. This equation is elliptic and highly non-linear.[1] It has been first studied by Lichnerowicz [2, 23] in the case $K = 0$ (Σ_0 maximal) and $\tilde{E} = 0$ (vacuum). Lichnerowicz has shown that given the value of Ψ at the boundary of a bounded domain of Σ_0 (Dirichlet problem), there exists at most one solution to Eq. (9.19). Besides, he showed the existence of a solution provided that $\hat{A}_{ij}\hat{A}^{ij}$ is not too large. These early results have been much improved since then. In particular Cantor [24] has shown that in the asymptotically flat case, still with $K = 0$ and $\tilde{E} = 0$, Eq. (9.19) is solvable if and only if the metric $\tilde{\boldsymbol{\gamma}}$ is conformal to a metric with vanishing scalar curvature (one says then that $\tilde{\boldsymbol{\gamma}}$ belongs to the ***positive Yamabe class***) (see also Ref. [25]). In the case of closed manifolds, the complete analysis of the CMC case has been achieved by Isenberg [26]. The non-CMC case is more tricky; see e.g. Refs. [27, 28] for recent progresses in this direction.

For more details and further references, we recommend the review articles by Choquet–Bruhat and York [14], Bartnik and Isenberg [17] and Chruściel, Galloway

---

[1] Although it is *quasi-linear* in the technical sense, i.e. linear with respect to the highest-order derivatives.

and Pollack [13], as well as Choquet–Bruhat's textbook [21]. Here we shall simply repeat the argument of York [8] to justify the rescaling (9.7) of E. This rescaling is indeed related to the uniqueness of solutions to the Lichnerowicz equation. Consider a solution Ψ_0 to Eq. (9.9) in the case $K = 0$, to which we restrict ourselves. Another solution close to Ψ_0 can be written $\Psi = \Psi_0 + \varepsilon$, with $|\varepsilon| \ll \Psi_0$:

$$\tilde{D}_i\tilde{D}^i(\Psi_0+\varepsilon) - \frac{1}{8}\tilde{R}(\Psi_0+\varepsilon) + \frac{1}{8}\hat{A}_{ij}\hat{A}^{ij}(\Psi_0+\varepsilon)^{-7} + 2\pi\tilde{E}(\Psi_0+\varepsilon)^{-3} = 0.$$

Expanding to the first order in ε/Ψ_0 leads to the following linear equation for ε:

$$\tilde{D}_i\tilde{D}^i\varepsilon - \alpha\varepsilon = 0, \tag{9.29}$$

with

$$\alpha := \frac{1}{8}\tilde{R} + \frac{7}{8}\hat{A}_{ij}\hat{A}^{ij}\Psi_0^{-8} + 6\pi\tilde{E}\Psi_0^{-4}. \tag{9.30}$$

Now, if $\alpha \geq 0$, one can show, by means of the maximum principle, that the solution of (9.29) which vanishes at spatial infinity is necessarily $\varepsilon = 0$ (see Ref. [29] or Sect. B.1 of Ref. [30]). We therefore conclude that the solution Ψ_0 to Eq. (9.19) is unique (at least locally) in this case. On the contrary, if $\alpha < 0$, non trivial oscillatory solutions of Eq. (9.29) exist, making the solution Ψ_0 not unique. The key point is that the scaling (9.7) of E yields the term $+6\pi\tilde{E}\Psi_0^{-4}$ in Eq. (9.30), which contributes to make α positive. If we had not rescaled *longitudinal part* E, i.e. had considered the original Hamiltonian constraint equation (7.94), the contribution to α would have been instead $-10\pi E\Psi_0^4$, i.e. would have been negative. Actually, any rescaling $\tilde{E} = \Psi^s E$ with $s > 5$ would have work to make α positive. The choice $s = 8$ in Eq. (9.7) is motivated by the fact that if the conformal data $(\tilde{E}, \tilde{p}^i)$ obey the "conformal" dominant energy condition (cf. Sect. 8.3.4)

$$\tilde{E} \geq \sqrt{\tilde{\gamma}_{ij}\tilde{p}^i\tilde{p}^j}, \tag{9.31}$$

then, via the scaling (9.8) of p^i, the reconstructed physical data (E, p^i) will automatically obey the dominant energy condition as stated by Eq. (8.38):

$$E \geq \sqrt{\gamma_{ij}p^ip^j}. \tag{9.32}$$

9.2.5 Conformally Flat and Momentarily Static Initial Data

In this section we search for asymptotically flat initial data $(\Sigma_0, \boldsymbol{\gamma}, \boldsymbol{K})$. Let us then consider the simplest case one may think of, namely choose the freely specifiable data $(\tilde{\gamma}_{ij}, \hat{A}^{ij}_{\rm TT}, K, \tilde{E}, \tilde{p}^i)$ to be a flat metric:

$$\tilde{\gamma}_{ij} = f_{ij}, \tag{9.33}$$

a vanishing transverse-traceless part of the extrinsic curvature:

$$\hat{A}^{ij}_{\mathrm{TT}} = 0, \tag{9.34}$$

a vanishing mean curvature (maximal hypersurface)

$$K = 0, \tag{9.35}$$

and a vacuum spacetime:

$$\tilde{E} = 0, \qquad \tilde{p}^i = 0. \tag{9.36}$$

Then $\tilde{D}_i = \bar{D}_i$, $\tilde{R} = 0$, $\tilde{\boldsymbol{L}} = \boldsymbol{L}$ [cf. Eq. (7.107)] and the constraint equations (9.19)–(9.20) reduce to

$$\Delta\Psi + \frac{1}{8}(LX)_{ij}(LX)^{ij}\Psi^{-7} = 0 \tag{9.37}$$

$$\Delta_L X^i = 0, \tag{9.38}$$

where Δ and Δ_L are respectively the scalar Laplacian and the conformal vector Laplacian associated with the flat metric $\boldsymbol{f}$:

$$\Delta := \bar{D}_i\bar{D}^i \tag{9.39}$$

and

$$\Delta_L X^i := \bar{D}_j\bar{D}^j X^i + \frac{1}{3}\bar{D}^i\bar{D}_j X^j. \tag{9.40}$$

Equations (9.37)–(9.38) must be solved with the boundary conditions

$$\Psi = 1 \quad \text{when} \quad r \to \infty \tag{9.41}$$

$$\boldsymbol{X} = 0 \quad \text{when} \quad r \to \infty, \tag{9.42}$$

which follow from the asymptotic flatness requirement. The solution depends on the topology of Σ_0, since the latter may introduce some inner boundary conditions in addition to (9.41)–(9.42).

Let us start with the simplest case: $\Sigma_0 = \mathbb{R}^3$. Then the solution of Eq. (9.38) subject to the boundary condition (9.42) is

$$\boldsymbol{X} = 0 \tag{9.43}$$

and there is no other solution (cf. Sect. A.2.4). Then obviously $(LX)^{ij} = 0$, so that Eq. (9.37) reduces to Laplace equation for Ψ:

$$\Delta\Psi = 0. \tag{9.44}$$

With the boundary condition (9.41), there is a unique regular solution on $\mathbb{R}^3$:

$$\Psi = 1. \tag{9.45}$$

The initial data reconstructed from Eqs. (9.23)–(9.24) is then

$$\boldsymbol{\gamma} = \boldsymbol{f} \tag{9.46}$$

$$\boldsymbol{K} = 0. \tag{9.47}$$

These data correspond to a spacelike hyperplane of Minkowski spacetime. Geometrically the condition $\boldsymbol{K} = 0$ is that of a *totally geodesic hypersurface* (cf. Sect. 3.4.3). Physical data with $\boldsymbol{K} = 0$ are said to be ***momentarily static*** or ***time symmetric***. Indeed, from Eq. (4.26),

$$\mathscr{L}_{\boldsymbol{m}}\boldsymbol{g} = -2N\boldsymbol{K} - 2\nabla_{\boldsymbol{n}}N\underline{\boldsymbol{n}} \otimes \underline{\boldsymbol{n}}.$$

So if $\boldsymbol{K} = 0$ and if moreover one chooses a geodesic slicing around Σ_0 (cf. Sect. 5.4.2), which yields $N = 1$ and $\nabla_{\boldsymbol{n}}N = 0$, then

$$\mathscr{L}_{\boldsymbol{m}}\boldsymbol{g} = 0. \tag{9.48}$$

This means that, locally (i.e. on Σ_0), the normal evolution vector $\boldsymbol{m}$ is a spacetime Killing vector. This vector being timelike, the configuration is then ***stationary***. Moreover, the Killing vector $\boldsymbol{m}$ being orthogonal to some hypersurface (i.e. Σ_0), the stationary configuration is called ***static***. Of course, this staticity properties holds a priori only on Σ_0 since there is no guarantee that the time development of Cauchy data with $\boldsymbol{K} = 0$ at $t = 0$ maintains $\boldsymbol{K} = 0$ at $t > 0$. Hence the qualifier *'momentarily'* in the expression *'momentarily static'* for data with $\boldsymbol{K} = 0$.

To get something less trivial than a hyperplane in Minkowski spacetime, let us consider a slightly more complicated topology for Σ_0, namely $\mathbb{R}^3$ minus a ball (cf. Fig. 9.1). The sphere $\mathscr{S}$ delimiting the ball is then the inner boundary of Σ_0 and we must provide boundary conditions for Ψ and $\boldsymbol{X}$ on $\mathscr{S}$ to solve Eqs. (9.37)–(9.38). For simplicity, let us choose

$$\boldsymbol{X}|_{\mathscr{S}} = 0. \tag{9.49}$$

Altogether with the outer boundary condition (9.42), this leads to $\boldsymbol{X}$ being identically zero as the unique solution of Eq. (9.38). So, again, the Hamiltonian constraint reduces to Laplace equation

$$\Delta\Psi = 0. \tag{9.50}$$

If we choose the boundary condition $\Psi|_{\mathscr{S}} = 1$, then the unique solution is $\Psi = 1$ and we are back to the previous example (slice of Minkowski spacetime). In order to have something non trivial, i.e. to ensure that the metric $\boldsymbol{\gamma}$ will not be flat, let us

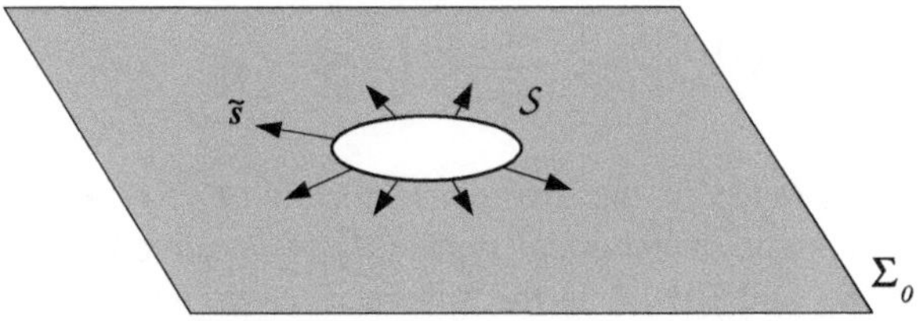

Fig. 9.1 Hypersurface Σ_0 as $\mathbb{R}^3$ minus a ball, displayed via an embedding diagram based on the metric $\tilde{\boldsymbol{\gamma}}$, which coincides with the Euclidean metric on $\mathbb{R}^3$. Hence Σ_0 appears to be flat. The unit normal of the inner boundary $\mathcal{S}$ with respect to the metric $\tilde{\boldsymbol{\gamma}}$ is $\tilde{\boldsymbol{s}}$. Notice that $\tilde{\boldsymbol{D}} \cdot \tilde{\boldsymbol{s}} > 0$

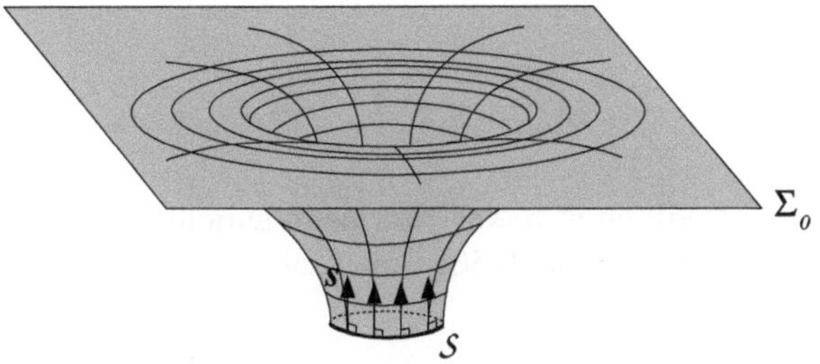

Fig. 9.2 Same hypersurface Σ_0 as in Fig. 9.1 but displayed via an embedding diagram based on the metric $\boldsymbol{\gamma}$ instead of $\tilde{\boldsymbol{\gamma}}$. The unit normal of the inner boundary $\mathcal{S}$ with respect to that metric is $\boldsymbol{s}$. Notice that $\boldsymbol{D} \cdot \boldsymbol{s} = 0$, which means that $\mathcal{S}$ is a minimal surface of $(\Sigma_0, \boldsymbol{\gamma})$

demand that $\boldsymbol{\gamma}$ admits a *closed minimal surface*, that we will choose to be $\mathcal{S}$. This will necessarily translate as a boundary condition for Ψ since all the information on the metric is encoded in Ψ (let us recall that from the choice (9.33), $\boldsymbol{\gamma} = \Psi^4 \boldsymbol{f}$). $\mathcal{S}$ is a ***minimal surface*** of $(\Sigma_0, \boldsymbol{\gamma})$ iff its mean curvature vanishes, or equivalently iff its unit normal $\boldsymbol{s}$ is divergence-free (cf. Fig. 9.2):

$$D_i s^i \Big|_{\mathcal{S}} = 0. \tag{9.51}$$

This is the analog of $\boldsymbol{\nabla} \cdot \boldsymbol{n} = 0$ for maximal hypersurfaces, the change from *minimal* to *maximal* being due to the change of signature, from the Riemannian to the Lorentzian one. By means of Eq. (7.34), condition (9.51) is equivalent to

$$\bar{D}_i (\Psi^6 s^i) \Big|_{\mathcal{S}} = 0, \tag{9.52}$$

where we have used $\tilde{D}_i = \bar{D}_i$, since $\tilde{\boldsymbol{\gamma}} = \boldsymbol{f}$. Let us rewrite this expression in terms of the unit vector $\tilde{\boldsymbol{s}}$ normal to $\mathcal{S}$ with respect to the metric $\tilde{\boldsymbol{\gamma}}$ (cf. Fig. 9.1); we have

$$\tilde{\boldsymbol{s}} = \Psi^2 \boldsymbol{s}, \tag{9.53}$$

since $\tilde{\boldsymbol{\gamma}}(\tilde{\boldsymbol{s}}, \tilde{\boldsymbol{s}}) = \Psi^4 \tilde{\boldsymbol{\gamma}}(\boldsymbol{s}, \boldsymbol{s}) = \boldsymbol{\gamma}(\boldsymbol{s}, \boldsymbol{s}) = 1$. Thus Eq. (9.52) becomes

$$\bar{D}_i(\Psi^4 \tilde{s}^i)\Big|_{\mathscr{S}} = \frac{1}{\sqrt{f}} \frac{\partial}{\partial x^i} \left(\sqrt{f} \Psi^4 \tilde{s}^i \right)\Big|_{\mathscr{S}} = 0, \tag{9.54}$$

where use has been made of the divergence formula (2.54). Let us introduce on Σ_0 a coordinate system of spherical type, $(x^i) = (r, \theta, \varphi)$, such that (i) $f_{ij} = \mathrm{diag}(1, r^2, r^2 \sin^2\theta)$ and (ii) $\mathscr{S}$ is the sphere $r = a$, where a is some positive constant. Since in these coordinates $\sqrt{f} = r^2 \sin\theta$ and $\tilde{s}^i = (1, 0, 0)$, the minimal surface condition (9.54) is written as

$$\frac{1}{r^2} \frac{\partial}{\partial r} \left(\Psi^4 r^2 \right)\Big|_{r=a} = 0,$$

i.e.

$$\left(\frac{\partial \Psi}{\partial r} + \frac{\Psi}{2r} \right)\Big|_{r=a} = 0 \tag{9.55}$$

This is a boundary condition of mixed Newmann/Dirichlet type for Ψ. The unique solution of the Laplace equation (9.50) which satisfies boundary conditions (9.41) and (9.55) is

$$\Psi = 1 + \frac{a}{r}. \tag{9.56}$$

The parameter a is then easily related to the ADM mass m of the hypersurface Σ_0. Indeed using formula (8.48), m is evaluated as

$$m = -\frac{1}{2\pi} \lim_{r \to \infty} \oint_{r=\mathrm{const}} \frac{\partial \Psi}{\partial r} r^2 \sin\theta \, \mathrm{d}\theta \, \mathrm{d}\varphi = -\frac{1}{2\pi} \lim_{r \to \infty} 4\pi r^2 \frac{\partial}{\partial r} \left(1 + \frac{a}{r} \right) = 2a. \tag{9.57}$$

Hence $a = m/2$ and we may write

$$\boxed{\Psi = 1 + \frac{m}{2r}}. \tag{9.58}$$

Therefore, in terms of the coordinates (r, θ, φ), the obtained initial data $(\boldsymbol{\gamma}, \boldsymbol{K})$ are

$$\gamma_{ij} = \left(1 + \frac{m}{2r} \right)^4 \mathrm{diag}(1, r^2, r^2 \sin\theta) \tag{9.59}$$

$$K_{ij} = 0. \tag{9.60}$$

So, as above, the initial data are momentarily static. Actually, we recognize on (9.59)–(9.60) a slice $t = \mathrm{const}$ of *Schwarzschild spacetime* in isotropic coordinates [compare with Eq. (7.23)].

The isotropic coordinates (r, θ, φ) covering the manifold Σ_0 are such that the range of r is $[m/2, +\infty)$. But thanks to the minimal character of the inner boundary

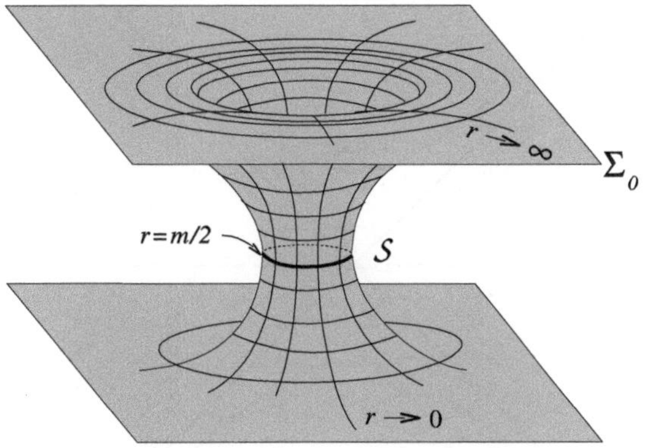

Fig. 9.3 Extended hypersurface Σ_0' obtained by gluing a copy of Σ_0 at the minimal surface $\mathscr{S}$ and defining an Einstein–Rosen bridge between two asymptotically flat regions

$\mathscr{S}$, we can extend $(\Sigma_0, \boldsymbol{\gamma})$ to a larger Riemannian manifold $(\Sigma_0', \boldsymbol{\gamma}')$ with $\boldsymbol{\gamma}'|_{\Sigma_0} = \boldsymbol{\gamma}$ and $\boldsymbol{\gamma}'$ smooth at $\mathscr{S}$. This is made possible by gluing a copy of Σ_0 at $\mathscr{S}$ (cf. Fig. 9.3). The topology of Σ_0' is $\mathbb{S}^2 \times \mathbb{R}$ and the range of r in Σ_0' is $(0, +\infty)$. The extended metric $\boldsymbol{\gamma}'$ keeps exactly the same form as (9.59):

$$\gamma'_{ij}\, \mathrm{d}x^i\, \mathrm{d}x^j = \left(1 + \frac{m}{2r}\right)^4 \left(\mathrm{d}r^2 + r^2 \mathrm{d}\theta^2 + r^2 \sin^2\theta \mathrm{d}\varphi^2\right).$$

By the change of variable

$$r \mapsto r' = \frac{m^2}{4r} \tag{9.61}$$

it is easily shown that the region $r \to 0$ does not correspond to some "center" but is actually a second asymptotically flat region (the lower one in Fig. 9.3). Moreover the transformation (9.61), with θ and φ kept fixed, is an isometry of $\boldsymbol{\gamma}'$. It maps a point p of Σ_0 to the point located at the vertical of p in Fig. 9.3. The minimal sphere $\mathscr{S}$ is invariant under this isometry. The region around $\mathscr{S}$ is called an ***Einstein–Rosen bridge***. $(\Sigma_0', \boldsymbol{\gamma}')$ is still a slice of Schwarzschild spacetime. It connects two asymptotically flat regions without entering below the event horizon, as shown in the Kruskal–Szekeres diagram of Fig. 9.4.

Remark 9.1 ***Kruskal–Szekeres diagrams*** are representations of the Schwarzschild spacetime based on the Kruskal–Szekeres coordinates, which have the nice features (i) to cover the entire Schwarzschild manifold, whose topology is $\mathbb{R}^2 \times \mathbb{S}^2$ and (ii) to be adapted to the null cones of the spacetime metric $\boldsymbol{g}$, so that they are depicted as $\pm 45°$ lines in Kruskal–Szekeres diagrams, as in Minkowski spacetime diagrams. We shall not recall here the construction of Kruskal–Szekeres coordinates and refer the reader to the textbooks [31–34]. Kruskal–Szekeres diagrams will be much used in Chap. 10.

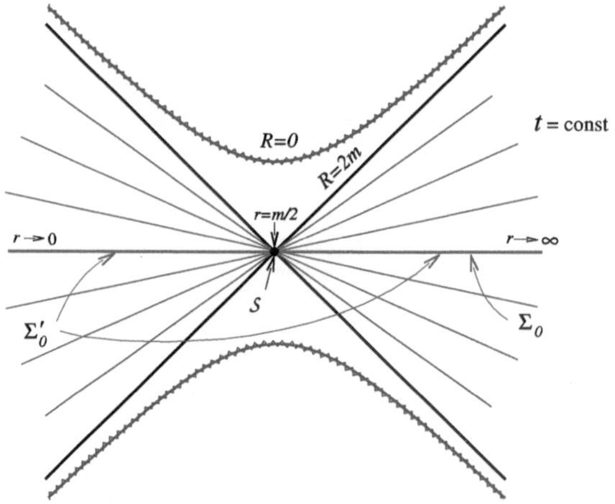

Fig. 9.4 Extended hypersurface Σ'_0 depicted in the Kruskal–Szekeres representation of Schwarzschild spacetime. R stands for Schwarzschild radial coordinate and r for the isotropic radial coordinate. $R = 0$ is the singularity and $R = 2m$ the event horizon. Σ'_0 is nothing but a hypersurface $t = \text{const}$, where t is the Schwarzschild time coordinate. In this diagram, these hypersurfaces are straight lines and the Einstein–Rosen bridge $\mathscr{S}$ is reduced to a point

9.2.6 Bowen–York Initial Data

Let us select the same simple free data as above, namely

$$\tilde{\gamma}_{ij} = f_{ij}, \quad \hat{A}^{ij}_{\rm TT} = 0, \quad K = 0, \quad \tilde{E} = 0 \quad \text{and} \quad \tilde{p}^i = 0. \tag{9.62}$$

For the hypersurface Σ_0, instead of $\mathbb{R}^3$ minus a ball, we choose $\mathbb{R}^3$ minus a point:

$$\Sigma_0 = \mathbb{R}^3 \backslash \{O\}. \tag{9.63}$$

The removed point O is called a ***puncture*** [35]. The topology of Σ_0 is $\mathbb{S}^2 \times \mathbb{R}$; it differs from the topology considered in Sect. 9.2.5 ($\mathbb{R}^3$ minus a ball); actually it is the same topology as that of the extended manifold Σ'_0 (cf. Fig. 9.3).

Thanks to the choice (9.62), the system to be solved is still (9.37)–(9.38). If we choose the trivial solution $\boldsymbol{X} = 0$ for Eq. (9.38), we are back to the slice of Schwarzschild spacetime considered in Sect. 9.2.5, except that now Σ_0 is the extended manifold previously denoted Σ'_0.

A simple non-trivial solution to Eq. (9.38) has been obtained by Bowen and York [36] (see also Ref. [37]). Given a Cartesian coordinate system $(x^i) = (x, y, z)$ on Σ_0 (i.e. a coordinate system such that $f_{ij} = \mathrm{diag}(1, 1, 1)$) with respect to which the coordinates of the puncture O are $(0, 0, 0)$, this solution writes

$$X^i = -\frac{1}{4r}\left(7f^{ij}P_j + \frac{P_j x^j x^i}{r^2}\right) - \frac{1}{r^3}\varepsilon^{ij}{}_k S_j x^k, \tag{9.64}$$

where $r := \sqrt{x^2+y^2+z^2}$, $\varepsilon^{ij}_{\ \ k}$ is the Levi–Civita tensor associated with the flat metric $\boldsymbol{f}$ (cf. Sect. 2.3.4) and $(P_i, S_j) = (P_1, P_2, P_3, S_1, S_2, S_3)$ are six real numbers, which constitute the six parameters of the Bowen–York solution. Notice that since $r \neq 0$ on Σ_0, the Bowen–York solution is a regular and smooth solution on the entire Σ_0.

Example 9.2 Choosing $P_i = (0, P, 0)$ and $S_i = (0, 0, S)$, where P and S are two real numbers, leads to the following expression of the Bowen–York solution:

$$\begin{cases} X^x = -\dfrac{P}{4}\dfrac{xy}{r^3} + S\dfrac{y}{r^3} \\ X^y = -\dfrac{P}{4r}\left(7 + \dfrac{y^2}{r^2}\right) - S\dfrac{x}{r^3} \\ X^z = -\dfrac{P}{4}\dfrac{xz}{r^3} \end{cases} \tag{9.65}$$

The conformal traceless extrinsic curvature corresponding to the solution (9.64) is deduced from formula (9.9), which in the present case reduces to $\hat{A}^{ij} = (LX)^{ij}$; one gets

$$\boxed{\hat{A}^{ij} = \frac{3}{2r^3}\left[x^i P^j + x^j P^i - \left(f^{ij} - \frac{x^i x^j}{r^2}\right)P_k x^k\right] + \frac{3}{r^5}\left(\varepsilon^{ik}{}_l S_k x^l x^j + \varepsilon^{jk}{}_l S_k x^l x^i\right)}, \tag{9.66}$$

where $P^i := f^{ij}P_j$. The tensor $\hat{A}^{ij}$ given by Eq. (9.66) is called the ***Bowen–York extrinsic curvature*** . Notice that the P_i part of $\hat{A}^{ij}$ decays asymptotically as $O(r^{-2})$, whereas the S_i part decays as $O(r^{-3})$.

Remark 9.2 Actually the expression of $\hat{A}^{ij}$ given in the original Bowen–York article [36] contains an additional term with respect to Eq. (9.66), but the role of this extra term is only to ensure that the solution is isometric through an inversion across some sphere. We are not interested by such a property here, so we have dropped this term. Therefore, strictly speaking, we should name expression (9.66) the *simplified* Bowen–York extrinsic curvature.

Example 9.3 Choosing $P_i = (0, P, 0)$ and $S_i = (0, 0, S)$ as in the previous example [Eq. (9.65)], we get

$$\hat{A}^{xx} = -\frac{3P}{2r^3}y\left(1 - \frac{x^2}{r^2}\right) - \frac{6S}{r^5}xy \tag{9.67}$$

$$\hat{A}^{xy} = \frac{3P}{2r^3}x\left(1 + \frac{y^2}{r^2}\right) + \frac{3S}{r^5}(x^2 - y^2) \tag{9.68}$$

$$\hat{A}^{xz} = \frac{3P}{2r^5}xyz - \frac{3S}{r^5}yz \tag{9.69}$$

$$\hat{A}^{yy} = \frac{3P}{2r^3}y\left(1 + \frac{y^2}{r^2}\right) + \frac{6S}{r^5}xy \tag{9.70}$$

$$\hat{A}^{yz} = \frac{3P}{2r^3}z\left(1 + \frac{y^2}{r^2}\right) + \frac{3S}{r^5}xz \tag{9.71}$$

$$\hat{A}^{zz} = -\frac{3P}{2r^3}y\left(1 - \frac{z^2}{r^2}\right). \tag{9.72}$$

In particular we verify that $\hat{A}^{ij}$ is traceless: $\tilde{\gamma}_{ij}\hat{A}^{ij} = f_{ij}\hat{A}^{ij} = \hat{A}^{xx} + \hat{A}^{yy} + \hat{A}^{zz} = 0$.

The Bowen–York extrinsic curvature provides an analytical solution of the momentum constraint (9.38) but there remains to solve the Hamiltonian constraint (9.37) for Ψ, with the asymptotic flatness boundary condition $\Psi = 1$ when $r \to \infty$. Since $\boldsymbol{X} \neq 0$, Eq. (9.37) is no longer a simple Laplace equation, as in Sect. 9.2.5, but a non-linear elliptic equation. There is no hope to get any analytical solution and one must solve Eq. (9.37) numerically to get Ψ and reconstruct the full initial data $(\boldsymbol{\gamma}, \boldsymbol{K})$ via Eqs. (9.23)–(9.24).

Let us now discuss the physical significance of the parameters (P_i, S_i) of the Bowen–York solution. First of all, the ADM momentum of the initial data $(\Sigma_0, \boldsymbol{\gamma}, \boldsymbol{K})$ is computed via formula (8.40). Taking into account that Ψ is asymptotically one and ***longitudinal part*** K vanishes, we can write

$$P_i^{\rm ADM} = \frac{1}{8\pi}\lim_{r\to\infty}\oint_{r=\rm const}\hat{A}_{ik}x^k r\sin\theta\,\mathrm{d}\theta\,\mathrm{d}\varphi, \qquad i \in \{1, 2, 3\},$$

where we have used the fact that, within the Cartesian coordinates $(x^i) = (x, y, z)$, $(\partial_i)^j = \delta^j_i$ and $s^k = x^k/r$. If we insert expression (9.66) for $\hat{A}_{jk}$ in this formula, we notice that the S_i part decays too fast to contribute to the integral; there remains only

$$P_i^{\mathrm{ADM}} = \frac{1}{8\pi} \lim_{r\to\infty} \oint_{r=\mathrm{const}} \frac{3}{2r^2} \Big[x_i P_j x^j + r^2 P_i - \underbrace{\left(x_i - \frac{x^i r^2}{r^2} \right)}_{0} P_k x^k \Big] \sin\theta \, \mathrm{d}\theta \, \mathrm{d}\varphi$$

$$= \frac{3}{16\pi} \Big(P_j \oint_{r=\mathrm{const}} \frac{x^i x^j}{r^2} \sin\theta \mathrm{d}\theta \, \mathrm{d}\varphi + P_i \underbrace{\oint_{r=\mathrm{const}} \sin\theta \, \mathrm{d}\theta \, \mathrm{d}\varphi}_{4\pi} \Big). \tag{9.73}$$

Now

$$\oint_{r=\mathrm{const}} \frac{x^i x^j}{r^2} \sin\theta \mathrm{d}\theta \, \mathrm{d}\varphi = \delta^{ij} \oint_{r=\mathrm{const}} \frac{(x^j)^2}{r^2} \sin\theta \mathrm{d}\theta \, \mathrm{d}\varphi$$
$$= \delta^{ij} \frac{1}{3} \oint_{r=\mathrm{const}} \frac{r^2}{r^2} \sin\theta \mathrm{d}\theta \, \mathrm{d}\varphi = \frac{4\pi}{3} \delta^{ij},$$

so that Eq. (9.73) becomes

$$P_i^{\mathrm{ADM}} = \frac{3}{16\pi} \left(\frac{4\pi}{3} + 4\pi \right) P_i,$$

i.e.

$$\boxed{P_i^{\mathrm{ADM}} = P_i}. \tag{9.74}$$

Hence the parameters P_i of the Bowen–York solution are nothing but the three components of the ADM linear momentum of the hypersurface Σ_0.

Regarding the angular momentum, we notice that since $\tilde{\gamma}_{ij} = f_{ij}$ in the present case, the Cartesian coordinates $(x^i) = (x, y, z)$ belong to the quasi-isotropic gauge introduced in Sect. 8.5.2 (condition (8.46) is trivially fulfilled). We may then use formula (8.45) to define the angular momentum of Bowen–York initial. Again, since $\Psi \to 1$ at spatial infinity and $K = 0$, we can write

$$J_i = \frac{1}{8\pi} \lim_{r\to\infty} \oint_{r=\mathrm{const}} \hat{A}_{jk} (\boldsymbol{\phi}_i)^j x^k r \sin\theta \, \mathrm{d}\theta \, \mathrm{d}\varphi, \qquad i \in \{1, 2, 3\}.$$

Substituting expression (9.66) for $\hat{A}_{jk}$ as well as expression (9.66) for $(\boldsymbol{\phi}_i)^j$, we get that only the S_i part contribute to this integral. After some computation, we find

$$\boxed{J_i = S_i}. \tag{9.75}$$

Hence the parameters S_i of the Bowen–York solution are nothing but the three components of the angular momentum of the hypersurface Σ_0.

Remark 9.3 The Bowen–York solution with $P^i = 0$ and $S^i = 0$ reduces to the momentarily static solution found in Sect. 9.2.5, i.e. is a slice $t = \mathrm{const}$ of the Schwarzschild spacetime (t being the Schwarzschild time coordinate). However Bowen–York initial data with $P^i = 0$ and $S^i \neq 0$ do not constitute a slice of a stationary black hole spacetime, i.e. a Kerr spacetime. Indeed, it has been shown

[38] (see also [39]) that there does not exist any foliation of Kerr spacetime by hypersurfaces which (i) are axisymmetric, (ii) smoothly reduce in the non-rotating limit to the hypersurfaces of constant Schwarzschild time and (iii) are conformally flat, i.e. have induced metric $\tilde{\boldsymbol{\gamma}} = \boldsymbol{f}$, as the Bowen–York hypersurfaces have. This means that a Bowen–York solution with $S^i \neq 0$ does represent initial data for a rotating black hole, but this black hole is not stationary: it is "surrounded" by gravitational radiation, as demonstrated by the time development of these initial data [40, 41].

9.3 Conformal Thin Sandwich Method

9.3.1 The Original Conformal Thin Sandwich Method

An alternative to the conformal transverse-traceless method for computing initial data has been introduced by York in 1999 [8]. It is motivated by expression (7.64) for the traceless part of the extrinsic curvature scaled with $\alpha = -4$:

$$\tilde{A}^{ij} = \frac{1}{2N}\left[\left(\frac{\partial}{\partial t} - \mathcal{L}_{\boldsymbol{\beta}}\right)\tilde{\gamma}^{ij} - \frac{2}{3}\tilde{D}_k\beta^k\tilde{\gamma}^{ij}\right]. \tag{9.76}$$

Noticing that [cf. Eq. (9.11)]

$$-\mathcal{L}_{\boldsymbol{\beta}}\tilde{\gamma}^{ij} = (\tilde{L}\beta)^{ij} + \frac{2}{3}\tilde{D}_k\beta^k, \tag{9.77}$$

and introducing the short-hand notation

$$\dot{\tilde{\gamma}}^{ij} := \frac{\partial}{\partial t}\tilde{\gamma}^{ij}, \tag{9.78}$$

we can rewrite Eq. (9.76) as

$$\tilde{A}^{ij} = \frac{1}{2N}\left[\dot{\tilde{\gamma}}^{ij} + (\tilde{L}\beta)^{ij}\right]. \tag{9.79}$$

The relation between $\tilde{A}^{ij}$ and $\hat{A}^{ij}$ is [cf. Eq. (7.85)]

$$\hat{A}^{ij} = \Psi^6\tilde{A}^{ij}. \tag{9.80}$$

Accordingly, Eq. (9.79) yields

$$\boxed{\hat{A}^{ij} = \frac{1}{2\tilde{N}}\left[\dot{\tilde{\gamma}}^{ij} + (\tilde{L}\beta)^{ij}\right]}, \tag{9.81}$$

where we have introduced the ***conformal lapse***

$$\boxed{\tilde{N} := \Psi^{-6}N}. \tag{9.82}$$

Equation (9.81) constitutes a decomposition of $\hat{A}^{ij}$ alternative to the longitudinal/transverse decomposition (9.9). Instead of expressing $\hat{A}^{ij}$ in terms of a vector $\boldsymbol{X}$ and a TT tensor $\hat{A}^{ij}_{\rm TT}$, it expresses it in terms of the shift vector $\boldsymbol{\beta}$, the time derivative of the conformal metric, $\dot{\tilde{\gamma}}^{ij}$, and the conformal lapse $\tilde{N}$.

The Hamiltonian constraint, written as the Lichnerowicz equation (9.81), takes the same form as before:

$$\boxed{\tilde{D}_i\tilde{D}^i\Psi - \frac{\tilde{R}}{8}\Psi + \frac{1}{8}\hat{A}_{ij}\hat{A}^{ij}\Psi^{-7} + 2\pi\tilde{E}\Psi^{-3} - \frac{K^2}{12}\Psi^5 = 0}, \qquad (9.83)$$

except that now $\hat{A}^{ij}$ is to be understood as the combination (9.81) of β^i, $\dot{\tilde{\gamma}}^{ij}$ and $\tilde{N}$. On the other side, the momentum constraint (9.6) becomes, once expression (9.81) is substituted for $\hat{A}^{ij}$,

$$\boxed{\tilde{D}_j\left(\frac{1}{\tilde{N}}(\tilde{L}\beta)^{ij}\right) + \tilde{D}_j\left(\frac{1}{\tilde{N}}\dot{\tilde{\gamma}}^{ij}\right) - \frac{4}{3}\Psi^6\tilde{D}^iK = 16\pi\tilde{p}^i}. \qquad (9.84)$$

In view of the system (9.83)–(9.84), the method to compute initial data consists in choosing freely $\tilde{\gamma}_{ij}$, $\dot{\tilde{\gamma}}^{ij}$, K, $\tilde{N}$, $\tilde{E}$ and $\tilde{p}^i$ on Σ_0 and solving (9.83)–(9.84) to get Ψ and β^i. This method is called ***conformal thin sandwich*** (*CTS*), because one input is the time derivative $\dot{\tilde{\gamma}}^{ij}$, which can be obtained from the value of the conformal metric on two neighbouring hypersurfaces Σ_t and $\Sigma_{t+\delta t}$ ("thin sandwich" view point).

Remark 9.4 The term "thin sandwich" originates from a previous method devised in the early sixties by Wheeler and his collaborators [42, 43]. Contrary to the methods exposed here, the thin sandwich method was not based on a conformal decomposition: it considered the constraint equations (9.1)–(9.2) as a system to be solved for the lapse N and the shift vector $\boldsymbol{\beta}$, given the metric $\boldsymbol{\gamma}$ and its time derivative. The extrinsic curvature which appears in (9.1)–(9.2) was then considered as the function of $\boldsymbol{\gamma}$, $\partial\boldsymbol{\gamma}/\partial t$, N and $\boldsymbol{\beta}$ given by Eq. (5.68). However, the thin sandwich system does have any solution except in special cases [44]. On the contrary the *conformal* thin sandwich method introduced by York [8] and exposed above was shown to be generic [30].

As for the conformal transverse-traceless method treated in Sect. 9.2, on CMC hypersurfaces, Eq. (9.84) decouples from Eq. (9.83) and becomes an elliptic linear equation for $\boldsymbol{\beta}$.

9.3.2 Extended Conformal Thin Sandwich Method

An input of the above method is the conformal lapse $\tilde{N}$. Considering the astrophysical problem stated in Sect. 9.1.1, it is not clear how to pick a relevant value for $\tilde{N}$. Instead of choosing an arbitrary value, Pfeiffer and York [9] have suggested to compute $\tilde{N}$

from the Einstein equation giving the time derivative of the trace K of the extrinsic curvature, i.e. Eq. (7.90):

$$\left(\frac{\partial}{\partial t} - \mathcal{L}_\beta\right) K = -\Psi^{-4}\left(\tilde{D}_i\tilde{D}^i N + 2\tilde{D}_i \ln\Psi \tilde{D}^i N\right) + N\left[4\pi(E+S) + \tilde{A}_{ij}\tilde{A}^{ij} + \frac{K^2}{3}\right]. \tag{9.85}$$

This amounts to add this equation to the initial data system. More precisely, Pfeiffer and York [9] suggested to combine Eq. (9.85) with the Hamiltonian constraint to get an equation involving the quantity $N\Psi = \tilde{N}\Psi^7$ and containing no scalar products of gradients as the $\tilde{D}_i \ln\Psi \tilde{D}^i N$ term in Eq. (9.85), thanks to the identity

$$\tilde{D}_i\tilde{D}^i N + 2\tilde{D}_i \ln\Psi \tilde{D}^i N = \Psi^{-1}\left[\tilde{D}_i\tilde{D}^i(N\Psi) + N\tilde{D}_i\tilde{D}^i\Psi\right].$$

Expressing the left-hand side of the above equation in terms of Eq. (9.85) and substituting $\tilde{D}_i\tilde{D}^i\Psi$ in the right-hand side by its expression deduced from Eq. (9.83), we get

$$\tilde{D}_i\tilde{D}^i(\tilde{N}\Psi^7) - (\tilde{N}\Psi^7)\left[\frac{1}{8}\tilde{R} + \frac{5}{12}K^2\Psi^4 + \frac{7}{8}\hat{A}_{ij}\hat{A}^{ij}\Psi^{-8} + 2\pi(\tilde{E}+2\tilde{S})\Psi^{-4}\right] + \left(\dot{K} - \beta^i\tilde{D}_i K\right)\Psi^5 = 0, \tag{9.86}$$

where we have used the short-hand notation

$$\dot{K} := \frac{\partial K}{\partial t} \tag{9.87}$$

and have set

$$\tilde{S} := \Psi^8 S. \tag{9.88}$$

Adding Eq. (9.86) to Eqs. (9.83) and (9.84), the initial data system becomes

$$\boxed{\tilde{D}_i\tilde{D}^i\Psi - \frac{\tilde{R}}{8}\Psi + \frac{1}{8}\hat{A}_{ij}\hat{A}^{ij}\Psi^{-7} + 2\pi\tilde{E}\Psi^{-3} - \frac{K^2}{12}\Psi^5 = 0} \tag{9.89}$$

$$\boxed{\tilde{D}_j\left(\frac{1}{\tilde{N}}(\tilde{L}\beta)^{ij}\right) + \tilde{D}_j\left(\frac{1}{\tilde{N}}\dot{\tilde{\gamma}}^{ij}\right) - \frac{4}{3}\Psi^6\tilde{D}^i K = 16\pi\tilde{p}^i} \tag{9.90}$$

$$\boxed{\begin{aligned}\tilde{D}_i\tilde{D}^i(\tilde{N}\Psi^7) - (\tilde{N}\Psi^7)\left[\tfrac{\tilde{R}}{8} + \tfrac{5}{12}K^2\Psi^4 + \tfrac{7}{8}\hat{A}_{ij}\hat{A}^{ij}\Psi^{-8} + 2\pi(\tilde{E}+2\tilde{S})\Psi^{-4}\right] \\ + \left(\dot{K} - \beta^i\tilde{D}_i K\right)\Psi^5 = 0\end{aligned}}, \tag{9.91}$$

where $\hat{A}^{ij}$ is the function of $\tilde{N}$, β^i, $\tilde{\gamma}_{ij}$ and $\dot{\tilde{\gamma}}^{ij}$ defined by Eq. (9.81). Equations (9.89)–(9.91) constitute the ***extended conformal thin sandwich*** (***XCTS***) system for the initial data problem. The free data are the conformal metric $\tilde{\boldsymbol{\gamma}}$, its coordinate time derivative $\dot{\tilde{\gamma}}$, the extrinsic curvature trace K, its coordinate time derivative $\dot{K}$, and the rescaled matter variables $\tilde{E}$, $\tilde{S}$ and $\tilde{p}^i$. The constrained data are the conformal factor Ψ, the conformal lapse $\tilde{N}$ and the shift vector $\boldsymbol{\beta}$.

Remark 9.5 The XCTS system (9.89)–(9.91) is a coupled system. Contrary to the CTT system (9.19)–(9.20), the assumption of constant mean curvature, and in particular of maximal slicing, does not allow to decouple it.

When solving the XCTS system (9.89)–(9.91) for black hole spacetimes, one may deal with a manifold Σ_0 with some non-trivial topology. For instance it can be $\mathbb{R}^3$ minus a ball (excised sphere), as in Fig. 9.1. Taking into account the elliptic nature of the XCTS system, one has to put boundary conditions on the excised sphere to get a unique solution. We shall not discuss these conditions here and refer the reader to [45–50].

9.3.3 XCTS at Work: Static Black Hole Example

Let us illustrate the extended conformal thin sandwich method on a simple example.

Example 9.4 Take for the hypersurface Σ_0 the punctured manifold considered in Sect. 9.2.6, namely

$$\Sigma_0 = \mathbb{R}^3 \backslash \{O\}. \tag{9.92}$$

For the free data, let us perform the simplest choice:

$$\tilde{\gamma}_{ij} = f_{ij}, \quad \dot{\tilde{\gamma}}^{ij} = 0, \quad K = 0, \quad \dot{K} = 0, \quad \tilde{E} = 0, \quad \tilde{S} = 0, \quad and \quad \tilde{p}^i = 0, \tag{9.93}$$

i.e. we are searching for vacuum initial data on a maximal and conformally flat hypersurface with all the freely specifiable time derivatives set to zero. Thanks to (9.93), the XCTS system (9.89)–(9.91) reduces to

$$\Delta\Psi + \frac{1}{8}\hat{A}_{ij}\hat{A}^{ij}\Psi^{-7} = 0 \tag{9.94}$$

$$\bar{D}_j\left(\frac{1}{\tilde{N}}(L\beta)^{ij}\right) = 0 \tag{9.95}$$

$$\Delta(\tilde{N}\Psi^7) - \frac{7}{8}\hat{A}_{ij}\hat{A}^{ij}\Psi^{-1}\tilde{N} = 0. \tag{9.96}$$

Aiming at finding the simplest solution, we notice that

$$\boldsymbol{\beta} = 0 \tag{9.97}$$

fulfills Eq. (9.95). Together with $\dot{\tilde{\gamma}}^{ij} = 0$, this leads to [cf. Eq. (9.81)]

$$\hat{A}^{ij} = 0. \tag{9.98}$$

The system (9.94)–(9.96) reduces then further:

$$\Delta\Psi = 0 \tag{9.99}$$

$$\Delta(\tilde{N}\Psi^7) = 0. \tag{9.100}$$

Hence we have only two Laplace equations to solve. Moreover Eq. (9.99) decouples from Eq. (9.100). For simplicity, let us assume spherical symmetry around the puncture O. We introduce an adapted spherical coordinate system $(x^i) = (r, \theta, \varphi)$ on Σ_0. The puncture O is then at $r = 0$. The simplest non-trivial solution of (9.99) which obeys the asymptotic flatness condition $\Psi \to 1$ as $r \to +\infty$ is

$$\Psi = 1 + \frac{m}{2r}, \tag{9.101}$$

where as in Sect. 9.2.5, the constant m is the ADM mass of Σ_0 [cf. Eq. (9.57)]. Notice that since $r = 0$ is excluded from Σ_0, Ψ is a perfectly regular solution on the entire manifold Σ_0. Let us recall that the Riemannian manifold $(\Sigma_0, \boldsymbol{\gamma})$ corresponding to this value of Ψ via $\boldsymbol{\gamma} = \Psi^4 \boldsymbol{f}$ is the Riemannian manifold denoted $(\Sigma_0', \boldsymbol{\gamma})$ in Sect. 9.2.5 and depicted in Fig. 9.3. In particular it has two asymptotically flat ends: $r \to +\infty$ and $r \to 0$ (the puncture).

As for Eq. (9.99), the simplest solution of Eq. (9.100) obeying the asymptotic flatness requirement $\tilde{N}\Psi^7 \to 1$ as $r \to +\infty$ is

$$\tilde{N}\Psi^7 = 1 + \frac{a}{r}, \tag{9.102}$$

where a is some constant. Let us determine a from the value of the lapse function at the second asymptotically flat end $r \to 0$. The lapse being related to $\tilde{N}$ via Eq. (9.82), Eq. (9.102) is equivalent to

$$N = \left(1 + \frac{a}{r}\right)\Psi^{-1} = \left(1 + \frac{a}{r}\right)\left(1 + \frac{m}{2r}\right)^{-1} = \frac{r + a}{r + m/2}. \tag{9.103}$$

Hence

$$\lim_{r\to 0} N = \frac{2a}{m}. \qquad (9.104)$$

There are two natural choices for $\lim_{r\to 0} N$. The first one is

$$\lim_{r\to 0} N = 1, \qquad (9.105)$$

yielding $a = m/2$. Then, from Eq. (9.103) N=1 everywhere on Σ_0. This value of N corresponds to a geodesic slicing (cf. Sect. 5.4.2). The second choice is

$$\lim_{r\to 0} N = -1. \qquad (9.106)$$

This choice is compatible with asymptotic flatness: it simply means that the coordinate time t is running "backward" near the asymptotic flat end $r \to 0$. This contradicts the assumption $N > 0$ in the definition of the lapse function given in Sect. 4.3.1. However, we shall generalize here this definition to allow for negative values: whereas the unit vector $\boldsymbol{n}$ is always future-oriented, the scalar field t is allowed to decrease towards the future. Such a situation has already been encountered for the part of the slices $t = \text{const}$ located on the left side of Fig. 9.4. Once reported into Eq. (9.104), the choice (9.106) yields $a = -m/2$, so that

$$N = \left(1 - \frac{m}{2r}\right)\left(1 + \frac{m}{2r}\right)^{-1}. \qquad (9.107)$$

Gathering relations (9.97), (9.101) and (9.107), we arrive at the following expression of the spacetime metric components:

$$g_{\mu\nu}\mathrm{d}x^\mu \mathrm{d}x^\nu = -\left(\frac{1 - \frac{m}{2r}}{1 + \frac{m}{2r}}\right)^2 \mathrm{d}t^2 + \left(1 + \frac{m}{2r}\right)^4 \left[\mathrm{d}r^2 + r^2(\mathrm{d}\theta^2 + \sin^2\theta \mathrm{d}\varphi^2)\right]. \qquad (9.108)$$

We recognize the line element of Schwarzschild spacetime in isotropic coordinates [cf. Eq. (7.23)]. Hence we recover the same initial data as in Sect. 9.2.5 and depicted in Figs. 9.3 and 9.4. The bonus is that we have the complete expression of the metric $\boldsymbol{g}$ on Σ_0, and not only the induced metric $\boldsymbol{\gamma}$.

Remark 9.6 The choices (9.105) and (9.106) for the asymptotic value of the lapse both lead to a momentarily static initial slice in Schwarzschild spacetime. The difference is that the time development corresponding to choice (9.105) (geodesic slicing) will depend on t, whereas the time development corresponding to choice (9.106) will not, since in the latter case $\boldsymbol{\partial}_t$ is a Killing vector.

9.3.4 Uniqueness Issue

Pfeiffer and York [51] have exhibited a choice of vacuum free data $(\tilde{\gamma}_{ij}, \dot{\tilde{\gamma}}^{ij}, K, \dot{K})$ for which the solution $(\Psi, \tilde{N}, \beta^i)$ to the XCTS system (9.89)–(9.91) is not unique (actually two solutions are found). The conformal metric $\tilde{\boldsymbol{\gamma}}$ is the flat metric plus a linearized quadrupolar gravitational wave, as obtained by Teukolsky [52], with a tunable amplitude. $\dot{\tilde{\gamma}}^{ij}$ corresponds to the time derivative of this wave, and both K and $\dot{K}$ are chosen to zero. On the contrary, for the same free data, with $\dot{K} = 0$ substituted by $\tilde{N} = 1$, Pfeiffer and York have shown that the original conformal thin sandwich method as described in Sect. 9.3.1 leads to a unique solution (or no solution at all if the amplitude of the wave is two large).

Baumgarte, Ó Murchadha and Pfeiffer [53] have argued that the lack of uniqueness for the XCTS system may be due to the term

$$-\frac{7}{8}(\tilde{N}\Psi^7)\hat{A}_{ij}\hat{A}^{ij}\Psi^{-8} = -\frac{7}{32}\Psi^6\tilde{\boldsymbol{\gamma}}_{ik}\tilde{\boldsymbol{\gamma}}_{jl}\left[\dot{\tilde{\gamma}}^{ij} + (\tilde{L}\beta)^{ij}\right]\left[\dot{\tilde{\gamma}}^{kl} + (\tilde{L}\beta)^{kl}\right](\tilde{N}\Psi^7)^{-1} \tag{9.109}$$

in Eq. (9.91). Indeed, if we proceed as for the analysis of Lichnerowicz equation in Sect. 9.2.4, we notice that this term, with the minus sign and the negative power of $(\tilde{N}\Psi^7)$, makes the linearization of Eq. (9.91) of the type $\tilde{D}_i\tilde{D}^i\varepsilon + \alpha\varepsilon = \sigma$, with $\alpha > 0$. This "wrong" sign of α prevents the application of the maximum principle to guarantee the uniqueness of the solution.

The non-uniqueness of solution of the XCTS system for certain choice of free data has been confirmed by Walsh [54] by means of bifurcation theory.

It turned out that for some highly relativistic systems, this non-uniqueness can be an issue and prevent numerical codes to converge [55]. A solution has been found [56, 55] and consists in solving an extra vector equation, of the type of Eq. (9.20), for the longitudinal part of $\hat{A}^{ij}$.

9.3.5 Comparing CTT, CTS and XCTS

The conformal transverse traceless (CTT) method exposed in Sect. 9.2 and the (extended) conformal thin sandwich (XCTS) method considered here differ by the choice of free data: whereas both methods use the conformal metric $\tilde{\boldsymbol{\gamma}}$ and the trace of the extrinsic curvature K as free data, CTT employs in addition $\hat{A}^{ij}_{\mathrm{TT}}$, whereas for CTS the additional free data are $\dot{\tilde{\gamma}}^{ij}$ and $\tilde{N}$. For XCTS, it is $\dot{\tilde{\gamma}}^{ij}$ and $\dot{K}$ instead. Since $\hat{A}^{ij}_{\mathrm{TT}}$ is directly related to the extrinsic curvature and the latter is linked to the canonical momentum of the gravitational field in the Hamiltonian formulation of general relativity (cf. Sect. 5.5), the CTT method can be considered as the approach to the initial data problem in the *Hamiltonian representation*. On the other side, $\dot{\tilde{\gamma}}^{ij}$ being the "velocity" of $\tilde{\gamma}^{ij}$, the (X)CTS method constitutes the approach in the *Lagrangian representation* [57].

Remark 9.7 The (X)CTS method assumes that the conformal metric is unimodular: $\det(\tilde{\gamma}_{ij}) = f$ [Eq. (7.18)] (since Eq. (9.81) follows from this assumption), whereas the CTT method can be applied with any conformal metric.

The advantage of CTT is that its mathematical theory is well developed, yielding existence and uniqueness theorems, at least for constant mean curvature (CMC) slices. The mathematical theory of CTS is very close to CTT. In particular, the momentum constraint decouples from the Hamiltonian constraint on CMC slices. On the contrary, XCTS has a much more involved mathematical structure. In particular the CMC slicing does not result in any decoupling in this case. The advantage of XCTS is then to be better suited to the description of quasi-stationary spacetimes, since $\dot{\tilde{\gamma}}^{ij} = 0$ and $\dot{K} = 0$ are necessary conditions for $\boldsymbol{\partial}_t$ to be a Killing vector. This makes XCTS the method to be used in order to prepare initial data in quasi-equilibrium. For instance, it has been shown [58, 59] that XCTS yields orbiting binary black hole configurations in much better agreement with post-Newtonian computations than the CTT treatment based on a superposition of two Bowen–York solutions.

A detailed comparison of CTT and XCTS for a single spinning or boosted black hole has been performed by Laguna [60].

9.4 Initial Data for Binary Systems

A major topic of contemporary numerical relativity is the computation of the merger of a binary system of black holes or neutron stars, for such systems are among the most promising sources of gravitational radiation for the interferometric detectors either ground-based (LIGO, VIRGO, GEO600, TAMA, LCGT) or in space (LISA/NGO). The problem of preparing initial data for these systems has therefore received a lot of attention in the past decade.

9.4.1 Helical Symmetry

Due to the gravitational-radiation reaction, a relativistic binary system has an inspiral motion, leading to the merger of the two components. However, when the two bodies are sufficiently far apart, one may approximate the spiraling orbits by closed ones. Moreover, it is well known that gravitational radiation circularizes the orbits very efficiently, at least for comparable mass systems [61]. We may then consider that the motion is described by a sequence of *closed circular orbits*.

The geometrical translation of this physical assumption is that the spacetime $(\mathscr{M}, \boldsymbol{g})$ is endowed with some symmetry, called ***helical symmetry***. Indeed exactly circular orbits imply the existence of a one-parameter symmetry group such that the associated Killing vector $\boldsymbol{\ell}$ obeys the following properties [62]: (i) $\boldsymbol{\ell}$ is timelike near

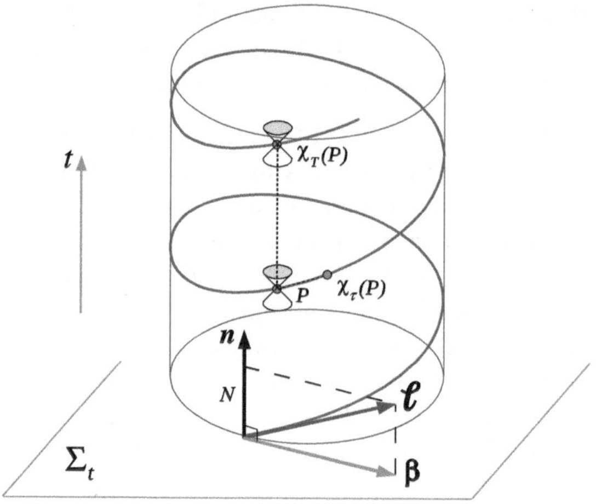

Fig. 9.5 Action of the helical symmetry group, with Killing vector $\boldsymbol{\ell}$. $\chi_\tau(P)$ is the displacement of the point P by the member of the symmetry group of parameter τ. N and $\boldsymbol{\beta}$ are respectively the lapse function and the shift vector associated with coordinates adapted to the symmetry, i.e. coordinates (t, x^i) such that $\partial_t = \boldsymbol{\ell}$

the system, (ii) far from it, $\boldsymbol{\ell}$ is spacelike but there exists a smallest number $T > 0$ such that the separation between any point P and its image $\chi_T(P)$ under the symmetry group is timelike (cf. Fig. 9.5). $\boldsymbol{\ell}$ is called a ***helical Killing vector***, its field lines in a spacetime diagram being helices (cf. Fig. 9.5).

Helical symmetry is exact in theories of gravity where gravitational radiation does not exist, namely:

- in Newtonian gravity,
- in post-Newtonian gravity, up to the second order,
- in the Isenberg-Wilson-Mathews approximation to general relativity discussed in Sect. 7.6.

Moreover helical symmetry can be exact in full general relativity for a non-axisymmetric system (such as a binary) with standing gravitational waves [63]. But notice that a spacetime with helical symmetry and standing gravitational waves cannot be asymptotically flat [64].

To treat helically symmetric spacetimes, it is natural to choose coordinates (t, x^i) that are adapted to the symmetry, i.e. such that

$$\partial_t = \boldsymbol{\ell}. \tag{9.110}$$

Then all the fields are independent of the coordinate t. In particular,

$$\dot{\tilde{\gamma}}^{ij} = 0 \quad \text{and} \quad \dot{K} = 0. \tag{9.111}$$

If we employ the XCTS formalism to compute initial data, we therefore get some definite prescription for the free data $\dot{\tilde{\gamma}}^{ij}$ and $\dot{K}$. On the contrary, the requirements (9.111) do not have any immediate translation in the CTT formalism.

Remark 9.8 Helical symmetry can also be useful to treat binary black holes outside the scope of the 3+1 formalism, as shown by Klein [65], who developed a quotient space formalism to reduce the problem to a 3-dimensional $\mathrm{SL}(2, \mathbb{R})/\mathrm{SO}(1, 1)$ sigma model.

Taking into account (9.111) and choosing maximal slicing ($K = 0$), the XCTS system (9.89)–(9.91) becomes

$$\tilde{D}_i \tilde{D}^i \Psi - \frac{\tilde{R}}{8}\Psi + \frac{1}{8}\hat{A}_{ij}\hat{A}^{ij}\Psi^{-7} + 2\pi \tilde{E}\Psi^{-3} = 0 \tag{9.112}$$

$$\tilde{D}_j \left(\frac{1}{\tilde{N}} (\tilde{L}\beta)^{ij} \right) - 16\pi \tilde{p}^i = 0 \tag{9.113}$$

$$\tilde{D}_i \tilde{D}^i (\tilde{N}\Psi^7) - (\tilde{N}\Psi^7) \left[\frac{\tilde{R}}{8} + \frac{7}{8}\hat{A}_{ij}\hat{A}^{ij}\Psi^{-8} + 2\pi(\tilde{E} + 2\tilde{S})\Psi^{-4} \right] = 0, \tag{9.114}$$

where [cf. Eq. (9.81)]

$$\hat{A}^{ij} = \frac{1}{2\tilde{N}} (\tilde{L}\beta)^{ij}. \tag{9.115}$$

9.4.2 Helical Symmetry and IWM Approximation

If we choose, as part of the free data, the conformal metric to be flat,

$$\tilde{\gamma}_{ij} = f_{ij}, \tag{9.116}$$

then the helically symmetric XCTS system (9.112)–(9.114) reduces to

$$\Delta \Psi + \frac{1}{8}\hat{A}_{ij}\hat{A}^{ij}\Psi^{-7} + 2\pi \tilde{E}\Psi^{-3} = 0 \tag{9.117}$$

$$\Delta \beta^i + \frac{1}{3}\bar{D}^i \bar{D}_j \beta^j - (L\beta)^{ij}\bar{D}_j \ln \tilde{N} = 16\pi \tilde{N}\tilde{p}^i \tag{9.118}$$

$$\Delta(\tilde{N}\Psi^7) - (\tilde{N}\Psi^7) \left[\frac{7}{8}\hat{A}_{ij}\hat{A}^{ij}\Psi^{-8} + 2\pi(\tilde{E} + 2\tilde{S})\Psi^{-4} \right] = 0, \tag{9.119}$$

where

$$\hat{A}^{ij} = \frac{1}{2\tilde{N}} (L\beta)^{ij} \tag{9.120}$$

and $\bar{\mathcal{D}}f$ is the connection associated with the flat metric $\boldsymbol{f}$, $\Delta := \bar{D}_i \bar{D}^i$ is the flat Laplacian [Eq. (9.39)], and $(L\beta)^{ij} := \bar{D}^i \beta^j + \bar{D}^j \beta^i - \frac{2}{3} \bar{D}_k \beta^k f^{ij}$ [Eq. (7.107)].

We remark that the system (9.117)–(9.119) is identical to the Isenberg–Wilson–Mathews (IWM) system (7.110)–(7.112) presented in Sect. 7.6: given that $\tilde{E} = \Psi^8 E$, $\tilde{p}^i = \Psi^{10} p^i$, $\tilde{N} = \Psi^{-6} N$, $\hat{A}^{ij} = \Psi^6 \tilde{A}^{ij}$ and $\hat{A}_{ij}\hat{A}^{ij} = \Psi^{12} \tilde{A}_{ij}\tilde{A}^{ij}$, Eq. (9.117) coincides with Eq. (7.111), Eq. (9.118) coincides with Eq. (7.112) and Eq. (9.119) is a combination of Eqs. (7.110) and (7.111). Hence, within helical symmetry, the XCTS system with the choice $K = 0$ and $\tilde{\boldsymbol{\gamma}} = \boldsymbol{f}$ is equivalent to the IWM system.

Remark 9.9 Contrary to IWM, XCTS is not some approximation to general relativity: it provides exact initial data. The only thing that may be questioned is the astrophysical relevance of the XCTS data with $\tilde{\boldsymbol{\gamma}} = \boldsymbol{f}$.

9.4.3 Initial Data for Orbiting Binary Black Holes

The concept of helical symmetry for generating orbiting binary black hole initial data has been introduced in 2002 by Gourgoulhon, Grandclément and Bonazzola [66, 58]. The system of equations that these authors have derived is equivalent to the XCTS system with $\tilde{\boldsymbol{\gamma}} = \boldsymbol{f}$, their work being anterior to the formulation of the XCTS method by Pfeiffer and York [9]. Since then other groups have combined XCTS with helical symmetry to compute binary black hole initial data [45, 67, 68, 47]. Since all these studies are using a flat conformal metric [choice (9.116)], the PDE system to be solved is (9.117)–(9.119), with the additional simplification $\tilde{E} = 0$ and $\tilde{p}^i = 0$ (vacuum). The initial data manifold Σ_0 is chosen to be $\mathbb{R}^3$ minus two balls:

$$\Sigma_0 = \mathbb{R}^3 \backslash (\mathscr{B}_1 \cup \mathscr{B}_2). \tag{9.121}$$

In addition to the asymptotic flatness conditions, some boundary conditions must be provided on the surfaces $\mathscr{S}_1$ and $\mathscr{S}_2$ of $\mathscr{B}_1$ and $\mathscr{B}_2$. One choose boundary conditions corresponding to a *non-expanding horizon*, since this concept characterizes black holes in equilibrium. We shall not detail these boundary conditions here; they can be found in Refs. [45, 69, 70, 48, 71]. The condition of non-expanding horizon provides 3 among the 5 required boundary conditions [for the 5 components $(\Psi, \tilde{N}, \beta^i)$]. The two remaining boundary conditions are given by (i) the choice of the foliation (choice of the value of N at $\mathscr{S}_1$ and $\mathscr{S}_2$) and (ii) the choice of the rotation state of each black hole ("individual spin"), as explained in Ref. [47].

Numerical codes for solving the above system have been constructed by

- Grandclément et al. [58] for corotating binary black holes;
- Cook et al. [45, 47] and Pfeiffer et al. [72] for corotating and irrotational binary black holes;
- Ansorg [67, 68] for corotating binary black holes;
- Grandclément [73] for irrotational binary black holes.

Detailed comparisons with post-Newtonian initial data (either from the standard post-Newtonian formalism [74] or from the Effective One-Body approach [75, 76]) have revealed a very good agreement, as shown in Refs. [59, 47]. More recently, XCTS initial data have been computed relaxing the hypothesis $\tilde{\boldsymbol{\gamma}} = \boldsymbol{f}$, using instead for $\tilde{\boldsymbol{\gamma}}$ a superposition of two boosted boosted Kerr–Schild metrics [77, 78].

An alternative to (9.121) for the initial data manifold would be to consider the twice-punctured $\mathbb{R}^3$:

$$\Sigma_0 = \mathbb{R}^3 \backslash \{O_1, O_2\}, \qquad (9.122)$$

where O_1 and O_2 are two points of $\mathbb{R}^3$. This would constitute some extension to the two bodies case of the punctured initial data discussed in Sect. 9.3.3. However, as shown by Hannam et al. [79], it is not possible to find a solution of the helically symmetric XCTS system with a regular lapse in this case.[2] For this reason, initial data based on the puncture manifold (9.122) are computed within the CTT framework discussed in Sect. 9.2. As already mentioned, there is no natural way to implement helical symmetry in this framework. One instead selects the free data $\hat{A}^{ij}_{\mathrm{TT}}$ to vanish identically, as in the single black hole case treated in Sect. 9.2.5 and 9.2.6. Then

$$\hat{A}^{ij} = (\tilde{L}X)^{ij}. \qquad (9.123)$$

The vector $\boldsymbol{X}$ must obey Eq. (9.38), which arises from the momentum constraint. Since this equation is linear, one may choose for $\boldsymbol{X}$ a linear superposition of two Bowen–York solutions (Sect. 9.2.6):

$$\boldsymbol{X} = \boldsymbol{X}_{(\boldsymbol{P}^{(1)}, \boldsymbol{S}^{(1)})} + \boldsymbol{X}_{(\boldsymbol{P}^{(2)}, \boldsymbol{S}^{(2)})}, \qquad (9.124)$$

where $\boldsymbol{X}_{(\boldsymbol{P}^{(a)}, \boldsymbol{S}^{(a)})}$ ($a = 1,2$) is the Bowen–York solution (9.64) centered on O_a. This method has been first implemented by Baumgarte in 2000 [81]. It has been since then used by Baker et al. [82] and Ansorg et al. [83]. The initial data hence obtained are close to helically symmetric XCTS initial data at large separation but deviate significantly from them, as well as from post-Newtonian initial data, when the two black holes are very close. This means that the Bowen–York extrinsic curvature is bad for close binary systems in quasi-equilibrium (see discussion in Ref. [59]).

Remark 9.10 Despite of this, CTT Bowen–York configurations have been often used as initial data for binary black hole inspiral and merger computations by Baker et al. [84–86] and Campanelli et al. [87–90]. Fortunately, these initial data had a relatively large separation, so that they differed only slightly from the helically symmetric XCTS ones.

Instead of choosing somewhat arbitrarily the free data of the CTT and XCTS methods, notably setting $\tilde{\boldsymbol{\gamma}} = \boldsymbol{f}$, one may deduce them from post-Newtonian results. This has been done for the binary black hole problem by Tichy et al. [91], who have

---

2 See however Ref. [80] for some attempt to circumvent this.

used the CTT method with the free data $(\tilde{\gamma}_{ij}, \hat{A}^{ij}_{\rm TT})$ given by the second order post-Newtonian (2PN) metric. This method has been improved by Mundim et al. [92]. In the same spirit, Nissanke [93] has provided 2PN free data for both the CTT and XCTS methods.

When evolved, all the binary black hole initial data discussed above show some small but spurious eccentricity, of the order of 1% [94, 72]. This originates from the circularity assumption (helical symmetry): due to the reaction to gravitational radiation, real orbits should not be exactly circular but slightly inspiraling. Various methods have been proposed to reduce the eccentricity so that the initial data are closer to circular orbits. We shall not discuss them here and refer the reader to [95] and references therein.

9.4.4 Initial Data for Orbiting Binary Neutron Stars

For computing initial data corresponding to orbiting binary neutron stars, one must solve equations for the fluid motion in addition to the Einstein constraints. Basically this amounts to solving $\nabla \cdot \vec{\boldsymbol{T}} = 0$ [Eq. (6.1)] in the context of helical symmetry. One can then show that a first integral of motion exists in two cases: (i) the stars are corotating, i.e. the fluid 4-velocity is colinear to the helical Killing vector (rigid motion), (ii) the stars are irrotational, i.e. the fluid vorticity vanishes. The most straightforward way to get the first integral of motion is by means of the Carter-Lichnerowicz formulation of relativistic hydrodynamics, as shown in Sect. 7 of Ref. [96]. Other derivations have been obtained in 1998 by Teukolsky [97] and Shibata [98].

From the astrophysical point of view, the irrotational motion is much more interesting than the corotating one, because the viscosity of neutron star matter is far too low to ensure the synchronization of the stellar spins with the orbital motion. The irrotational state is a very good approximation for neutron stars that are not millisecond rotators. Indeed, for these stars the spin frequency is much lower than the orbital frequency at the late stages of the inspiral and thus can be neglected.

The first initial data for binary neutron stars on circular orbits have been computed by Baumgarte et al. in 1997 [99, 100] in the corotating case, and by Bonazzola et al. in 1999 [101] in the irrotational one. These results were based on a polytropic equation of state. Since then configurations in the irrotational regime have been obtained

- for a polytropic equation of state [102–108];
- for nuclear matter equations of state issued from recent nuclear physics computations [108–110];
- for strange quark matter [111, 112].

All these computations are based on a flat conformal metric [choice (9.116)], by solving the helically symmetric XCTS system (9.117)–(9.119), supplemented by an elliptic equation for the velocity potential. Configurations based on a non-flat conformal metric have been obtained by Uryu, Limousin, Friedman, Gourgoulhon and Shibata [113, 114]. The conformal metric is then deduced from a waveless approximation developed by Shibata, Uryu and Friedman [115] and which goes beyond the IWM approximation (see also [116]).

9.4.5 Initial Data for Black Hole: Neutron Star Binaries

Let us mention briefly that initial data for a mixed binary system, i.e. a system composed of a black hole and a neutron star, have been obtained by Grandclément [117] and Taniguchi, Baumgarte, Faber and Shapiro [118, 119]. Codes aiming at computing such systems have also been presented by Ansorg [68] and Tsokaros and Uryu [120].

References

1. Fourès-Bruhat, Y., Choquet-Bruhat, Y.: Sur l'Intégration des Équations de la Relativité Générale. J. Rational Mech. Anal. **5**, 951 (1956)
2. Lichnerowicz, A.: L'intégration des équations de la gravitation relativiste et le problème des n corps, J. Math. Pures Appl. **23**, 37 (1944); reprinted in A. Lichnerowicz : Choix d'œuvres mathématiques, Hermann, Paris (1982), p. 4
3. Choquet-Bruhat, Y.: New elliptic system and global solutions for the constraints equations in general relativity. Commun. Math. Phys. **21**, 211 (1971)
4. York, J.W.: Mapping onto Solutions of the Gravitational Initial Value Problem. J. Math. Phys. **13**, 125 (1972)
5. York, J.W.: Conformally invariant orthogonal decomposition of symmetric tensors on Riemannian manifolds and the initial-value problem of general relativity. J. Math. Phys. **14**, 456 (1973)
6. Ó Murchadha, N., York, J.W.: Initial-value problem of general relativity. I. General formulation and physical interpretation. Phys. Rev. D **10**, 428 (1974)
7. York, J.W.: Kinematics and dynamics of general relativity. In: Smarr, L.L. (eds) Sources of Gravitational Radiation. pp. 83. Cambridge University Press, Cambridge (1979)
8. York, J.W.: Conformal "thin-sandwich" data for the initial-value problem of general relativity. Phys. Rev. Lett. **82**, 1350 (1999)
9. Pfeiffer, H.P., York, J.W.: Extrinsic curvature and the Einstein constraints. Phys. Rev. D **67**, 044022 (2003)
10. Bartnik, R.: Quasi-spherical metrics and prescribed scalar curvature. J. Diff. Geom. **37**, 31 (1993)
11. Corvino, J.: Scalar curvature deformation and a gluing construction for the Einstein constraint equations. Commun. Math. Phys. **214**, 137 (2000)
12. Isenberg, J., Mazzeo, R., Pollack, D.: Gluing and wormholes for the Einstein constraint equations. Commun. Math. Phys. **231**, 529 (2002)
13. Chruściel, P.T., Galloway, G.J., Pollack, D.: Mathematical general relativity: A sampler. Bull. Amer. Math. Soc. **47**, 567 (2010)
14. Choquet-Bruhat, Y., York, J.W.: The Cauchy Problem. In: Held, A. (eds) General Relativity and Gravitation, one hundred Years after the Birth of Albert Einstein, Vol. 1, pp. 99. Plenum Press, New York (1980)

15. Cook, G.B.: Initial data for numerical relativity. Living Rev. Relat **3**, 5 (2000); http://www.livingreviews.org/lrr-2000-5
16. Pfeiffer, H.P.: The initial value problem in numerical relativity. In: Proceedings Miami Waves Conference 2004. [preprint gr-qc/0412002].
17. Bartnik, R., Isenberg, J.: The Constraint Equations, in Ref. [121], p. 1.
18. Gourgoulhon, E.: Construction of initial data for 3+1 numerical relativity. In: Proceedings of the VII Mexican School on Gravitation and Mathematical Physics, held in Playa del Carmen, Mexico (Nov. 26 - Dec. 2, 2006), J. Phys.: Conf. Ser. **91**, 012001 (2007).
19. Alcubierre, M.: Introduction to 3+1 Numerical Relativity. Oxford University Press, Oxford (2008)
20. Baumgarte, T.W., Shapiro, S.L.: Numerical relativity. Solving Einstein's Equations on the Computer. Cambridge University Press, Cambridge (2010)
21. Choquet-Bruhat, Y.: General Relativity and Einstein's Equations. Oxford University Press, New York (2009)
22. York, J.W.: Covariant decompositions of symmetric tensors in the theory of gravitation. Ann. Inst. Henri Poincaré A **21**, 319 (1974); available at http://www.numdam.org/item?id=AIHPA_1974__21_4_319_0
23. Lichnerowicz, A.: Sur les équations relativistes de la gravitation, Bulletin de la S.M.F. **80**, 237 (1952); available at http://www.numdam.org/item?id=BSMF_1952__80__237_0
24. Cantor, M.: The existence of non-trivial asymptotically flat initial data for vacuum spacetimes. Commun. Math. Phys. **57**, 83 (1977)
25. Maxwell, D.: Initial data for black holes and rough spacetimes. PhD Thesis, University of Washington (2004)
26. Isenberg, J.: Constant mean curvature solutions of the Einstein constraint equations on closed manifolds. Class. Quantum Grav. **12**, 2249 (1995)
27. Dahl, M., Gicquaud, R., Humbert, E.: A limit equation associated to the solvability of the vacuum Einstein constraint equations using the conformal method. preprint arXiv:1012.2188
28. R. Gicquaud and A. Sakovich : A large class of non constant mean curvature solutions of the Einstein constraint equations on an asymptotically hyperbolic manifold, preprint arXiv:1012.2246
29. Choquet-Bruhat, Y., Christodoulou, D.: Elliptic systems of $H_{s,\delta}$ spaces on manifolds which are Euclidean at infinity. Acta Math. **146**, 129 (1981)
30. Choquet-Bruhat, Y., Isenberg, J., York, J.W.: Einstein constraints on asymptotically Euclidean manifolds. Phys. Rev. D **61**, 084034 (2000)
31. S.M. Carroll : Spacetime and Geometry: An Introduction to General Relativity. Addison Wesley (Pearson Education), San Fransisco (2004) http://preposterousuniverse.com/spacetimeandgeometry/
32. Misner, C.W., Thorne, K.S., Wheeler, J.A.: Gravitation. Freeman, New York (1973)
33. Straumann, N.: General relavity, with applications to astrophysics. Springer, Berlin (2004)
34. Wald, R.M.: General relativity. University of Chicago Press, Chicago (1984)
35. Brandt, S., Brügmann, B.: A simple construction of initial data for multiple black holes. Phys. Rev. Lett. **78**, 3606 (1997)
36. Bowen, J.M., York, J.W.: Time-asymmetric initial data for black holes and black-hole collisions. Phys. Rev. D **21**, 2047 (1980)
37. Beig, R., Krammer, W.: Bowen–York tensors. Class. Quantum Grav. **21**, S73 (2004)
38. Garat, A., Price, R.H.: Nonexistence of conformally flat slices of the Kerr spacetime. Phys. Rev. D **61**, 124011 (2000)
39. Valiente Kroon, J.A.: Nonexistence of conformally flat slices in Kerr and other stationary spacetimes. Phys. Rev. Lett. **92**, 041101 (2004)
40. Brandt, S.R., Seidel, E.: Evolution of distorted rotating black holes. II. Dynamics and analysis. Phys. Rev. D **52**, 870 (1995)
41. Gleiser, R.J., Nicasio, C.O., Price, R.H., Pullin, J.: Evolving the Bowen–York initial data for spinning black holes. Phys. Rev. D **57**, 3401 (1998)

42. Baierlein, R.F., Sharp, D.H., Wheeler, J.A.: Three-dimensional geometry as carrier of information about time. Phys. Rev. **126**, 1864 (1962)
43. Wheeler, J.A.: Geometrodynamics and the issue of the final state. In: DeWitt, C., DeWitt, B.S. (eds) Relativity, Groups and Topology., pp. 316. Gordon and Breach, New York (1964)
44. Bartnik, R., Fodor, G.: On the restricted validity of the thin sandwich conjecture. Phys. Rev. D **48**, 3596 (1993)
45. Cook, G.B., Pfeiffer, H.P.: Excision boundary conditions for black-hole initial data. Phys. Rev. D **70**, 104016 (2004)
46. Jaramillo, J.L., Gourgoulhon, E., Mena Marugán, G.A.: Inner boundary conditions for black hole initial data derived from isolated horizons. Phys. Rev. D **70**, 124036 (2004)
47. Caudill, M., Cook, G.B., Grigsby, J.D., Pfeiffer, H.P.: Circular orbits and spin in black-hole initial data. Phys. Rev. D **74**, 064011 (2006)
48. Gourgoulhon, E., Jaramillo, J.L.: A 3+1 perspective on null hypersurfaces and isolated horizons. Phys. Rep. **423**, 159 (2006)
49. Matera, K., Baumgarte, T.W., Gourgoulhon, E.: Shells around black holes: the effect of freely specifiable quantities in Einstein's constraint equations. Phys. Rev. D **77**, 024049 (2008)
50. Vasset, N., Novak, J., Jaramillo, J.L.: Excised black hole spacetimes: Quasilocal horizon formalism applied to the Kerr example. Phys. Rev. D **79**, 124010 (2009)
51. Pfeiffer, H.P., York, J.W.: Uniqueness and Nonuniqueness in the Einstein Constraints. Phys. Rev. Lett. **95**, 091101 (2005)
52. Teukolsky, S.A.: Linearized quadrupole waves in general relativity and the motion of test particles. Phys. Rev. D **26**, 745 (1982)
53. Baumgarte, T.W., Ó Murchadha, N., Pfeiffer, H.P.: Einstein constraints: Uniqueness and non-uniqueness in the conformal thin sandwich approach. Phys. Rev. D **75**, 044009 (2007)
54. Walsh, D.: Non-uniqueness in conformal formulations of the Einstein Constraints. Class. Quantum Grav. **24**, 1911 (2007)
55. Cordero-Carrión, I., Cerdá-Durán, P., Dimmelmeier, H., Jaramillo, J.L., Novak, J., Gourgoulhon, E.: Improved constrained scheme for the Einstein equations: An approach to the uniqueness issue. Phys. Rev. D **79**, 024017 (2009)
56. Shibata, M., Uryu, K.: Merger of black hole-neutron star binaries: Nonspinning black hole case. Phys. Rev. D **74**, 121503(R) (2006)
57. York, J.W.: Velocities and momenta in an extended elliptic form of the initial value conditions. Nuovo Cim. B **119**, 823 (2004)
58. Grandclément, P., Gourgoulhon, E., Bonazzola, S.: Binary black holes in circular orbits. II. Numerical methods and first results. Phys. Rev. D **65**, 044021 (2002)
59. Damour, T., Gourgoulhon, E., Grandclément, P.: Circular orbits of corotating binary black holes: comparison between analytical and numerical results. Phys. Rev. D **66**, 024007 (2002)
60. Laguna, P.: Conformal-thin-sandwich initial data for a single boosted or spinning black hole puncture. Phys. Rev. D **69**, 104020 (2004)
61. Blanchet, L.: Gravitational radiation from post-newtonian sources and inspiralling compact binaries. Living Rev. Relat. **9**, 4 (2006); http://www.livingreviews.org/lrr-2006-4
62. Friedman, J.L., Uryu, K., Shibata, M.: Thermodynamics of binary black holes and neutron stars. Phys. Rev. D **65**, 064035 (2002); erratum in Phys. Rev. D **70**, 129904(E) (2004).
63. Detweiler, S.: Periodic solutions of the Einstein equations for binary systems. Phys. Rev. D **50**, 4929 (1994)
64. Gibbons, G.W., Stewart, J.M.: Absence of asymptotically flat solutions of Einstein's equations which are periodic and empty near infinity. In: Bonnor, W.B., Islam, J.N., MacCallum, M.A.H. (eds) Classical General Relativity., pp. 77. Cambridge University Press, Cambridge (1983)
65. Klein, C.: Binary black hole spacetimes with a helical Killing vector. Phys. Rev. D **70**, 124026 (2004)
66. Gourgoulhon, E., Grandclément, P., Bonazzola, S.: Binary black holes in circular orbits. I. A global spacetime approach. Phys. Rev. D **65**, 044020 (2002)

67. Ansorg, M.: Double-domain spectral method for black hole excision data. Phys. Rev. D **72**, 024018 (2005)
68. Ansorg, M.: Multi-Domain spectral method for initial data of arbitrary binaries in general relativity. Class. Quantum Grav. **24**, S1 (2007)
69. Dain, S.: Trapped surfaces as boundaries for the constraint equations. Class. Quantum Grav. **21**, 555 (2004); errata in Class. Quantum Grav. **22**, 769 (2005)
70. Dain, S., Jaramillo, J.L., Krishnan, B.: On the existence of initial data containing isolated black holes. Phys.Rev. D **71**, 064003 (2005)
71. Jaramillo, J.L., Ansorg, M., Limousin, F.: Numerical implementation of isolated horizon boundary conditions. Phys. Rev. D **75**, 024019 (2007)
72. Pfeiffer, H.P., Brown, D.A., Kidder, L.E., Lindblom, L., Lovelace, G., Scheel, M.A.: Reducing orbital eccentricity in binary black hole simulations. Class. Quantum Grav. **24**, S59 (2007)
73. Grandclément, P. : KADATH: a spectral solver for theoretical physics. J. Comput. Phys. **229**, 3334 (2010)
74. Blanchet, L.: Innermost circular orbit of binary black holes at the third post-Newtonian approximation. Phys. Rev. D **65**, 124009 (2002)
75. Buonanno, A., Damour, T.: Effective one-body approach to general relativistic two-body dynamics. Phys. Rev. D **59**, 084006 (1999)
76. Damour, T.: Coalescence of two spinning black holes: An effective one-body approach. Phys. Rev. D **64**, 124013 (2001)
77. Lovelace, G.: Reducing spurious gravitational radiation in binary-black-hole simulations by using conformally curved initial data. Class. Quantum Grav. **26**, 114002 (2009)
78. Lovelace, G., Owen, R., Pfeiffer, H.P., Chu, T.: Binary-black-hole initial data with nearly extremal spins. Phys. Rev. D **78**, 084017 (2008)
79. Hannam, M.D., Evans, C.R., Cook, G.B., Baumgarte, T.W.: Can a combination of the conformal thin-sandwich and puncture methods yield binary black hole solutions in quasiequilibrium?. Phys. Rev. D **68**, 064003 (2003)
80. Hannam, M.D.: Quasicircular orbits of conformal thin-sandwich puncture binary black holes. Phys. Rev. D **72**, 044025 (2005)
81. Baumgarte, T.W.: Innermost stable circular orbit of binary black holes. Phys. Rev. D **62**, 024018 (2000)
82. Baker, J.G., Campanelli, M., Lousto, C.O., Takahashi, R.: Modeling gravitational radiation from coalescing binary black holes. Phys. Rev. D **65**, 124012 (2002)
83. Ansorg, M., Brügmann, B., Tichy, W.: Single-domain spectral method for black hole puncture data. Phys. Rev. D **70**, 064011 (2004)
84. Baker, J.G., Centrella, J., Choi, D.-I., Koppitz, M., van Meter, J.: Gravitational-Wave extraction from an inspiraling configuration of merging black holes. Phys. Rev. Lett. **96**, 111102 (2006)
85. Baker, J.G., Centrella, J., Choi, D.-I., Koppitz, M., van Meter, J.: Binary black hole merger dynamics and waveforms. Phys. Rev. D **73**, 104002 (2006)
86. van Meter, J.R., Baker, J.G., Koppitz, M., Choi, D.I.: How to move a black hole without excision: gauge conditions for the numerical evolution of a moving puncture. Phys. Rev. D **73**, 124011 (2006)
87. Campanelli, M., Lousto, C.O., Marronetti, P., Zlochower, Y.: Accurate evolutions of orbiting black-hole binaries without excision. Phys. Rev. Lett. **96**, 111101 (2006)
88. Campanelli, M., Lousto, C.O., Zlochower, Y.: Last orbit of binary black holes. Phys. Rev. D **73**, 061501(R) (2006)
89. Campanelli, M., Lousto, C.O., Zlochower, Y.: Spinning-black-hole binaries: The orbital hang-up. Phys. Rev. D **74**, 041501(R) (2006)
90. Campanelli, M., Lousto, C.O., Zlochower, Y.: Spin-orbit interactions in black-hole binaries. Phys. Rev. D **74**, 084023 (2006)
91. Tichy, W., Brügmann, B., Campanelli, M., Diener, P.: Binary black hole initial data for numerical general relativity based on post-Newtonian data. Phys. Rev. D **67**, 064008 (2003)

92. Mundim, B.C., Kelly, B.J., Zlochower, Y., Nakano, H., Campanelli, M.: Hybrid black-hole binary initial data. Class. Quantum Grav. **28**, 134003 (2011)
93. Nissanke, S.: Post-Newtonian freely specifiable initial data for binary black holes in numerical relativity. Phys. Rev. D **73**, 124002 (2006)
94. Buonanno, A., Cook, G.B., Pretorius, F.: Inspiral, merger, and ring-down of equal-mass black-hole binaries. Phys. Rev. D **75**, 124018 (2007)
95. Buonanno, A., Kidder, L.E., Mroué, A.H., Pfeiffer, H.P., Taracchini, A.: Reducing orbital eccentricity of precessing black-hole binaries. Phys. Rev. D **83**, 104034 (2011)
96. Gourgoulhon, E.: An introduction to relativistic hydrodynamics, in Stellar Fluid Dynamics and Numerical Simulations: From the Sun to Neutron Stars. edited by M. Rieutord & B. Dubrulle, EAS Publications Series **21**, EDP Sciences, Les Ulis (2006), p. 43; available at http://arxiv.org/abs/gr-qc/0603009
97. Teukolsky, S.A.: Irrotational binary neutron stars in quasi-equilibrium in general relativity. Astrophys. J. **504**, 442 (1998)
98. Shibata, M.: Relativistic formalism for computation of irrotational binary stars in quasiequilibrium states. Phys. Rev. D **58**, 024012 (1998)
99. Baumgarte, T.W., Cook, G.B., Scheel, M.A., Shapiro, S.L., Teukolsky, S.A.: Binary neutron stars in general relativity: Quasiequilibrium models. Phys. Rev. Lett. **79**, 1182 (1997)
100. Baumgarte, T.W., Cook, G.B., Scheel, M.A., Shapiro, S.L., Teukolsky, S.A.: General relativistic models of binary neutron stars in quasiequilibrium. Phys. Rev. D **57**, 7299 (1998)
101. Bonazzola, S., Gourgoulhon, E., Marck, J.-A.: Numerical models of irrotational binary neutron stars in general relativity. Phys. Rev. Lett. **82**, 892 (1999)
102. Marronetti, P., Mathews, G.J., Wilson, J.R.: Irrotational binary neutron stars in quasiequilibrium. Phys. Rev. D **60**, 087301 (1999)
103. Uryu, K., Eriguchi, Y.: New numerical method for constructing quasiequilibrium sequences of irrotational binary neutron stars in general relativity. Phys. Rev. D **61**, 124023 (2000)
104. Uryu, K., Shibata, M., Eriguchi, Y.: Properties of general relativistic, irrotational binary neutron stars in close quasiequilibrium orbits: Polytropic equations of state. Phys. Rev. D **62**, 104015 (2000)
105. Gourgoulhon, E., Grandclément, P., Taniguchi, K., Marck, J.-A., Bonazzola, S.: Quasiequilibrium sequences of synchronized and irrotational binary neutron stars in general relativity: Method and tests. Phys. Rev. D **63**, 064029 (2001)
106. Taniguchi, K., Gourgoulhon, E.: Quasiequilibrium sequences of synchronized and irrotational binary neutron stars in general relativity. III. Identical and different mass stars with $\gamma = 2$. Phys. Rev. D **66**, 104019 (2002)
107. Taniguchi, K., Gourgoulhon, E.: Various features of quasiequilibrium sequences of binary neutron stars in general relativity. Phys. Rev. D **68**, 124025 (2003)
108. Taniguchi, K., Shibata, M.: Binary neutron stars in quasi-equilibrium. Astrophys. J. Suppl. Ser. **188**, 187 (2010)
109. Bejger, M., Gondek-Rosińska, D., Gourgoulhon, E., Haensel, P., Taniguchi, K., Zdunik, J.L.: Impact of the nuclear equation of state on the last orbits of binary neutron stars. Astron. Astrophys. **431**, 297–306 (2005)
110. Oechslin, R., Janka, H.-T., Marek, A.: Relativistic neutron star merger simulations with non-zero temperature equations of state I. Variation of binary parameters and equation of state. Astron. Astrophys. **467**, 395 (2007)
111. Oechslin, R., Uryu, K., Poghosyan, G., Thielemann, F.K.: The influence of quark matter at high densities on binary neutron star mergers. Mon. Not. Roy. Astron. Soc. **349**, 1469 (2004)
112. Limousin, F., Gondek-Rosińska, D., Gourgoulhon, E.: Last orbits of binary strange quark stars . Phys Rev. D **71**, 064012 (2005)
113. Uryu, K., Limousin, F., Friedman, J.L., Gourgoulhon, E., Shibata, M.: Binary neutron stars: Equilibrium models beyond spatial conformal flatness. Phys. Rev. Lett. **97**, 171101 (2006)
114. Uryu, K., Limousin, F., Friedman, J.L., Gourgoulhon, E., Shibata, M.: Nonconformally flat initial data for binary compact objects. Phys. Rev. D **80**, 124004 (2009)

115. Shibata, M., Uryu, K., Friedman, J.L.: Deriving formulations for numerical computation of binary neutron stars in quasicircular orbits. Phys. Rev. D **70**, 044044 (2004); errata in Phys. Rev. D **70**, 129901(E) (2004)
116. Damour, T., Nagar, A.: Effective one body description of tidal effects in inspiralling compact binaries. Phys. Rev. D **81**, 084016 (2010)
117. Grandclément, P.: Accurate and realistic initial data for black hole-neutron star binaries. Phys. Rev. D **74**, 124002 (2006); erratum in Phys. Rev. D **75**, 129903(E) (2007).
118. Taniguchi, K., Baumgarte, T.W., Faber, J.A., Shapiro, S.L.: Quasiequilibrium sequences of black-hole-neutron-star binaries in general relativity. Phys. Rev. D **74**, 041502(R) (2006)
119. Taniguchi, K., Baumgarte, T.W., Faber, J.A., Shapiro, S.L.: Quasiequilibrium black hole-neutron star binaries in general relativity. Phys. Rev. D **75**, 084005 (2007)
120. Tsokaros, A.A., Uryu, K.: Numerical method for binary black hole/neutron star initial data: Code test. Phys. Rev. D **75**, 044026 (2007)
121. Chruściel, P.T., Friedrich, H. (eds): The Einstein equations and the large scale behavior of gravitational fields—50 years of the Cauchy problem in general relativity. Birkhäuser Verlag, Basel (2004)

Chapter 10
Choice of Foliation and Spatial Coordinates

Abstract We discuss here the choice of spacetime coordinates, from a 3 + 1 point of view. This amounts to discuss first the choice of foliation, via the lapse function: the geodesic, maximal, harmonic, and 1+log slicings are presented here. In a second stage, we focus on of the propagation of the spatial coordinates from slice to slice, via the choice of the shift vector. We introduce the concepts of normal coordinates, minimal distortion and variants of it, Gamma freezing coordinates and Gamma drivers. Finally we discuss choices that fix fully the spatial coordinates on a given slice: spatial harmonic coordinates and Dirac gauge.

10.1 Introduction

Having investigated the initial data problem in the preceding chapter, the next logical step is to discuss the evolution problem, i.e. the development $(\Sigma_t, \boldsymbol{\gamma})$ of initial data $(\Sigma_0, \boldsymbol{\gamma}, \boldsymbol{K})$. This constitutes the integration of the Cauchy problem introduced in Sect. 5.4. As discussed in Sect. 5.4.1, a key feature of this problem is the freedom of choice for the lapse function N and the shift vector $\boldsymbol{\beta}$, reflecting respectively the choice of foliation $(\Sigma_t)_{t\in\mathbb{R}}$ and the choice of coordinates (x^i) on each leaf Σ_t of the foliation. These choices are crucial because they determine the specific form of the 3+1 Einstein system (5.68)–(5.71) that one has actually to deal with. In particular, depending of the choice of $(N, \boldsymbol{\beta})$, this system can be made more hyperbolic or more elliptic.

Extensive discussions about the various possible choices of foliations and spatial coordinates can be found in the seminal articles by Smarr and York [1, 2] as well as in the review articles by Alcubierre [3], Baumgarte and Shapiro [4], and Lehner [5], or the textbooks by Alcubierre [6], Baumgarte and Shapiro [7] and Choquet–Bruhat [8].

É. Gourgoulhon, *3+1 Formalism in General Relativity*, Lecture Notes in Physics 846, DOI: 10.1007/978-3-642-24525-1_10, © Springer-Verlag Berlin Heidelberg 2012

10.2 Choice of Foliation

10.2.1 Geodesic Slicing

The simplest choice of foliation one might think about is the ***geodesic slicing***, for it corresponds to a unit lapse:

$$\boxed{N = 1}. \tag{10.1}$$

Since the 4-acceleration $\boldsymbol{a}$ of the Eulerian observers is nothing but the spatial gradient of In N (logarithm of N) [cf. Eq. (4.19)], the choice (10.1) implies $\boldsymbol{a} = 0$, i.e. the worldlines of the Eulerian observers are geodesics, hence the name *geodesic slicing*. Moreover the choice (10.1) implies that the proper time along these worldlines coincides with the coordinate time t.

We have already used the geodesic slicing to discuss the basics feature of the 3+1 Einstein system in Sect. 5.4.2. We have also argued there that, due to the tendency of timelike geodesics without vorticity (as the worldlines of the Eulerian observers are) to focus and eventually cross, this type of foliation can become pathological within a finite range of t.

Example 10.1 A simple example of geodesic slicing is provided by the Painlevé–Gullstrand slicing of Schwarzschild spacetime discussed in Example 8.2. Indeed we have seen that for this slicing, $N = 1$ [Eq. (8.22a)].

Example 10.2 Another example of geodesic slicing, still in Schwarzschild spacetime, is provided by the time development with $N = 1$ of the initial data constructed in Sects. 9.2.5 and 9.3.3, namely the momentarily static slice $t_S = 0$ of Schwarzschild spacetime (t_S standing for the standard Schwarzschild time coordinate), with topology $\mathbb{R} \times \mathbb{S}^2$ (Einstein–Rosen bridge, cf. Sect. 9.2.5). The resulting foliation is depicted in Fig. 10.1. It hits the singularity at $t = \pi m$, reflecting the bad behavior of geodesic slicing.

In numerical relativity, geodesic slicings have been used by Nakamura, Oohara and Kojima to perform in 1987 the first 3D evolutions of vacuum spacetimes with gravitational waves [9]. However, as discussed in Ref. [10], the evolution was possible only for a pretty limited range of t, because of the focusing property mentioned above.

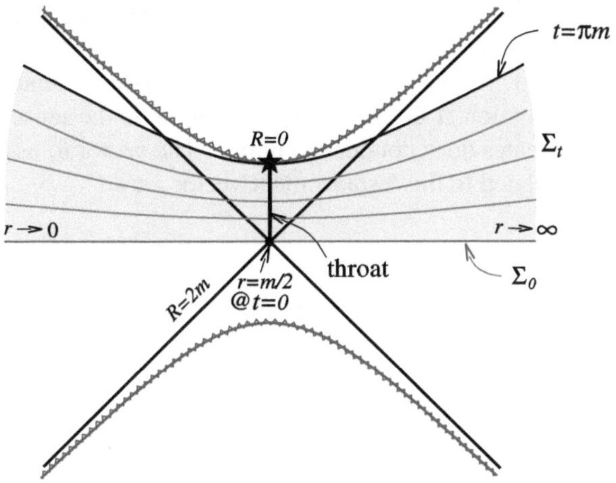

Fig. 10.1 Geodesic-slicing evolution from the initial slice $t = t_S = 0$ of Schwarzschild spacetime depicted in a Kruskal–Szekeres diagram. R stands for Schwarzschild radial coordinate (areal radius), so that $R = 0$ is the singularity and $R = 2m$ is the event horizon (figure adapted from Fig. 2a of [1])

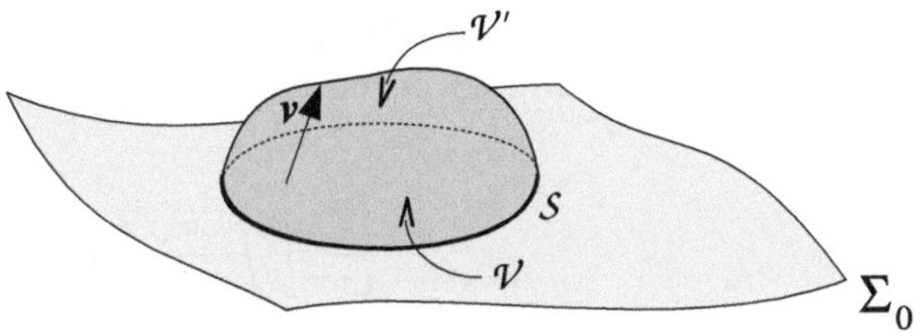

Fig. 10.2 Deformation of a volume $\mathscr{V}$ delimited by the surface $\mathscr{S}$ in the hypersurface Σ_0

10.2.2 Maximal Slicing

A very famous type of foliation is maximal slicing, already encountered in Sect. 7.6 and in Chap. 9, where it plays a great role in decoupling the constraint equations. The ***maximal slicing*** corresponds to the vanishing of the mean curvature of the hypersurfaces Σ_t:

$$\boxed{K = 0}. \tag{10.2}$$

The fact that this condition leads to hypersurfaces of *maximal volume* can be seen as follows. Consider some hypersurface Σ_0 and a closed two-dimensional surface $\mathscr{S}$ lying in Σ_0 (cf. Fig. 10.2). The volume of the domain $\mathscr{V}$ enclosed in $\mathscr{S}$ is

$$V = \int_{\mathscr{V}} \sqrt{\gamma}\, \mathrm{d}^3x, \tag{10.3}$$

where $\gamma = \det \gamma_{ij}$ is the determinant of the metric $\boldsymbol{\gamma}$ with respect to some coordinates (x^i) used in Σ_t. Let us consider a small deformation $\mathscr{V}'$ of $\mathscr{V}$ that keeps the boundary $\mathscr{S}$ fixed. $\mathscr{V}'$ is generated by a small displacement along a vector field $\boldsymbol{v}$ of every point of $\mathscr{V}$, such that $\boldsymbol{v}|_{\mathscr{S}} = 0$. Without any loss of generality, we may consider that $\mathscr{V}'$ lies in a hypersurface $\Sigma_{\partial t}$ that is a member of some "foliation" $(\Sigma_t)_{t\in\mathbb{R}}$ such that

$\Sigma_{t=0} = \Sigma_0$. The hypersurfaces Σ_t intersect each other at $\mathscr{S}$, which violates condition (4.2) in the definition of a foliation given in Sect. 4.2.2, hence the quotes around the word "foliation". Let us consider a 3+1 coordinate system (t, x^i) associated with the "foliation" $(\Sigma_t)_{t\in\mathbb{R}}$ and adapted to $\mathscr{S}$ in the sense that the position of $\mathscr{S}$ in these coordinates does not depend upon t. The vector $\boldsymbol{\partial}_t$ associated to these coordinates is then related to the displacement vector $\boldsymbol{v}$ by

$$\boldsymbol{v} = \delta t \boldsymbol{\partial}_t. \tag{10.4}$$

Introducing the lapse function N and shift vector $\boldsymbol{\beta}$ associated with the coordinates (t, x^i), the above relation becomes [cf. Eq. (5.29)] $\boldsymbol{v} = \delta t(N\boldsymbol{n} + \boldsymbol{\beta})$. Accordingly, the condition $\boldsymbol{v}|_{\mathscr{S}} = 0$ implies

$$N|_{\mathscr{S}} = 0 \quad \text{and} \quad \boldsymbol{\beta}|_{\mathscr{S}} = 0. \tag{10.5}$$

Let us define $V(t)$ as the volume of the domain $\mathscr{V}_t$ delimited by $\mathscr{S}$ in Σ_t. It is given by a formula identical to Eq. (10.3), except of course that the integration domain has to be replaced by $\mathscr{V}_t$. Moreover, the domain $\mathscr{V}_t$ lying at fixed values of the coordinates (x^i), we have

$$\frac{\mathrm{d}V}{\mathrm{d}t} = \int_{\mathscr{V}_t} \frac{\partial\sqrt{\gamma}}{\partial t}\, \mathrm{d}^3x.$$

Replacing the integrand by means of formula (6.66), we get

$$\frac{\mathrm{d}V}{\mathrm{d}t} = \int_{\mathscr{V}_t} \left[-NK + D_i\beta^i\right]\sqrt{\gamma}\, \mathrm{d}^3x. \tag{10.6}$$

Now from the Gauss–Ostrogradsky theorem,

$$\int_{\mathscr{V}_t} D_i\beta^i\sqrt{\gamma}\, \mathrm{d}^3x = \oint_{\mathscr{S}} \beta^i s_i\sqrt{q}\, \mathrm{d}^2y,$$

where $\boldsymbol{s}$ is the unit normal to $\mathscr{S}$ lying in Σ_t, $\boldsymbol{q}$ is the induced metric on $\mathscr{S}$, (y^a) are coordinates on $\mathscr{S}$ and $q = \det q_{ab}$. Since $\boldsymbol{\beta}$ vanishes on $\mathscr{S}$ [property (10.5)], the above integral is identically zero and Eq. (10.6) reduces to

$$\boxed{\frac{\mathrm{d}V}{\mathrm{d}t} = -\int_{\mathscr{V}_t} NK\sqrt{\gamma}\, \mathrm{d}^3x}\,. \tag{10.7}$$

We conclude that if $K = 0$ on Σ_0, the volume V enclosed in $\mathscr{S}$ is extremal with respect to variations of the domain delimited by $\mathscr{S}$, provided that the boundary of the domain remains $\mathscr{S}$. In the Euclidean space, such an extremum would define a *minimal surface*, the corresponding variation problem being a ***Plateau problem*** [named after the Belgian physicist Joseph Plateau (1801–1883)]: given a closed contour $\mathscr{S}$ (wire loop), find the surface $\mathscr{V}$ (soap film) of minimal area (minimal surface tension energy) bounded by $\mathscr{S}$. However, in the present case of a Lorentzian metric, it can be shown that the extremum is actually a maximum, hence the name *maximal slicing*. For the same reason, a timelike geodesic between two points in spacetime is the curve of *maximum* length joining these two points.

Demanding that the maximal slicing condition (10.2) holds for all hypersurfaces Σ_t, once combined with the evolution equation (7.74) for K, yields the following elliptic equation for the lapse function:

$$\boxed{D_i D^i N = N\left[4\pi(E+S) + K_{ij}K^{ij}\right]}. \tag{10.8}$$

Remark 10.1 We have already noticed that at the Newtonian limit, Eq. (10.8) reduces to the Poisson equation for the gravitational potential Φ (cf. Sect. 7.5.1). Therefore the maximal slicing can be considered as a natural generalization to the relativistic case of the canonical slicing of Newtonian spacetime by hypersurfaces of constant absolute time. In this respect, let us notice that the "beyond Newtonian" approximation of general relativity constituted by the Isenberg–Wilson–Mathews approach discussed in Sect. 7.6 is also based on maximal slicing.

Example 10.3 In Schwarzschild spacetime, the standard Schwarzschild time coordinate t defines maximal hypersurfaces Σ_t, which are spacelike for $R > 2m$ (R being Schwarzschild radial coordinate). Indeed these hypersurfaces are totally geodesic: $\boldsymbol{K} = 0$ (cf. Sect. 3.4.3), so that, in particular, $K = \mathrm{tr}_\gamma \boldsymbol{K} = 0$. This maximal slicing is shown in Fig. 10.3. The corresponding lapse function expressed in terms of the isotropic radial coordinate r is

$$N = \left(1 - \frac{m}{2r}\right)\left(1 + \frac{m}{2r}\right)^{-1}. \tag{10.9}$$

As shown in Sect. 9.3.3, the above expression can be derived by means of the XCTS formalism. Notice that the foliation $(\Sigma_t)_{t\in\mathbb{R}}$ does not penetrate under the event horizon ($R = 2m$) and that the lapse is negative for $r < m/2$ (cf. discussion in Sect. 9.3.3 about negative lapse values).

Besides its nice geometrical definition, an interesting property of maximal slicing is ***singularity avoidance***. This is related to the fact that the set of Eulerian observers of a maximal foliation define an *incompressible flow*: indeed, thanks to Eq. (3.66), the condition $K = 0$ is equivalent to the incompressibility condition

$$\boldsymbol{\nabla} \cdot \boldsymbol{n} = 0 \tag{10.10}$$

for the 4-velocity field $\boldsymbol{n}$ of the Eulerian observers. If we compare with the Eulerian observers of geodesic slicings (Sect. 10.2.1), who have the tendency to squeeze, we may say that maximal-slicing Eulerian observers do not converge because they are

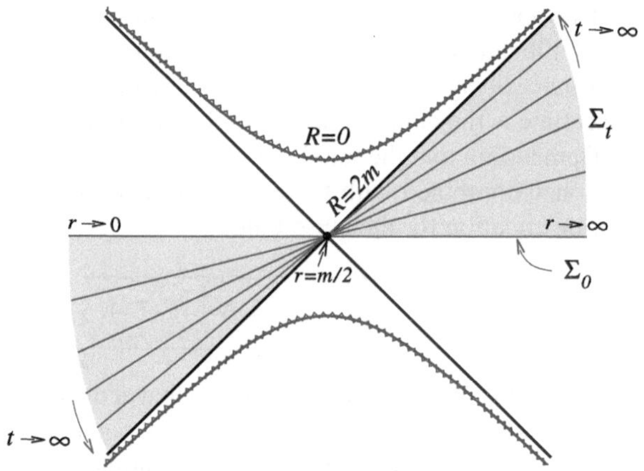

Fig. 10.3 Kruskal–Szekeres diagram showing the maximal slicing of Schwarzschild spacetime defined by the standard Schwarzschild time coordinate t. As for Fig. 10.1, R stands for Schwarzschild radial coordinate (areal radius), so that $R = 0$ is the singularity and $R = 2m$ is the event horizon, whereas r stands for the isotropic radial coordinate [cf. Eq. (9.108)]

accelerating ($\boldsymbol{D}N \neq 0$) in order to balance the focusing effect of gravity. Loosely speaking, the incompressibility prevents the Eulerian observers from converging towards the central singularity if the latter forms during the time evolution. This is illustrated by the following example in Schwarzschild spacetime. A more extended discussion of the singularity avoidance feature of maximal slicing can be found in Baumgarte and Shapiro's textbook [7].

Example 10.4 Let us consider the time development of the initial data constructed in Sects. 9.2.5 and 9.3.3, namely the momentarily static slice $t_S = 0$ of Schwarzschild spacetime (with the Einstein–Rosen bridge). A first maximal slicing development of these initial data is that based on Schwarzschild time coordinate t_S and discussed in Example 10.3. The corresponding lapse function is given by Eq. (10.9) and is *antisymmetric* about the minimal surface $r = m/2$ (throat) (cf. Fig. 10.3). There exists a second maximal-slicing development of the same initial data but with a lapse which is *symmetric* about the throat. It has been found in 1973 by Estabrook et al. [11], as well as Reinhart [12]. The corresponding time coordinate t is different from Schwarzschild time coordinate t_S, except for $t = 0$ (initial slice $t_S = 0$). In the coordinates $(x^\alpha) = (t, R, \theta, \varphi)$, where R is Schwarzschild radial coordinate, the metric components obtained by Estabrook et al. [11] (see also Refs. [13–15]) take the form

$$g_{\mu\nu}\mathrm{d}x^\mu \mathrm{d}x^\nu = -N^2\mathrm{d}t^2 + \left(1 - \frac{2m}{R} + \frac{C(t)^2}{R^4}\right)^{-1}\left(\mathrm{d}R + \frac{C(t)}{R^2}N\mathrm{d}t\right)^2 + R^2(\mathrm{d}\theta^2 + \sin^2\theta\mathrm{d}\varphi^2), \tag{10.11}$$

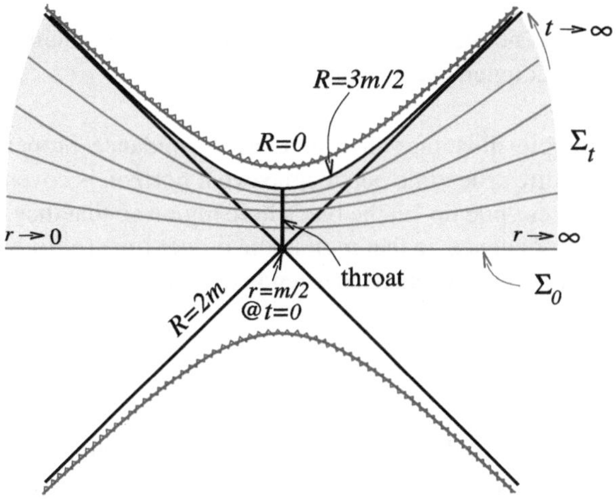

Fig. 10.4 Kruskal–Szekeres diagram depicting the maximal slicing of Schwarzschild spacetime defined by the Reinhart/Estabrook et al. time function t [cf. Eq. (10.11)]. As for Fig. 10.1 and Fig. 10.3, R stands for Schwarzschild radial coordinate (areal radius), so that $R = 0$ is the singularity and $R = 2m$ is the event horizon, whereas r stands for the isotropic radial coordinate. At the throat (minimal surface), $R = R_C$ where R_C is the function of t defined below Eq. (10.13) (figure adapted from Fig. 1 of Ref. [11])

where

$$N = N(R,t) = \sqrt{1 - \frac{2m}{R} + \frac{C(t)^2}{R^4}} \left\{ 1 + \frac{\mathrm{d}C}{\mathrm{d}t} \int_R^{+\infty} \frac{x^4 \, \mathrm{d}x}{\left[x^4 - 2mx^3 + C(t)^2\right]^{3/2}} \right\}, \tag{10.12}$$

and $C(t)$ is the function of t defined implicitly by

$$t = -C \int_{R_C}^{+\infty} \frac{\mathrm{d}x}{(1 - 2m/x)\sqrt{x^4 - 2mx^3 + C^2}}, \tag{10.13}$$

R_C being the unique root of the polynomial $P_C(x) := x^4 - 2mx^3 + C^2$ in the interval $(3m/2, 2m]$. $C(t)$ varies from 0 at $t = 0$ to $C_\infty := (3\sqrt{3}/4)m^2$ as $t \to +\infty$. Accordingly, R_C decays from $2m$ ($t = 0$) to $3m/2$ ($t \to +\infty$). Actually, for $C = C(t)$, R_C represents the smallest value of the radial coordinate R in the slice Σ_t. This maximal slicing of Schwarzschild spacetime is represented in Fig. 10.4. We notice that, as $t \to +\infty$, the slices Σ_t accumulate on a limiting hypersurface: the hypersurface $R = 3m/2$ (let us recall that for $R < 2m$, the hypersurfaces $R =$ const are spacelike and are thus eligible for a 3+1 foliation). Actually, it can be seen that the hypersurface $R = 3m/2$ is the only hypersurface $R =$ const which is spacelike and maximal [14]. If we compare with Fig. 10.1, we notice that, contrary to the geodesic slicing, the present foliation never encounters the singularity.

The above example illustrates the singularity-avoidance property of maximal slicing: while the entire spacetime outside the event horizon is covered by the foliation, the hypersurfaces "pile up" in the black hole region so that they never reach the singularity. As a consequence, in that region, the proper time (of Eulerian observers) between two neighbouring hypersurfaces tends to zero as t increases. According to Eq. (4.17), this implies

$$N \to 0 \quad \text{as} \quad t \to +\infty. \tag{10.14}$$

This "phenomenon" is called ***collapse of the lapse***. Beyond the Schwarzschild case discussed above, the collapse of the lapse is a generic feature of maximal slicing of spacetimes describing black hole formation via gravitational collapse. For instance, it occurs in the analytic solution obtained by Petrich et al. [16] for the maximal slicing of the Oppenheimer-Snyder spacetime (gravitational collapse of a spherically symmetric homogeneous ball of pressureless matter).

In numerical relativity, maximal slicing has been used in the computation of the (axisymmetric) head-on collision of two black holes by Smarr et al. in the seventies [17, 18], as well as in computations of axisymmetric gravitational collapse by Nakamura and Sato (1981) [19, 20], Stark and Piran (1985) [21] and Evans (1986) [22]. Actually Stark and Piran used a mixed type of foliation introduced by Bardeen and Piran [23]: maximal slicing near the origin ($r = 0$) and polar slicing far from it. The ***polar slicing*** is defined in spherical-type coordinates $(x^i) = (r, \theta, \varphi)$ by

$$K^\theta{}_\theta + K^\varphi{}_\varphi = 0, \tag{10.15}$$

instead of $K^r{}_r + K^\theta{}_\theta + K^\varphi{}_\varphi = 0$ for maximal slicing.

Whereas maximal slicing is a nice choice of foliation, with a clear geometrical meaning, a natural Newtonian limit and a singularity-avoidance feature, it has not been much used in 3D (no spatial symmetry) numerical relativity. The reason is a technical one: imposing maximal slicing requires to solve the elliptic equation (10.8) for the lapse and elliptic equations are usually CPU-time consuming, except if one make uses of fast elliptic solvers [24, 25]. For this reason, most of the recent computations of binary black hole inspiral and merger have been performed with the 1+log slicing, to be discussed in Sect. 10.2.4. Nevertheless, it is worth to note that maximal slicing has been used for the first grazing collisions of binary black holes, computed by Brügmann [26].

To avoid solving an elliptic equation while preserving most of the good properties of maximal slicing, an ***approximate maximal slicing*** has been introduced by Shibata [27]. It consists in transforming Eq. (10.8) into a *parabolic* equation by adding a term of the type $\partial N/\partial\lambda$ in the right-hand side and to compute the "λ-evolution" for some range of the parameter λ. This amounts to resolve a heat like equation. Generically the solution converges towards a stationary one, so that $\partial N/\partial\lambda \to 0$ and the original elliptic equation (10.8) is solved. The approximate maximal slicing has been used by Shibata et al. to compute the merger of binary neutron stars [28–33], as well as by Shibata and Sekiguchi for 2D (axisymmetric) gravitational collapses [34–36] or 3D ones [37].

10.2.3 Harmonic Slicing

Another important category of time slicing is deduced from the standard ***harmonic*** or ***de Donder*** condition for the spacetime coordinates (x^α):

$$\Box_g x^\alpha = 0, \tag{10.16}$$

where $\Box_g := \nabla_\mu \nabla^\mu$ is the d'Alembertian associated with the metric $\boldsymbol{g}$ and each coordinate x^α is considered as a scalar field on $\mathscr{M}$. Harmonic coordinates have been introduced by De Donder in 1921 [38] and have played an important role in theoretical developments, notably in Choquet–Bruhat's demonstration [39] of the well-posedness of the Cauchy problem for 3+1 Einstein equations (cf. Sect. 5.4.4).

The ***harmonic slicing*** is defined by requiring that the harmonic condition holds for the $x^0 = t$ coordinate, but not necessarily for the other coordinates, leaving the freedom to choose any coordinate (x^i) in each hypersurface Σ_t:

$$\boxed{\Box_g t = 0}\,. \tag{10.17}$$

Since $\Box_g t = \nabla_\mu \nabla^\mu t$ is the divergence of the vector field $\vec{\nabla} t$, formula (2.65) gives

$$\frac{1}{\sqrt{-g}}\frac{\partial}{\partial x^\mu}\left(\sqrt{-g}g^{\mu\nu}\underbrace{\frac{\partial t}{\partial x^\nu}}_{\delta^0{}_\nu}\right) = 0,$$

i.e.

$$\frac{\partial}{\partial x^\mu}\left(\sqrt{-g}g^{\mu 0}\right) = 0.$$

Thanks to the relation $\sqrt{-g} = N\sqrt{\gamma}$ [Eq. (5.55)], this equation becomes

$$\frac{\partial}{\partial t}\left(N\sqrt{\gamma}g^{00}\right) + \frac{\partial}{\partial x^i}\left(N\sqrt{\gamma}g^{i0}\right) = 0.$$

Now, one reads on (5.51) that $g^{00} = -1/N^2$ and $g^{i0} = \beta^i/N^2$. Thus

$$-\frac{\partial}{\partial t}\left(\frac{\sqrt{\gamma}}{N}\right) + \frac{\partial}{\partial x^i}\left(\frac{\sqrt{\gamma}}{N}\beta^i\right) = 0. \tag{10.18}$$

Expanding and reordering gives

$$\frac{\partial N}{\partial t} - \beta^i \frac{\partial N}{\partial x^i} - N \Bigg[\frac{1}{\sqrt{\gamma}} \frac{\partial \sqrt{\gamma}}{\partial t} - \underbrace{\frac{1}{\sqrt{\gamma}} \frac{\partial}{\partial x^i} \left(\sqrt{\gamma} \beta^i \right)}_{D_i \beta^i} \Bigg] = 0.$$

Thanks to Eq. (6.66), the term in brackets can be replaced by $-NK$, so that the harmonic slicing condition becomes

$$\boxed{\left(\frac{\partial}{\partial t} - \mathcal{L}_\beta \right) N = -KN^2}. \tag{10.19}$$

Thus we get an *evolution* equation for the lapse function. This contrasts with Eq. (10.1) for geodesic slicing and Eq. (10.8) for maximal slicing.

The harmonic slicing has been introduced by Choquet–Bruhat and Ruggeri (1983) [40] as a way to put the 3+1 Einstein system in a hyperbolic form. It has been considered more specifically in the context of numerical relativity by Bona and Masso (1988) [41]. For a review and more references see Ref. [42].

Remark 10.2 The harmonic slicing equation (10.19) was already laid out by Smarr and York in 1978 [43], as a part of the expression of de Donder coordinate condition in terms of 3+1 variables.

Example 10.5 In Schwarzschild spacetime, the hypersurfaces of constant standard Schwarzschild time coordinate $t = t_S$ and depicted in Fig. 10.3 constitute some harmonic slicing, in addition to being maximal (cf. Sect. 10.2.2). Indeed, using Schwarzschild coordinates (t, R, θ, φ) or isotropic coordinates (t, r, θ, φ), we have $\partial N/\partial t = 0$ and $\boldsymbol{\beta} = 0$. Since $K = 0$ for these hypersurfaces, we conclude that the harmonic slicing condition (10.19) is satisfied.

Example 10.6 The above slicing does not penetrate under the event horizon. A harmonic slicing of Schwarzschild spacetime (and more generally Kerr-Newman spacetime) which passes smoothly through the event horizon has been found by Bona and Massó [41], as well as Cook and Scheel [44]. It is given by a time coordinate t that is related to Schwarzschild time t_S by

$$t = t_S + 2m \ln \left| 1 - \frac{2m}{R} \right|, \tag{10.20}$$

where R is Schwarzschild radial coordinate (areal radius). The corresponding expression of Schwarzschild metric is [44]

$$g_{\mu\nu}\mathrm{d}x^\mu \mathrm{d}x^\nu = -N^2\mathrm{d}t^2 + \frac{1}{N^2}\left(\mathrm{d}R + \frac{4m^2}{R^2}N^2\mathrm{d}t\right)^2 + R^2(\mathrm{d}\theta^2 + \sin^2\theta \mathrm{d}\varphi^2), \tag{10.21}$$

where

$$N = \left[\left(1 + \frac{2m}{R}\right)\left(1 + \frac{4m^2}{R^2}\right)\right]^{-1/2}. \tag{10.22}$$

Notice that all metric coefficients are regular at the event horizon ($R = 2m$). This harmonic slicing is represented in a Kruskal–Szekeres diagram in Fig. 1 of Ref. [44]. It is clear from that figure that the hypersurfaces Σ_t never hit the singularity (contrary to those of the geodesic slicing shown in Fig. 10.1), but they come arbitrary close to it as $t \to +\infty$.

We infer from the above example that the harmonic slicing has some singularity avoidance feature, but weaker than that of maximal slicing: for the latter, the hypersurfaces Σ_t never come close to the singularity as $t \to +\infty$ (cf. Fig. 10.4). This has been confirmed by means of numerical computations by Shibata and Nakamura [10].

Remark 10.3 If one uses normal coordinates, i.e. spatial coordinates (x^i) such that $\boldsymbol{\beta} = 0$, then the harmonic slicing condition in the form (10.23) is easily integrated to

$$N = C(x^i)\sqrt{\gamma}, \tag{10.23}$$

where $C(x^i)$ is an arbitrary function of the spatial coordinates, which does not depend upon t. Equation (10.23) is as easy to implement as the geodesic slicing condition ($N = 1$). It is related to the ***conformal time slicing*** introduced by Shibata and Nakamura [45].

10.2.4 1+log Slicing

Bona, Massó, Seidel and Stela (1995) [46] have generalized the harmonic slicing condition (10.19) to

$$\left(\frac{\partial}{\partial t} - \mathcal{L}_{\boldsymbol{\beta}}\right) N = -KN^2 f(N), \tag{10.24}$$

where f is an arbitrary function. The harmonic slicing corresponds to $f(N) = 1$. The geodesic slicing also fulfills this relation with $f(N) = 0$. The choice $f(N) = 2/N$ leads to

$$\boxed{\left(\frac{\partial}{\partial t} - \mathcal{L}_{\boldsymbol{\beta}}\right) N = -2KN}. \tag{10.25}$$

Substituting Eq. (6.66) for $-KN$, we obtain

$$\left(\frac{\partial}{\partial t} - \mathscr{L}_{\boldsymbol{\beta}}\right) N = \frac{\partial}{\partial t} \ln \gamma - 2 D_i \beta^i. \tag{10.26}$$

If normal coordinates are used, $\boldsymbol{\beta} = 0$ and the above equation reduces to

$$\frac{\partial N}{\partial t} = \frac{\partial}{\partial t} \ln \gamma,$$

a solution of which is

$$N = 1 + \ln \gamma. \tag{10.27}$$

For this reason, a foliation whose lapse function obeys Eq. (10.25) is called a ***1+log slicing***. The original 1+log condition (10.27) has been introduced by Bernstein (1993) [47] and Anninos et al. (1995) [48] (see also Ref. [49]). Notice that, even when $\boldsymbol{\beta} \neq 0$, we still define the 1+log slicing by condition (10.25), although the "1+log" relation (10.27) does no longer hold.

Remark 10.4 As for the geodesic slicing [Eq. (10.1)], the harmonic slicing with zero shift [Eq. (10.23)], the original 1+log slicing with zero shift [Eq. (10.27)] belongs to the family of ***algebraic slicings*** [4, 50]: the determination of the lapse function does not require to solve any equation. It is therefore very easy to implement.

Example 10.7 Examples of 1+log slicings of Schwarzschild spacetime can be found in Refs. [7, 51–53], to which we refer the reader. In particular these references discuss the so-called *trumpet hypersurface* to which some 1+log slicings converge and which has a direct connection with the moving-puncture approach to black holes in numerical relativity.

The 1+log slicing has stronger singularity avoidance properties than harmonic slicing: it has been found to "mimic" maximal slicing [48].

Alcubierre has shown in 1997 [54] that for any slicing belonging to the family (10.24), and in particular for the harmonic and 1+log slicings, some smooth initial data (Σ_0, γ) can be found such that the foliation (Σ_t) become singular for a finite value of t.

Remark 10.5 The above finding does not contradict the well-posedness of the Cauchy problem established by Choquet–Bruhat in 1952 [39] for generic smooth initial data by means of harmonic coordinates (which define a harmonic slicing) (cf. Sect. 5.4.4). Indeed it must be remembered that Choquet-Bruhat's theorem is a *local* one, whereas the pathologies found by Alcubierre develop for a finite value of time. Moreover, these pathologies are far from being generic, as the tremendous successes of the 1+log slicing in numerical relativity have shown (see below).

The 1+log slicing has been used the 3D investigations of the dynamics of relativistic stars by Font et al. in 2002 [55]. It has also been used in most of the computations of binary black hole inspiral and merger: Baker et al. [56–58], Campanelli et al. [59–62], Sperhake [63], Diener et al. [64], Brügmann et al. [65, 66], and Herrmann et al. [67, 68]. The first three groups employ exactly Eq. (10.25), whereas the last two groups are using a modified ("zero-shift") version:

$$\frac{\partial N}{\partial t} = -2KN. \tag{10.28}$$

This version has also been employed by Sekiguchi and Shibata for computing gravitational collapse of a stellar core to a black hole [69]. The 3D gravitational collapse calculations of Baiotti et al. [70–72] are based on a slight modification of the 1+log slicing: instead of Eq. (10.25), these authors have used

$$\left(\frac{\partial}{\partial t} - \mathcal{L}_{\boldsymbol{\beta}}\right) N = -2N(K - K_0), \tag{10.29}$$

where K_0 is the value of K at $t = 0$. The same choice is adopted in the binary neutron star merger computations of Baiotti, Giacomazzo and Rezzolla [73, 74].

The original 1+log prescription (10.25) has also been employed in computations of (i) neutron star—neutron star mergers by Kiuchi et al. [75] and (ii) neutron star—black hole mergers by Kyutoku et al. [76].

Remark 10.6 There is a basic difference between maximal slicing and the other types of foliations presented above (geodesic, harmonic and 1+log slicings): the property of being maximal is applicable to a *single hypersurface* Σ_0, whereas the property of being geodesic, harmonic or 1+log are meaningful only for a *foliation* $(\Sigma_t)_{t\in\mathbb{R}}$. This is reflected in the basic definition of these slicings: the maximal slicing is defined from the extrinsic curvature tensor only ($K = 0$), which characterizes a single hypersurface (cf. Chap. 3), whereas the definitions of geodesic, harmonic and 1+log slicings all involve the lapse function N, which of course makes sense only for a foliation (cf. Chap. 4).

10.3 Evolution of Spatial Coordinates

Having discussed the choice of the foliation $(\Sigma_t)_{t\in\mathbb{R}}$, let us turn now to the choice of the coordinates (x^i) in each hypersurface Σ_t. As discussed in Sect. 5.2.4, this is done via the shift vector $\boldsymbol{\beta}$. More precisely, once some coordinates (x^i) are set in the initial slice Σ_0, the shift vector governs the propagation of these coordinates to all the slices Σ_t.

10.3.1 Normal Coordinates

As for the lapse choice $N = 1$ (geodesic slicing, Sect. 10.2.1), the simplest choice for the shift vector is to set it to zero:

$$\boxed{\boldsymbol{\beta} = 0}. \tag{10.30}$$

For this choice, the lines $x^i = \text{const}$ are normal to the hypersurfaces Σ_t (cf. Fig. 5.1), hence the name ***normal coordinates***. The alternative name is ***Eulerian coordinates***, defining the so-called ***Eulerian gauge*** [77]. This is of course justified by the fact that the lines $x^i = \text{const}$ are then the worldlines of the Eulerian observers introduced in Sect. 4.3.3.

Besides their simplicity, an advantage of normal coordinates is to be as regular as the foliation itself: they cannot introduce some pathology per themselves. On the other hand, the major drawback of these coordinates is that they may lead to a large coordinate shear, resulting in large values of the metric coefficients γ_{ij}. This is specially true if rotation is present. For instance, in Kerr or rotating star spacetimes, the field lines of the stationary Killing vector $\boldsymbol{\xi}$ are not orthogonal to the hypersurfaces $t = \text{const}$. Therefore, if one wishes to have coordinates adapted to stationarity, i.e. to have $\boldsymbol{\partial}_t = \boldsymbol{\xi}$, one must allow for $\boldsymbol{\beta} \neq 0$.

Despite of the shear problem mentioned above, normal coordinates have been used because of their simplicity in early treatments of two famous axisymmetric problems in numerical relativity: the head-on collision of black holes by Smarr et al. in 1976–1977 [17, 18] and the gravitational collapse of a rotating star by Nakamura in 1981 [19, 20]. Normal coordinates have also been used in the 3D evolution of gravitational waves performed by Shibata and Nakamura [10] and Baumgarte and Shapiro [78], as well as in the 3D grazing collisions of binary black holes computed by Brügmann [26] and Alcubierre et al. [79].

10.3.2 Minimal Distortion

A very well motivated choice of spatial coordinates has been introduced in 1978 by Smarr and York [1, 43] (see also Ref. [2]). As discussed in Sect. 7.1, the physical degrees of freedom of the gravitational field are carried by the conformal 3-metric $\tilde{\boldsymbol{\gamma}}$. The evolution of the latter with respect to the coordinates (t, x^i) is given by the derivative $\dot{\tilde{\boldsymbol{\gamma}}} := \mathcal{L}_{\boldsymbol{\partial}_t} \tilde{\boldsymbol{\gamma}}$, the components of which are

$$\dot{\tilde{\gamma}}_{ij} = \frac{\partial \tilde{\gamma}_{ij}}{\partial t}. \tag{10.31}$$

Given a foliation $(\Sigma_t)_{t\in\mathbb{R}}$, the idea of Smarr and York is to choose the coordinates (x^i), and hence the vector $\boldsymbol{\partial}_t$, in order to minimize this time derivative. There is not a unique way to minimize $\dot{\tilde{\gamma}}_{ij}$; this can be realized by counting the degrees of

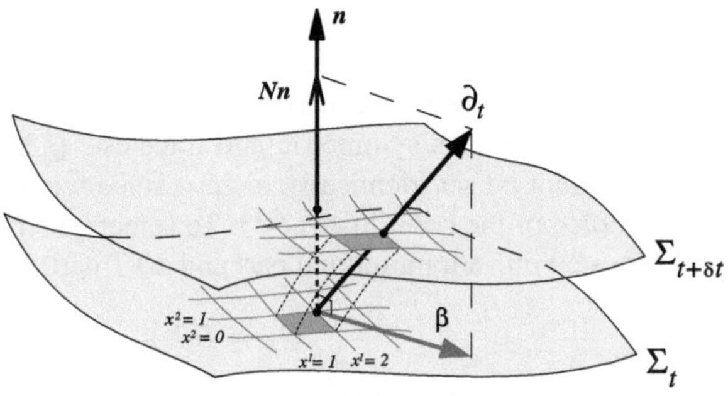

Fig. 10.5 Distortion of a spatial domain defined by fixed values of the coordinates (x^i)

freedom: $\dot{\tilde{\gamma}}_{ij}$ has 5 independent components[1] and, for a given foliation, only 3 degrees of freedom can be controlled via the 3 coordinates (x^i). One then proceeds as follows. First one notices that $\dot{\tilde{\boldsymbol{\gamma}}}$ is related to the ***distortion tensor*** $\boldsymbol{Q}$, the latter being defined as the trace-free part of the time derivative of the physical metric $\boldsymbol{\gamma}$:

$$\boldsymbol{Q} := \mathcal{L}_{\partial_t}\boldsymbol{\gamma} - \frac{1}{3}\left(\mathrm{tr}_{\gamma}\mathcal{L}_{\partial_t}\boldsymbol{\gamma}\right)\boldsymbol{\gamma}, \tag{10.32}$$

or in components,

$$Q_{ij} = \frac{\partial \gamma_{ij}}{\partial t} - \frac{1}{3}\gamma^{kl}\frac{\partial \gamma_{kl}}{\partial t}\gamma_{ij}. \tag{10.33}$$

$\boldsymbol{Q}$ measures the change in shape from Σ_t to $\Sigma_{t+\partial t}$ of any spatial domain $\mathcal{V}$ which lies at fixed values of the coordinates (x^i) (the evolution of $\mathcal{V}$ is then along the vector $\boldsymbol{\partial}_t$, cf. Fig. 10.5). Thanks to the trace removal, $\boldsymbol{Q}$ does not take into account the change of volume, but only the change in shape (shear). From the law (2.64) of variation of a determinant,

$$\gamma^{kl}\frac{\partial \gamma_{kl}}{\partial t} = \frac{\partial}{\partial t}\ln\gamma = 12\frac{\partial}{\partial t}\ln\Psi + \underbrace{\frac{\partial}{\partial t}\ln f}_{0} = 12\frac{\partial}{\partial t}\ln\Psi,$$

where we have used the relation (7.15) between the determinant γ and the conformal factor Ψ, as well as the property (7.7). Thus we may rewrite Eq. (10.33) as

$$Q_{ij} = \frac{\partial \gamma_{ij}}{\partial t} - 4\frac{\partial}{\partial t}\ln\Psi\,\gamma_{ij} = \frac{\partial}{\partial t}(\Psi^4\tilde{\gamma}_{ij}) - 4\Psi^3\frac{\partial\Psi}{\partial t}\tilde{\gamma}_{ij} = \Psi^4\frac{\partial\tilde{\gamma}_{ij}}{\partial t}.$$

Hence the relation between the distortion tensor and the time derivative of the conformal metric:

[1] As a symmetric 3 × 3 matrix, $\dot{\tilde{\gamma}}_{ij}$ has a priori six components, but one degree of freedom is lost in the demand $\det\tilde{\gamma}_{ij} = \det f_{ij}$ [Eq. (7.18)], which implies $\det\dot{\tilde{\gamma}}_{ij} = 0$ via Eq. (7.7).

$$\boldsymbol{Q} = \Psi^4 \dot{\tilde{\boldsymbol{\gamma}}}. \tag{10.34}$$

The rough idea would be to choose the coordinates (x^i) in order to minimize $\boldsymbol{Q}$. Taking into account that it is symmetric and traceless, $\boldsymbol{Q}$ has 5 independent components. Thus it cannot be set identically to zero since we have only 3 degrees of freedom in the choice of the coordinates (x^i). To select which part of $\boldsymbol{Q}$ to set to zero, let us decompose it into a longitudinal part and a TT part, in a manner similar to Eq. (9.9):

$$Q_{ij} = (LX)_{ij} + Q_{ij}^{\rm TT}. \tag{10.35}$$

LX denotes the conformal Killing operator associated with the metric $\boldsymbol{\gamma}$ and acting on some vector field $\boldsymbol{X}$ (cf. Appendix A)[2]:

$$(LX)_{ij} := D_i X_j + D_j X_i - \frac{2}{3} D_k X^k \gamma_{ij} \tag{10.36}$$

and $Q_{ij}^{\rm TT}$ is both traceless and transverse (i.e. divergence-free) with respect to the metric $\boldsymbol{\gamma}$: $D^j Q_{ij}^{\rm TT} = 0$. $\boldsymbol{X}$ is then related to the divergence of $\boldsymbol{Q}$ by $D^j (LX)_{ij} = D^j Q_{ij}$. It is legitimate to relate the TT part to the dynamics of the gravitational field and to attribute the longitudinal part to the change in γ_{ij} which arises because of the variation of coordinates from Σ_t to $\Sigma_{t+\delta t}$. This longitudinal part has 3 degrees of freedom (the three components of the vector $\boldsymbol{X}$) and we might set it to zero by some judicious choice of the coordinates (x^i). The ***minimal distortion*** coordinates are thus defined by the requirement $\boldsymbol{X} = 0$ or

$$Q_{ij} = Q_{ij}^{\rm TT}, \tag{10.37}$$

i.e.

$$\boxed{D^j Q_{ij} = 0}. \tag{10.38}$$

Let us now express $\boldsymbol{Q}$ in terms of the shift vector to turn the above condition into an equation for the evolution of spatial coordinates. By means of Eqs. (5.68) and (6.65), Eq. (10.33) becomes

$$Q_{ij} = -2NK_{ij} + \mathcal{L}_{\boldsymbol{\beta}} \gamma_{ij} - \frac{1}{3}\left(-2NK + 2D_k\beta^k\right)\gamma_{ij},$$

i.e. (since $\mathcal{L}_{\boldsymbol{\beta}} \gamma_{ij} = D_i\beta_j + D_j\beta_i$)

$$Q_{ij} = -2NA_{ij} + (L\beta)_{ij}, \tag{10.39}$$

where we let appear the trace-free part $\boldsymbol{A}$ of the extrinsic curvature $\boldsymbol{K}$ [Eq. (7.46)]. If we insert this expression into the minimal distortion requirement (10.38), we get

[2] In Sect. 7.6, we have also used the notation L for the conformal Killing operator associated with the flat metric $\boldsymbol{f}$, but no confusion should arise in the present context.

$$-2ND_jA^{ij} - 2A^{ij}D_jN + D_j(L\beta)^{ij} = 0.$$

Let us then use the momentum constraint (5.71) to express the divergence of $\boldsymbol{A}$ as

$$D_jA^{ij} = 8\pi p^i + \frac{2}{3}D^iK.$$

Besides, we recognize in $D_j(L\beta)^{ij}$ the conformal vector Laplacian associated with the metric $\boldsymbol{\gamma}$, so that we can write [cf. Eq. (A.8)]

$$D_j(L\beta)^{ij} = D_jD^j\beta^i + \frac{1}{3}D^iD_j\beta^j + R^i{}_j\beta^j,$$

where $\boldsymbol{R}$ is the Ricci tensor associated with $\boldsymbol{\gamma}$. Thus we arrive at

$$\boxed{D_jD^j\beta^i + \frac{1}{3}D^iD_j\beta^j + R^i{}_j\,\beta^j = 16\pi Np^i + \frac{4}{3}ND^iK + 2A^{ij}D_jN}\,. \tag{10.40}$$

This is the elliptic equation on the shift vector that one has to solve in order to enforce minimal distortion.

Remark 10.7 For a constant mean curvature (CMC) slicing, and in particular for a maximal slicing, the term D^iK vanishes and the above equation is slightly simplified. Incidentally, this is the form originally derived by Smarr and York (Eq. (3.27) in Ref. [43]).

Another way to introduce minimal distortion amounts to minimizing the integral

$$S = \int_{\Sigma_t} Q_{ij}Q^{ij}\sqrt{\gamma}\,\mathrm{d}^3x \tag{10.41}$$

with respect to the shift vector $\boldsymbol{\beta}$, keeping the slicing fixed (i.e. fixing $\boldsymbol{\gamma}, \boldsymbol{K}$ and N). Indeed, if we replace $\boldsymbol{Q}$ by its expression (10.39), we get

$$S = \int_{\Sigma_t}\left[4N^2A_{ij}A^{ij} - 4NA_{ij}(L\beta)^{ij} + (L\beta)_{ij}(L\beta)^{ij}\right]\sqrt{\gamma}\,\mathrm{d}^3x. \tag{10.42}$$

At fixed values of $\boldsymbol{\gamma}, \boldsymbol{K}$ and N, $\delta N = 0$, $\delta A_{ij} = 0$ and $\delta(L\beta)^{ij} = (L\delta\beta)^{ij}$, so that the variation of S with respect to $\boldsymbol{\beta}$ is

$$\delta S = \int_{\Sigma_t}\left[-4NA_{ij}(L\delta\beta)^{ij} + 2(L\beta)_{ij}(L\delta\beta)^{ij}\right]\sqrt{\gamma}\,\mathrm{d}^3x = 2\int_{\Sigma_t} Q_{ij}(L\delta\beta)^{ij}\sqrt{\gamma}\,\mathrm{d}^3x.$$

Now, since $\boldsymbol{Q}$ is symmetric and traceless, $Q_{ij}(L\delta\beta)^{ij} = Q_{ij}(D^i\delta\beta^j + D^j\delta\beta^i - 2/3D_k\delta\beta^k\gamma^{ij}) = Q_{ij}(D^i\delta\beta^j + D^j\delta\beta^i) = 2Q_{ij}D^i\delta\beta^j$. Hence

$$\begin{aligned}\delta S &= 4\int_{\Sigma_t} Q_{ij}D^i\delta\beta^j\sqrt{\gamma}\,\mathrm{d}^3x \\ &= 4\int_{\Sigma_t}\left[D^i\left(Q_{ij}\delta\beta^j\right) - D^iQ_{ij}\delta\beta^j\right]\sqrt{\gamma}\mathrm{d}^3x \\ &= 4\oint_{\partial\Sigma_t} Q_{ij}\delta\beta^j s^i\sqrt{q}\,\mathrm{d}^2y - 4\int_{\Sigma_t} D^iQ_{ij}\delta\beta^j\sqrt{\gamma}\mathrm{d}^3x\end{aligned}$$

Assuming that $\delta\beta^i = 0$ at the boundaries of Σ_t (for instance at spatial infinity), we deduce from the above relation that $\delta S = 0$ for any variation of the shift vector if and only if $D^i Q_{ij} = 0$. Hence we recover condition (10.38).

In stationary spacetime, an important property of the minimal distortion gauge is to be fulfilled by coordinates adapted to the stationarity (i.e. such that $\boldsymbol{\partial}_t$ is a Killing vector): it is immediate from Eq. (10.32) that $\boldsymbol{Q} = 0$ when $\boldsymbol{\partial}_t$ is a symmetry generator, so that condition (10.38) is trivially satisfied. Another nice feature of the minimal distortion gauge is that in the weak field region (radiative zone), it includes the standard TT gauge of linearized gravity [43]. Actually Smarr and York [43] have advocated for maximal slicing combined with minimal distortion as a very good coordinate choice for radiative spacetimes, calling such choice the ***radiation gauge***.

Remark 10.8 A "new minimal distortion" gauge has been introduced in 2006 by Jantzen and York [80]. It corrects the time derivative of $\tilde{\boldsymbol{\gamma}}$ in the original minimal distortion condition by the lapse function N [cf. relation (4.17) between the coordinate time t and the Eulerian observer's proper time τ], i.e. one requires

$$D^j \left(\frac{1}{N} Q_{ij} \right) = 0 \tag{10.43}$$

instead of (10.38). This amounts to minimizing the integral

$$S' = \int_{\Sigma_t} (N^{-1} Q_{ij})(N^{-1} Q^{ij}) \sqrt{-g}\, \mathrm{d}^3 x \tag{10.44}$$

with respect to the shift vector. Notice the spacetime measure $\sqrt{-g} = N\sqrt{\gamma}$ instead of the spatial measure $\sqrt{\gamma}$ in Eq. (10.41).

The minimal distortion condition can be expressed in terms of the time derivative of the conformal metric by combining Eqs. (10.34) and (10.38):

$$\boxed{D^j (\Psi^4 \dot{\tilde{\gamma}}_{ij}) = 0}\,. \tag{10.45}$$

Let us write this relation in terms of the connection $\tilde{\boldsymbol{D}}$ (associated with the metric $\tilde{\boldsymbol{\gamma}}$) instead of the connection $\boldsymbol{D}$ (associated with the metric $\boldsymbol{\gamma}$). To this purpose, let us use Eq. (7.66) which relates the $\boldsymbol{D}$-divergence of a traceless symmetric tensor, such as Q^{ij}, to its $\tilde{\boldsymbol{D}}$-divergence:

$$D_j Q^{ij} = \Psi^{-10} \tilde{D}_j \left(\Psi^{10} Q^{ij} \right).$$

Now $Q^{ij} = \gamma^{ik}\gamma^{jl} Q_{kl} = \Psi^{-8} \tilde{\gamma}^{ik} \tilde{\gamma}^{jl} Q_{kl} = \Psi^{-4} \tilde{\gamma}^{ik} \tilde{\gamma}^{jl} \dot{\tilde{\gamma}}_{kl}$; hence

$$D_j Q^{ij} = \Psi^{-10} \tilde{D}_j \left(\Psi^6 \tilde{\gamma}^{ik} \tilde{\gamma}^{jl} \dot{\tilde{\gamma}}_{kl} \right) = \Psi^{-10} \tilde{\gamma}^{ik} \tilde{D}^l \left(\Psi^6 \dot{\tilde{\gamma}}_{kl} \right).$$

The minimal distortion condition is therefore

$$\boxed{\tilde{D}^j (\Psi^6 \dot{\tilde{\gamma}}_{ij}) = 0}\,. \tag{10.46}$$

10.3.3 Approximate Minimal Distortion

In view of Eq. (10.46), it is natural to consider the simpler condition

$$\tilde{D}^j \dot{\tilde{\gamma}}_{ij} = 0, \tag{10.47}$$

which of course differs from the true minimal distortion (10.46) by a term $6\dot{\gamma}_{ij}\tilde{D}^j \ln \Psi$. Nakamura (1994) [81, 82] has then introduced the ***pseudo-minimal distortion*** condition by replacing (10.47) by

$$\boxed{\bar{D}^j \dot{\tilde{\gamma}}_{ij} = 0}, \tag{10.48}$$

where $\bar{\boldsymbol{D}}$ is the connection associated with the flat metric $\boldsymbol{f}$.

An alternative has been introduced by Shibata (1999) [83] as follows. Starting from Eq. (10.47), let us express $\dot{\tilde{\boldsymbol{\gamma}}}_{ij}$ in terms of $\boldsymbol{A}$ and $\boldsymbol{\beta}$: from Eq. (9.79), we deduce that

$$\begin{aligned} 2N\tilde{A}_{ij} &= \tilde{\gamma}_{ik}\tilde{\gamma}_{jl}\left[\dot{\tilde{\gamma}}^{kl} + (\tilde{L}\beta)^{kl}\right] = \tilde{\gamma}_{jl}\Bigg[\frac{\partial}{\partial t}\underbrace{(\tilde{\gamma}_{ik}\tilde{\gamma}^{kl})}_{\delta^l{}_i} - \tilde{\gamma}^{kl}\frac{\partial \tilde{\gamma}_{ik}}{\partial t} + \tilde{\gamma}_{ik}(\tilde{L}\beta)^{kl}\Bigg] \\ &= -\dot{\tilde{\gamma}}_{ij} + \tilde{\gamma}_{ik}\tilde{\gamma}_{jl}(\tilde{L}\beta)^{kl}, \end{aligned}$$

where $\tilde{A}_{ij} := \tilde{\gamma}_{ik}\tilde{\gamma}_{jl}\tilde{A}^{kl} = \Psi^{-4}A_{ij}$. Equation (10.47) becomes then

$$\tilde{D}^j\left[\tilde{\gamma}_{ik}\tilde{\gamma}_{jl}(\tilde{L}\beta)^{kl} - 2N\tilde{A}_{ij}\right] = 0,$$

or equivalently (cf. Sect. A.2.1),

$$\tilde{D}_j\tilde{D}^j\beta^i + \frac{1}{3}\tilde{D}^i\tilde{D}_j\beta^j + \tilde{R}^i_j\beta^j - 2\tilde{A}^{ij}\tilde{D}_jN - 2N\tilde{D}_j\tilde{A}^{ij} = 0.$$

We can express $\tilde{D}_j\tilde{A}^{ij}$ via the momentum constraint (7.93) and get

$$\tilde{D}_j\tilde{D}^j\beta^i + \frac{1}{3}\tilde{D}^i\tilde{D}_j\beta^j + \tilde{R}^i{}_j\beta^j - 2\tilde{A}^{ij}\tilde{D}_jN + 4N\left[3\tilde{A}^{ij}\tilde{D}_j\ln\Psi - \frac{1}{3}\tilde{D}^iK - 4\pi\Psi^4p^i\right] = 0. \tag{10.49}$$

At this stage, Eq. (10.49) is nothing but a rewriting of Eq. (10.47) as an elliptic equation for the shift vector. Shibata [83] then proposes to replace in this equation the conformal vector Laplacian relative to $\tilde{\boldsymbol{\gamma}}$ and acting on $\boldsymbol{\beta}$ by the conformal vector Laplacian relative to the flat metric $\boldsymbol{f}$, thereby writing

$$\bar{D}_j\bar{D}^j\beta^i + \frac{1}{3}\bar{D}^i\bar{D}_j\beta^j - 2\tilde{A}^{ij}\tilde{D}_jN + 4N\left[3\tilde{A}^{ij}\tilde{D}_j\ln\Psi - \frac{1}{3}\tilde{D}^iK - 4\pi\Psi^4p^i\right] = 0. \tag{10.50}$$

The choice of coordinates defined by solving Eq. (10.50) instead of (10.40) is called ***approximate minimal distortion***.

The approximate minimal distortion has been used by Shibata and Uryu [29, 30] for their first computations of the merger of binary neutron stars, as well as by Shibata, Baumgarte and Shapiro for computing the collapse of supramassive neutron stars at the mass-shedding limit (Keplerian angular velocity) [84] and for studying the dynamical bar-mode instability in differentially rotating neutron stars [85]. It has also been used by Shibata [34] to devise a 2D (axisymmetric) code to compute the long-term evolution of rotating neutron stars and gravitational collapse.

10.3.4 Gamma Freezing

The ***Gamma freezing*** prescription for the evolution of spatial coordinates is very much related to Nakamura's pseudo-minimal distortion (10.48): it differs from it only in the replacement of $\bar{D}^j$ by $\bar{D}_j$ and $\dot{\tilde{\gamma}}_{ij}$ by $\dot{\tilde{\gamma}}^{ij} := \partial\tilde{\gamma}^{ij}/\partial t$:

$$\boxed{\bar{D}_j \dot{\tilde{\gamma}}^{ij} = 0}. \tag{10.51}$$

The name *Gamma freezing* is justified as follows: since $\partial/\partial t$ and $\bar{\boldsymbol{D}}$ commute [as a consequence of (7.7)], Eq. (10.51) is equivalent to

$$\frac{\partial}{\partial t}\left(\bar{D}_j \tilde{\gamma}^{ij}\right) = 0. \tag{10.52}$$

Now, expressing the covariant derivative $\bar{D}_j$ in terms of the Christoffel symbols $\bar{\Gamma}^i{}_{jk}$ of the metric $\boldsymbol{f}$ with respect to the coordinates (x^i), we get

$$\begin{aligned}
D_j\tilde{\gamma}^{ij} &= \frac{\partial\tilde{\gamma}^{ij}}{\partial x^j} + \bar{\Gamma}^i{}_{jk}\tilde{\gamma}^{kj} + \underbrace{\bar{\Gamma}^j{}_{jk}}_{\frac{1}{2}\frac{\partial}{\partial x^k}\ln f}\ \tilde{\gamma}^{ik} \\
&= \frac{\partial\tilde{\gamma}^{ij}}{\partial x^j} + \tilde{\Gamma}^i{}_{jk}\tilde{\gamma}^{kj} + \left(\bar{\Gamma}^i{}_{jk} - \tilde{\Gamma}^i{}_{jk}\right)\tilde{\gamma}^{kj} + \underbrace{\frac{1}{2}\frac{\partial}{\partial x^k}\ln\tilde{\gamma}}_{\tilde{\Gamma}^j{}_{jk}}\ \tilde{\gamma}^{ik} \\
&= \underbrace{\frac{\partial\tilde{\gamma}^{ij}}{\partial x^j} + \tilde{\Gamma}^i_{jk}\tilde{\gamma}^{kj} + \tilde{\Gamma}^j_{jk}\tilde{\gamma}^{ik}}_{\tilde{D}_j\tilde{\gamma}^{ij}=0} + \left(\bar{\Gamma}^i_{jk} - \tilde{\Gamma}^i_{jk}\right)\tilde{\gamma}^{kj} \\
&= \tilde{\gamma}^{jk}\left(\bar{\Gamma}^i{}_{jk} - \tilde{\Gamma}^i{}_{jk}\right),
\end{aligned}$$

where $\tilde{\Gamma}^i{}_{jk}$ denote the Christoffel symbols of the metric $\tilde{\boldsymbol{\gamma}}$ with respect to the coordinates (x^i) and we have used $\tilde{\gamma} = f$ [Eq. (7.18)] to write the second line. If we introduce the notation

$$\boxed{\tilde{\Gamma}^i := \tilde{\gamma}^{jk}\left(\tilde{\Gamma}^i{}_{jk} - \bar{\Gamma}^i{}_{jk}\right)}, \tag{10.53}$$

then the above relation becomes

$$\boxed{\bar{D}_j \tilde{\gamma}^{ij} = -\tilde{\Gamma}^i}. \tag{10.54}$$

Remark 10.9 If one uses Cartesian-type coordinates, then $\bar{\Gamma}^i{}_{jk} = 0$ and the $\tilde{\Gamma}^i$'s reduce to the contracted Christoffel symbols introduced by Baumgarte and Shapiro [78] [cf. their Eq. (21)]. In the present case, the $\tilde{\Gamma}^i$'s are the components of a *vector field* $\tilde{\boldsymbol{\Gamma}}$ on Σ_t, as it is clear from relation (10.54), or from expression (10.53) if one remembers that, although the Christoffel symbols are not the components of any tensor field, the differences between two sets of them are. Of course the vector field $\tilde{\boldsymbol{\Gamma}}$ depends on the choice of the background metric $\boldsymbol{f}$.

By combining Eqs. (10.54) and (10.52), we see that the Gamma freezing condition is equivalent to

$$\boxed{\frac{\partial \tilde{\Gamma}^i}{\partial t} = 0}, \tag{10.55}$$

hence the name *Gamma freezing*: for such a choice, the vector $\tilde{\boldsymbol{\Gamma}}$ does not evolve, in the sense that $\mathcal{L}_{\partial_t} \tilde{\boldsymbol{\Gamma}} = 0$. The Gamma freezing prescription has been introduced by Alcubierre and Brügmann in 2001 [86], in the form of Eq. (10.55).

Let us now derive the equation that the shift vector must obey in order to enforce the Gamma freezing condition. If we express the Lie derivative in the evolution equation (7.89) for $\tilde{\gamma}^{ij}$ in terms of the covariant derivative $\bar{\boldsymbol{D}}$ [cf. Eq. (2.90)], we get

$$\dot{\tilde{\gamma}}^{ij} = 2N\tilde{A}^{ij} + \beta^k \bar{D}_k \tilde{\gamma}^{ij} - \tilde{\gamma}^{kj}\bar{D}_k \beta^i - \tilde{\gamma}^{ik}\bar{D}_k\beta^j + \frac{2}{3}\bar{D}_k\beta^k \tilde{\gamma}^{ij}.$$

Taking the flat-divergence of this relation and using relation (10.54) (with the commutation property of $\partial/\partial t$ and $\bar{\boldsymbol{D}}$) yields

$$\begin{aligned}\frac{\partial \tilde{\Gamma}^i}{\partial t} ==& -2N\bar{D}_j\tilde{A}^{ij} - 2A^{ij}\bar{D}_j N + \beta^k\bar{D}_k\tilde{\Gamma}^i - \tilde{\Gamma}^k\bar{D}_k\beta^i + \frac{2}{3}\tilde{\Gamma}^i\bar{D}_k\beta^k \\ & \tilde{\gamma}^{jk}\bar{D}_j\bar{D}_k\beta^i + \frac{1}{3}\tilde{\gamma}^{ij}\bar{D}_j\bar{D}_k\beta^k.\end{aligned} \tag{10.56}$$

Now, we may use the momentum constraint (7.93) to express $\bar{D}_j\tilde{A}^{ij}$:

$$\tilde{D}_j\tilde{A}^{ij} = -6\tilde{A}^{ij}\tilde{D}_j \ln\Psi + \frac{2}{3}\tilde{D}^i K + 8\pi\Psi^4 p^i,$$

with

$$\tilde{D}_j\tilde{A}^{ij} = \bar{D}_j\tilde{A}^{ij} + \left(\tilde{\Gamma}^i{}_{jk} - \bar{\Gamma}^i{}_{jk}\right)\tilde{A}^{kj} + \underbrace{\left(\tilde{\Gamma}^j{}_{jk} - \bar{\Gamma}^j{}_{jk}\right)}_{0}\tilde{A}^{ik},$$

where the "$=0$" results from the fact that $2\tilde{\Gamma}^j{}_{jk} = \partial \ln \tilde{\gamma}/\partial x^k$ and $2\bar{\Gamma}^j{}_{jk} = \partial \ln f/\partial x^k$, with $\tilde{\gamma} := \det \tilde{\gamma}_{ij} = \det f_{ij} =: f$ [Eq. (7.18)]. Thus Eq. (10.56) becomes

$$\begin{aligned}\frac{\partial \tilde{\Gamma}^i}{\partial t} =& \tilde{\gamma}^{jk}\bar{D}_j\bar{D}_k\beta^i + \frac{1}{3}\tilde{\gamma}^{ij}\bar{D}_j\bar{D}_k\beta^k + \frac{2}{3}\tilde{\Gamma}^i\bar{D}_k\beta^k - \tilde{\Gamma}^k\bar{D}_k\beta^i + \beta^k\bar{D}_k\tilde{\Gamma}^i \\ & - 2N\left[8\pi\Psi^4 p^i - \tilde{A}^{jk}\left(\tilde{\Gamma}^i_{jk} - \bar{\Gamma}^i_{jk}\right) - 6\tilde{A}^{ij}\bar{D}_j\ln\Psi + \frac{2}{3}\tilde{\gamma}^{ij}\bar{D}_j K\right] - 2\tilde{A}^{ij}\bar{D}_j N.\end{aligned} \tag{10.57}$$

We conclude that the Gamma freezing condition (10.55) is equivalent to

$$\boxed{\begin{aligned}&\tilde{\gamma}^{jk}\bar{D}_j\bar{D}_k\beta^i + \tfrac{1}{3}\tilde{\gamma}^{ij}\bar{D}_j\bar{D}_k\beta^k + \tfrac{2}{3}\tilde{\Gamma}^i\bar{D}_k\beta^k - \tilde{\Gamma}^k\bar{D}_k\beta^i + \beta^k\bar{D}_k\tilde{\Gamma}^i = \\ &\quad 2N\left[8\pi\Psi^4 p^i - \tilde{A}^{jk}\left(\tilde{\Gamma}^i{}_{jk} - \bar{\Gamma}^i{}_{jk}\right) - 6\tilde{A}^{ij}\bar{D}_j\ln\Psi + \tfrac{2}{3}\tilde{\gamma}^{ij}\bar{D}_j K\right] + 2\tilde{A}^{ij}\bar{D}_j N\end{aligned}}. \tag{10.58}$$

This is an elliptic equation for the shift vector, which bears some resemblance with Shibata's approximate minimal distortion, Eq. (10.50).

10.3.5 Gamma Drivers

As seen above the Gamma freezing condition (10.55) yields to the elliptic equation (10.58) for the shift vector. Alcubierre and Brügmann [86] have proposed to turn it into a parabolic equation by considering, instead of Eq. (10.55), the relation

$$\boxed{\frac{\partial \beta^i}{\partial t} = k\frac{\partial \tilde{\Gamma}^i}{\partial t}}, \tag{10.59}$$

where K is a positive function. The resulting coordinate choice is called a ***parabolic Gamma driver***. Indeed, if we inject Eq. (10.59) into Eq. (10.57), we clearly get a parabolic equation for the shift vector, of the type $\partial\beta^i/\partial t = k\left[\tilde{\gamma}^{jk}\bar{D}_j\bar{D}_k\beta^i + \frac{1}{3}\tilde{\gamma}^{ij}\bar{D}_j\bar{D}_k\beta^k + \cdots\right]$.

An alternative has been introduced in 2003 by Alcubierre et al. [87] (see also Refs. [88] and [89]); it requires

$$\boxed{\frac{\partial^2 \beta^i}{\partial t^2} = k\frac{\partial \tilde{\Gamma}^i}{\partial t} - \left(\eta - \frac{\partial}{\partial t}\ln k\right)\frac{\partial \beta^i}{\partial t}}, \tag{10.60}$$

where k and η are two positive functions. The prescription (10.60) is called a ***hyperbolic Gamma driver*** [87–89]. Indeed, thanks to Eq. (10.57), it is equivalent to

$$\frac{\partial^2 \beta^i}{\partial t^2} + \left(\eta - \frac{\partial}{\partial t}\ln k\right)\frac{\partial \beta^i}{\partial t} = k\left\{\tilde{\gamma}^{jk}\bar{D}_j\bar{D}_k\beta^i + \frac{1}{3}\tilde{\gamma}^{ij}\bar{D}_j\bar{D}_k\beta^k + \frac{2}{3}\tilde{\Gamma}^i\bar{D}_k\beta^k - \tilde{\Gamma}^k\bar{D}_k\beta^i \right.$$
$$+ \beta^k\bar{D}_k\tilde{\Gamma}^i - 2N\left[8\pi\Psi^4 p^i - \tilde{A}^{jk}\left(\tilde{\Gamma}^i_{jk} - \bar{\Gamma}^i_{jk}\right) - 6\tilde{A}^{ij}\bar{D}_j\ln\Psi + \frac{2}{3}\tilde{\gamma}^{ij}\bar{D}_j K\right]$$
$$\left. - 2\tilde{A}^{ij}\bar{D}_j N\right\}, \tag{10.61}$$

which is a hyperbolic equation for the shift vector, of the type of the telegrapher's equation. The term with the coefficient η is a dissipation term. It has been found by Alcubierre et al. [87] crucial to add it in order to avoid strong oscillations in the shift.

The hyperbolic Gamma driver condition (10.60) is equivalent to the following first order system

$$\begin{cases} \dfrac{\partial \beta^i}{\partial t} = kB^i \\ \dfrac{\partial B^i}{\partial t} = \dfrac{\partial \tilde{\Gamma}^i}{\partial t} - \eta B^i. \end{cases} \tag{10.62}$$

Remark 10.10 In the case where K does not depend on t, the Gamma driver condition (10.60) reduces to a previous hyperbolic condition proposed by Alcubierre et al. [90], namely

$$\frac{\partial^2 \beta^i}{\partial t^2} = k\frac{\partial \tilde{\Gamma}^i}{\partial t} - \eta\frac{\partial \beta^i}{\partial t}. \tag{10.63}$$

Hyperbolic Gamma driver conditions have been employed in many numerical computations:

- 3D gravitational collapse calculations by Baiotti et al. (2005, 2006) [70, 72], with $k = 3/4$ and $\eta = 3/M$, where M is the ADM mass;
- the first evolution of a binary black hole system lasting for about one orbit by Brügmann, Tichy and Jansen (2004) [91], with $k = 3/4N\Psi^{-2}$ and $\eta = 2/M$;
- binary black hole mergers by
  - Campanelli et al. (2006) [59–62], with $k = 3/4$;
  - Baker et al. (2006) [56, 57], with $k = 3N/4$ and a slightly modified version of Eq. (10.62), namely $\partial\tilde{\Gamma}^i/\partial t$ replaced by $\partial\tilde{\Gamma}^i/\partial t - \beta^j\partial\tilde{\Gamma}^i/\partial x^j$ in the second equation;
  - Sperhake [63], with $k = 1$ and $\eta = 1/M$.

van Meter et al. [58] and Brügmann et al. [65] have considered a modified version of Eq. (10.62), by replacing all the derivatives $\partial/\partial t$ by $\partial/\partial t - \beta^j\partial/\partial x^j$, i.e. writing

$$\begin{cases} \dfrac{\partial \beta^i}{\partial t} - \beta^j \dfrac{\partial \beta^i}{\partial x^j} = kB^i \\ \dfrac{\partial B^i}{\partial t} - \beta^j \dfrac{\partial B^i}{\partial x^j} = \dfrac{\partial \tilde{\Gamma}^i}{\partial t} - \beta^j \dfrac{\partial \tilde{\Gamma}^i}{\partial x^j} - \eta B^i. \end{cases} \tag{10.64}$$

In particular, Brügmann et al. [65, 66] have computed binary black hole mergers using (10.64) with $k = 3/4$ and η ranging from 0 to $3.5/M$, whereas Herrmann et al. [68] have used (10.64) with $k = 3/4$ and $\eta = 2/M$. The choice (10.64) has also been used by Baiotti, Giacomazzo and Rezzolla [73, 74] to compute binary neutron star mergers (with $k = 3/4$ and $\eta = 1$), as well as by Kyutoku et al. [76] for neutron star - black hole mergers (with $k = 3/4$ and η equal to the ratio of a solar mass to the irreducible mass of the black hole). Finally let us mention the recent work by Alic et al. [92] which discuss the shortcomings of the Gamma driver for long-term evolution of unequal-mass binary systems and propose some improvement to it.

10.3.6 Other Dynamical Shift Gauges

Shibata [35] has introduced a spatial gauge that is closely related to the hyperbolic Gamma driver: it is defined by the requirement

$$\frac{\partial \beta^i}{\partial t} = \tilde{\gamma}^{ij} \left(F_j + \delta t \frac{\partial F_j}{\partial t} \right), \tag{10.65}$$

where δt is the time step used in the numerical computation and[3]

$$F_i := \bar{D}^j \tilde{\gamma}_{ij}. \tag{10.66}$$

From the definition of the inverse metric $\tilde{\gamma}^{ij}$, namely the identity $\tilde{\gamma}^{ik}\tilde{\gamma}_{kj} = \delta^i{}_j$, and relation (10.54), it is easy to show that F_i is related to $\tilde{\Gamma}^i$ by

$$F_i = \tilde{\gamma}_{ij}\tilde{\Gamma}^j - \left(\tilde{\gamma}^{jk} - f^{jk}\right) \bar{D}_k \tilde{\gamma}_{ij}. \tag{10.67}$$

Notice that in the weak field region, i.e. where $\tilde{\gamma}^{ij} = f^{ij} + h^{ij}$ with $f_{ik} f_{jl} h^{kl} h^{ij} \ll 1$, the second term in Eq. (10.67) is of second order in $\boldsymbol{h}$, so that at first order in $\boldsymbol{h}$, Eq. (10.67) reduces to $F_i \simeq \tilde{\gamma}_{ij}\tilde{\Gamma}^j$. Accordingly Shibata's prescription (10.65) becomes

$$\frac{\partial \beta^i}{\partial t} \simeq \tilde{\Gamma}^i + \tilde{\gamma}^{ij} \delta t \frac{\partial F_j}{\partial t}. \tag{10.68}$$

If we disregard the δt term in the right-hand side and take the time derivative of this equation, we obtain the Gamma-driver condition (10.60) with $k = 1$ and $\eta = 0$.

---

[3] Let us recall that $\bar{D}^i := f^{ij} \bar{D}_j$.

The term in δt has been introduced by Shibata [35] in order to stabilize the numerical code.

The spatial gauge (10.65) has been used by Shibata [35] and Sekiguchi and Shibata [36, 39] to compute axisymmetric gravitational collapse to back hole of rapidly rotating neutron stars [36] and stellar cores [69], as well as by Shibata and Sekiguchi [37] to compute 3D gravitational collapses, allowing for the development of nonaxisymmetric instabilities. It has also been used by Shibata, Taniguchi and Uryu [31, 32, 33] and Kiuchi, Sekiguchi, Shibata, and Taniguchi [75, 93] to compute the merger of binary neutron stars, while their preceding computations [29, 30] rely on the approximate minimal distortion gauge (Sect. 10.3.3).

10.4 Full Spatial Coordinate-Fixing Choices

The spatial coordinate choices discussed in Sect. 10.3, namely vanishing shift, minimal distortion, Gamma freezing, Gamma driver and related prescriptions, are relative to the *propagation* of the coordinates (x^i) away from the initial hypersurface Σ_0. They do not restrict at all the choice of coordinates in Σ_0. Here we discuss some coordinate choices that fix completely the coordinate freedom, including in the initial hypersurface.

10.4.1 Spatial Harmonic Coordinates

The first full coordinate-fixing choice we shall discuss is that of ***spatial harmonic coordinates***. They are defined by

$$\boxed{D_j D^j x^i = 0}\,, \tag{10.69}$$

in full analogy with the spacetime harmonic coordinates [cf. Eq. (10.16)]. The above condition is equivalent to

$$\frac{1}{\sqrt{\gamma}}\frac{\partial}{\partial x^j}\Big(\sqrt{\gamma}\gamma^{jk}\underbrace{\frac{\partial x^i}{\partial x^k}}_{\delta^i{}_k}\Big) = 0,$$

i.e.

$$\frac{\partial}{\partial x^j}\left(\sqrt{\gamma}\gamma^{ij}\right) = 0. \tag{10.70}$$

This relation restricts the coordinates to be of Cartesian type. Notably, it forbids the use of spherical-type coordinates, even in flat space, for it is violated by $\gamma_{ij} =$

diag$(1, r^2, r^2 \sin^2\theta)$. To allow for any type of coordinates, let us rewrite condition (10.70) in terms of a background flat metric $\boldsymbol{f}$ (cf. discussion in Sect. 7.2.2), as

$$\boxed{\bar{D}_j \left[\left(\frac{\gamma}{f} \right)^{1/2} \gamma^{ij} \right] = 0}, \qquad (10.71)$$

where $\bar{\boldsymbol{D}}$ is the connection associated with $\boldsymbol{f}$ and $f := \det f_{ij}$ is the determinant of $\boldsymbol{f}$ with respect to the coordinates (x^i).

Spatial harmonic coordinates have been considered by Čadež [94] for binary black holes and by Andersson and Moncrief [95] in order to put the 3+1 Einstein system into an elliptic-hyperbolic form and to show that the corresponding Cauchy problem is well posed.

Remark 10.11 The spatial harmonic coordinates discussed above should not be confused with *spacetime* harmonic coordinates; the latter would be defined by $\Box_{\boldsymbol{g}} x^i = 0$ [spatial part of Eq. (10.16)] instead of (10.69). Spacetime harmonic coordinates, as well as some generalizations, are considered e.g. in Ref. [96].

10.4.2 Dirac Gauge

As a natural way to fix the coordinates in his Hamiltonian formulation of general relativity (cf. Sect. 5.5), Dirac [97] has introduced in 1959 the following condition:

$$\frac{\partial}{\partial x^j} \left(\gamma^{1/3} \gamma^{ij} \right) = 0. \qquad (10.72)$$

It differs from the definition (10.70) of spatial harmonic coordinates only by the power of the determinant γ. Similarly, we may rewrite it more covariantly in terms of the background flat metric $\boldsymbol{f}$ as [25]

$$\boxed{\bar{D}_j \left[\left(\frac{\gamma}{f} \right)^{1/3} \gamma^{ij} \right] = 0}. \qquad (10.73)$$

We recognize in this equation the inverse conformal metric [cf. Eqs. (7.15) and (7.21)], so that we may write:

$$\boxed{\bar{D}_j \tilde{\gamma}^{ij} = 0}. \qquad (10.74)$$

We call this condition the ***Dirac gauge***. It has been first discussed in the context of numerical relativity in 1978 by Smarr and York [43] but disregarded in profit of the minimal distortion gauge (Sect. 10.3.2), for the latter leaves the freedom to choose the coordinates in the initial hypersurface. In terms of the vector $\tilde{\boldsymbol{\Gamma}}$ introduced in Sect. 10.3.4, the Dirac gauge has a simple expression, thanks to relation (10.54):

$$\boxed{\tilde{\Gamma}^i = 0}. \tag{10.75}$$

It is clear that if the coordinates (x^i) obey the Dirac gauge at all times t, then they belong to the Gamma freezing class discussed in Sect. 10.3.4, for Eq. (10.75) implies Eq. (10.55). Accordingly, the shift vector of Dirac-gauge coordinates has to satisfy the Gamma freezing elliptic equation (10.58), with the additional simplification $\tilde{\Gamma}^i = 0$:

$$\boxed{\begin{aligned}\tilde{\gamma}^{jk}\bar{D}_j\bar{D}_k\beta^i + \frac{1}{3}\tilde{\gamma}^{ij}\bar{D}_j\bar{D}_k\beta^k =& 2N\Big[8\pi\Psi^4 p^i - \tilde{A}^{jk}\left(\tilde{\Gamma}^i{}_{jk} - \bar{\Gamma}^i{}_{jk}\right) - 6\tilde{A}^{ij}\bar{D}_j \ln\Psi \\ & + \frac{2}{3}\tilde{\gamma}^{ij}\bar{D}_j K\Big] + 2\tilde{A}^{ij}\bar{D}_j N.\end{aligned}} \tag{10.76}$$

The Dirac gauge, along with maximal slicing, has been employed by Bonazzola et al. [25] to devise a constrained scheme[4] (see also [98, 99, 100]), which has been applied to evolutions of gravitational waves in spacetimes with matter content [101]. It has also been used by Shibata, Uryu and Friedman [102] to formulate waveless approximations of general relativity that go beyond the IWM approximation discussed in Sect. 7.6. Such a formulation has been employed to compute quasi-equilibrium configurations of binary neutron stars [103, 104]. Since Dirac gauge is a full coordinate-fixing gauge, the initial data must fulfill it. Recently, Lin and Novak [105] have computed equilibrium configurations of rapidly rotating stars within the Dirac gauge, which may serve as initial data for gravitational collapse.

References

1. Smarr, L., York, J.W.: Kinematical conditions in the construction of spacetime. Phys. Rev. D **17**, 2529 (1978)
2. York, J.W.: Kinematics and dynamics of general relativity. In: Smarr, L.L.(ed.) Sources of Gravitational Radiation. Cambridge University Press, Cambridge, **83**, 83 (1979)
3. Alcubierre. M.: The status of numerical relativity in general relativity and gravitation. In: Florides, P., Nolan B., Ottewill A. (eds.) Proceedings of the 17th International Conference, Dublin, 18–23 July 2004. World Scientific (2005)
4. Baumgarte, T.W., Shapiro, S.L.: Numerical relativity and compact binaries. Phys. Rep. **376**, 41 (2003)
5. Lehner, L.: Numerical relativity: a review. Class. Quantum Grav. **18**, R25 (2001)
6. Alcubierre, M.: Introduction to 3+1 Numerical Relativity. Oxford University Press, Oxford (2008)
7. Baumgarte, T.W., Shapiro, S.L.: Numerical Relativity. Solving Einstein's Equations on the Computer. Cambridge University Press, Cambridge (2010)

[4] The concept of *constrained scheme* will be discussed in Sect. 11.2.

8. Choquet-Bruhat, Y.: General Relativity and Einstein's Equations. Oxford University Press, New York (2009)
9. Nakamura, T., Oohara, K., Kojima, Y.: General relativistic collapse to black holes and gravitational waves from black holes. Prog. Theor. Phys. Suppl. **90**, 1 (1987)
10. Shibata, M., Nakamura, T.: Evolution of three-dimensional gravitational waves: Harmonic slicing case. Phys. Rev. D **52**, 5428 (1995)
11. Estabrook, F., Wahlquist, H., Christensen, S., DeWitt, B., Smarr, L., Tsiang, E.: Maximally Slicing a Black Hole. Phys. Rev. D **7**, 2814 (1973)
12. Reinhart, B.L.: Maximal foliations of extended Schwarzschild space. J. Math. Phys. **14**, 719 (1973)
13. Beig, R., O'Murchadha, N.: Late time behavior of the maximal slicing a the Schwarzschild black hole. Phys. Rev. D **57**, 4728 (1998)
14. Beig, R.: The maximal slicing of a Schwarzschild black hole. Ann. Phys. (Leipzig) **11**, 507 (2000)
15. Reimann, B., Brügmann, B.: Maximal slicing for puncture evolutions of Schwarzschild and Reissner-Nordström black holes. Phys. Rev. D **69**, 044006 (2004)
16. Petrich, L.I., Shapiro, S.L, Teukolsky, S.A.: Oppenheimer-Snyder collapse with maximal time slicing and isotropic coordinates. Phys. Rev. D **31**, 2459 (1985)
17. Smarr, L., Čadež, A., DeWitt, B., Eppley, K.: Collision of two black holes: Theoretical framework. Phys. Rev. D **14**, 002443 (1976)
18. Smarr, L.: Gauge Conditions, Radiation Formulae and the Two Black Hole Collisions. In: Smarr, L.L. (ed.) Sources of Gravitational Radiation. Cambridge University Press, Cambridge, 245 (1979)
19. Nakamura, T.: General relativistic collapse of axially symmetric stars leading to the formation of rotating black holes. Prog. Theor. Phys. **65**, 1876 (1981)
20. Nakamura, T., Sato, H.: General relativistic colaapse of rotating supermassive stars. Prog. Theor. Phys. **66**, 2038 (1981)
21. Stark, R.F., Piran, T.: Gravitational-wave emission from rotating gravitational collapse. Phys. Rev. Lett. **55**, 891 (1985)
22. Evans C.R.: An approach for calculating axisymmetric gravitational collapse. In: Centrella, J. (ed) Dynamical spacetimes and numerical relativity. Cambridge University Press, Cambridge, p. 3 (1986)
23. Bardeen, J.M., Piran.: General relativistic axisymmetric rotating systems: coordinates and equations. Phys. Rep. **96**, 206 (1983)
24. Grandclément, P., Bonazzola, S., Gourgoulhon, E., Marck, J.-.A.: A multi-domain spectral method for scalar and vectorial Poisson equations with non-compact sources. J. Comput. Phys. **170**, 231 (2001)
25. Bonazzola, S., Gourgoulhon, E., Grandclément, P., Novak, J.: Constrained scheme for the Einstein equations based on the Dirac gauge and spherical coordinates. Phys. Rev. D **70**, 104007 (2004)
26. Brügmann, B.: Binary black hole mergers in 3d numerical relativity. Int. J. Mod. Phys. D **8**, 85 (1999)
27. Shibata, M.: 3D numerical simulations of black hole formation using collisionless particles. Prog. Theor. Phys. **101**, 251 (1999)
28. Shibata, M.: Fully general relativistic simulation of coalescing binary neutron stars: Preparatory tests. Phys. Rev. D **60**, 104052 (1999)
29. Shibata, M., Uryu, K.: Simulation of merging binary neutron stars in full general relativity: $\Gamma = 2$ case. Phys. Rev. D **61**, 064001 (2000)
30. Shibata, M., Uryu, K.: Gravitational waves from the merger of binary neutron stars in a fully general relativistic simulation. Prog. Theor. Phys. **107**, 265 (2002)
31. Shibata, M., Taniguchi, K., Uryu, K.: Merger of binary neutron stars of unequal mass in full general relativity. Phys. Rev. D **68**, 084020 (2003)

32. Shibata, M., Taniguchi, K., Uryu, K.: Merger of binary neutron stars with realistic equations of state in full general relativity. Phys. Rev. D **71**, 084021 (2005)
33. Shibata, M., Taniguchi, K.: Merger of binary neutron stars to a black hole: Disk mass, short gamma-ray bursts, and quasinormal mode ringing. Phys. Rev. D **73**, 064027 (2006)
34. Shibata, M.: Axisymmetric general relativistic hydrodynamics: Long-term evolution of neutron stars and stellar collapse to neutron stars and black holes. Phys. Rev. D **67**, 024033 (2003)
35. Shibata, M.: Collapse of rotating supramassive neutron stars to black holes: fully general relativistic simulations. Astrophys. J. **595**, 992 (2003)
36. Sekiguchi, Y., Shibata, M.: Axisymmetric collapse simulations of rotating massive stellar cores in full general relativity: Numerical study for prompt black hole formation. Phys. Rev. D **71**, 084013 (2005)
37. Shibata, M., Sekiguchi, Y.: Three-dimensional simulations of stellar core collapse in full general relativity: Nonaxisymmetric dynamical instabilities. Phys. Rev. D **71**, 024014 (2005)
38. De Donder, T.: La Gravifique einsteinienne. Gauthier-Villars, Paris (1921). A related downloadable article is http://www.numdam.org/item?id=AIHP_1930__1_2_77_0
39. Fourès-Bruhat, Y. (Choquet-Bruhat, Y.): Théorème d'existence pour certains systèmes d'équations aux dérivées partielles non linéaires. Acta Mathematica **88**, 141 (1952); available at http://fanfreluche.math.univ-tours.fr
40. Choquet-Bruhat, Y., Ruggeri, T.: Hyperbolicity of the 3+1 system of Einstein equations. Commun. Math. Phys. **89**, 269 (1983)
41. Bona, C., Massó, J.: Harmonic synchronizations of spacetime. Phys. Rev. D **38**, 2419 (1988)
42. Reula, O.A., Hyperbolic methods for Einstein's equations. Living Rev. Relativity 1:3 (1998); http://www.livingreviews.org/lrr-1998-3
43. Smarr, L., York, J.W.: Radiation gauge in general relativity. Phys. Rev. D **17**, 1945 (1978)
44. Cook, G.B., Scheel, M.A.: Well-behaved harmonic time slices of a charged, rotating, boosted black hole. Phys. Rev. D **56**, 4775 (1997)
45. Shibata, M., Nakamura, T.: Conformal time slicing condition in Three Dimensional numerical relativity. Prog. Theor. Phys. **88**, 317 (1992)
46. Bona, C., Massó, J., Seidel, E., Stela, J.: New formalism for numerical relativity. Phys. Rev. Lett. **75**, 600 (1995)
47. Bernstein, D.H.: A Numerical Study of the Black Hole Plus Brill Wave Spacetime. PhD Thesis, Department of Physics, University of Illinois at Urbana-Champaign (1993)
48. Anninos, P., Massó, J.,Seidel, E.,Suen, W.-M., Towns, J.: Three-dimensional numerical relativity: The evolution of black holes. Phys. Rev. D 52:2059 (1995)
49. Bona, C., Massó, J., Seidel, E., Stela, J.: First order hyperbolic formalism for numerical relativity. Phys. Rev. D **56**, 3405 (1997)
50. Piran, T.: Methods of Numerical Relativity. In: Deruelle, N., Piran T. (eds.) Rayonnement gravitationnel Gravitation Radiation. North Holland (1983)
51. Hannam, M., Husa, S., Ohme, F., Brügmann, B., Ó Murchadha, N.: Wormholes and trumpets: Schwarzschild spacetime for the moving-puncture generation. Phys. Rev. D **78**, 064020 (2008)
52. Baumgarte, T.W., Naculich, S.G.: Analytical representation of a black hole puncture solution. Phys. Rev. D **75**, 067502 (2007)
53. Brown, J.D.: Probing the puncture for black hole simulations. Phys. Rev. D **80**, 084042 (2009)
54. Alcubierre, M.: Appearance of coordinate shocks in hyperbolic formalisms of general relativity. Phys. Rev. D **55**, 5981 (1997)
55. Font, J.A., Goodale, T., Iyer, S., Miller, M., Rezzolla, L., Seidel, E., Stergioulas, N., Suen, W.-.M., Tobias, M.: Three-dimensional numerical general relativistic hydrodynamics. II. Long-term dynamics of single relativistic stars. Phys. Rev. D **65**, 084024 (2002)
56. Baker, J.G., Centrella, J., Choi, D.-.I., Koppitz, M., van Meter, J.: Gravitational-wave extraction from an inspiraling configuration of merging black holes. Phys. Rev. Lett. **96**, 111102 (2006)

57. Baker, J.G., Centrella, J., Choi, D.-.I., Koppitz, M., van Meter, J.: Binary black hole merger dynamics and waveforms. Phys. Rev. D **73**, 104002 (2006)
58. van Meter, J.R., Baker, J.G., Koppitz, M., Choi, D.I.: How to move a black hole without excision: gauge conditions for the numerical evolution of a moving puncture. Phys. Rev. D **73**, 124011 (2006)
59. Campanelli, M., Lousto, C.O., Marronetti, P., Zlochower, Y.: Accurate evolutions of orbiting black-hole binaries without excision. Phys. Rev. Lett. **96**, 111101 (2006)
60. Campanelli, M., Lousto, C.O., Zlochower, Y.: Last orbit of binary black holes. Phys. Rev. D **73**, 061501(R) (2006)
61. Campanelli, M., Lousto, C.O., Zlochower, Y.: Spinning-black-hole binaries: The orbital hang-up. Phys. Rev. D(R) **74**, 041501 (2006)
62. Campanelli, M., Lousto, C.O., Zlochower, Y.: Spin-orbit interactions in black-hole binaries. Phys. Rev. D **74**, 084023 (2006)
63. Sperhake, U.: Binary black-hole evolutions of excision and puncture data. Phys. Rev. D **76**, 104015 (2007)
64. Diener, P., Herrmann, F., Pollney, D., Schnetter, E., Seidel, E., Takahashi, R., Thornburg, J., Ventrella, J.: Accurate evolution of orbiting binary black holes. Phys. Rev. Lett. **96**, 121101 (2006)
65. Brügmann, B., González, J.A., Hannam, M., Husa, S., Sperhake, U., Tichy, W.: Calibration of moving puncture simulations. Phys. Rev. D **77**, 024027 (2008)
66. Marronetti, P., Tichy, W., Brügmann, B., González, J., Hannam, M., Husa, S., Sperhake, U.: Binary black holes on a budget: simulations using workstations. Class. Quantum Grav. **24**, S43 (2007)
67. Herrmann, F., Hinder, I., Shoemaker, D., Laguna, P.: Unequal mass binary black hole plunges and gravitational recoil. Class. Quantum Grav. **24**, S33 (2007)
68. Herrmann, F., Hinder, I., Shoemaker, D., Laguna, P., Matzner, R.A: Gravitational Recoil from Spinning Binary Black Hole Mergers. Astrophys. J. **661**, 430 (2007)
69. Sekiguchi, Y., Shibata, M.: Formation of black hole and accretion disk in a massive high-entropy stellar core collapse. Astrophys. J. **737**, 6, (2011)
70. Baiotti, L., Hawke, I., Montero, P.J., Löffler, F., Rezzolla, L., Stergioulas, N., Font, J.A., Seidel E.: Three-dimensional relativistic simulations of rotating neutron-star collapse to a Kerr black hole: Phys. Rev. D 71. 024035 (2005)
71. Baiotti, L., Hawke, I., Rezzolla, L., Schnetter, E.: Gravitational-wave emission from rotating gravitational collapse in Three Dimensions. (2005)
72. Baiotti, L., Rezzolla, L.: Challenging the paradigm of singularity excision in gravitational collapse. Phys. Rev. Lett. **97**, 141101 (2006)
73. Baiotti, L., Giacomazzo, B., Rezzolla, L.: Accurate evolutions of inspiralling neutron-star binaries: Prompt and delayed collapse to a black hole. Phys. Rev. D **78**, 084033 (2008)
74. Giacomazzo, B., Rezzolla, L., Baioti, L.: Accurate evolutions of inspiralling and magnetized neutron stars: Equal-mass binaries. Phys. Rev. D **83**, 044014 (2011)
75. Kiuchi, K., Sekiguchi, Y., Shibata, M., Taniguchi, K.: Long-term general relativistic simulation of binary neutron stars collapsing to a black hole. Phys. Rev. D **80**, 064037 (2009)
76. Kyutoku, K., Shibata, M., Taniguchi, K.: Gravitational waves from nonspinning black hole-neutron star binaries: Dependence on equations of state. Phys. Rev. D **82**, 044049 (2010)
77. Bardeen, J.M.: Gauge and radiation conditions in numerical relativity. In: Deruelle, N., Piran, T. (eds.) Rayonnement gravitationnel Gravitation Radiation. North Holland, Amsterdam, p. 433 (1983)
78. Baumgarte, T.W., Shapiro, S.L.: Numerical integration of Einstein's field equations. Phys. Rev. D **59**, 024007 (1999)
79. Alcubierre, M., Benger, W., Brügmann, B., Lanfermann, G., Nerger, L., Seidel, E., Takahashi, R.: 3D Grazing collision of two black holes. Phys. Rev. Lett. **87**, 271103 (2001)
80. Jantzen, R.T., York, J.W.: New minimal distortion shift gauge. Phys. Rev. D **73**, 104008 (2006)

81. Nakamura, T.: 3D Numerical Relativity. In: Sasaki, M. (ed.) Relativistic Cosmology, Proceedings of the 8th Nishinomiya-Yukawa Memorial Symposium. Universal Academy Press, Tokyo, p. 155 (1994)
82. Oohara, K., Nakamura, T., Shibata, M.: A way to 3D numerical relativity. Prog. Theor. Phys. Suppl. **128**, 183 (1997)
83. Shibata, M.: Fully general relativistic simulation of merging binary clusters—spatial gauge condition. Prog. Theor. Phys. **101**, 1199 (1999)
84. Shibata, M., Baumgarte, T.W., Shapiro, S.L.: Stability and collapse of rapidly rotating, supramassive neutron stars: 3D simulations in general relativity. Phys. Rev. D **61**, 044012 (2000)
85. Shibata, M., Baumgarte, T.W., Shapiro, S.L.: The bar-mode instability in differentially rotating neutron stars: simulations in full general relativity. Astrophys. J. **542**, 453 (2000)
86. Alcubierre, M., Brügmann, B.: Simple excision of a black hole in 3+1 numerical relativity. Phys. Rev. D **63**, 104006 (2001)
87. Alcubierre, M., Brügmann, B., Diener, P., Koppitz, M., Pollney, D., Seidel, E., Takahashi, R.: Gauge conditions for long-term numerical black hole evolutions without excision. Phys. Rev. D **67**, 084023 (2003)
88. Lindblom, L., Scheel, M.A.: Dynamical gauge conditions for the Einstein evolution equations. Phys. Rev. D **67**, 124005 (2003)
89. Bona, C., Lehner, L., Palenzuela-Luque, C.: Geometrically motivated hyperbolic coordinate conditions for numerical relativity: Analysis, issues and implementations. Phys. Rev. D **72**, 104009 (2005)
90. Alcubierre, M., Brügmann, B., Pollney, D., Seidel, E., Takahashi, R.: Black hole excision for dynamic black holes. Phys. Rev. D **64**, 061501 (2001)
91. Brügmann, B., Tichy, W., Jansen, N.: Numerical simulation of orbiting black holes. Phys. Rev. Lett. **92**, 211101 (2004)
92. Alic, D., Rezzolla, L., Hinder, I., Mösta, P.: Dynamical damping terms for symmetry-seeking shift conditions. Class. Quantum Grav. **27**, 245023 (2010)
93. Sekiguchi, Y., Kiuchi, K., Kyutoku, K., Shibata, M.: Gravitational waves and neutrino emission from the merger of binary neutron stars. Phys. Rev. Lett. **107**, 051102 (2011)
94. Čadež, A.: Some remarks on the two-body-problem in geometrodynamics: Ann. Phys. N.Y. **91**, 58 (1975)
95. Andersson, L., Moncrief, V.: Elliptic-Hyperbolic Systems and the Einstein Equations. Ann. Henri Poincaré **4**, 1 (2003)
96. Alcubierre, M., Corichi, A., González, J.A., Núnez, D., Reimann, B., Salgado, M.: Generalized harmonic spatial coordinates and hyperbolic shift conditions. Phys. Rev. D **72**, 124018 (2005)
97. Dirac, P.A.M.: Fixation of coordinates in the Hamiltonian theory of gravitation. Phys. Rev. **114**, 924 (1959)
98. Cordero-Carrión, I., Cerdá-Durán, P., Dimmelmeier, H., Jaramillo, J.L, Novak, J., Gourgoulhon, E.: Improved constrained scheme for the Einstein equations: An approach to the uniqueness issue. Phys. Rev. D **79**, 024017 (2009)
99. Cordero-Carrión, I., Cerdá-Durán, P., Ibánez, J.M.: Dynamical spacetimes and gravitational radiation in a Fully Constrained Formulation. J. Phys.: Conf. Ser. **228**, 012055 (2010)
100. Cordero-Carrión, I., Ibánez, J.M., Gourgoulhon, E., Jaramillo, J.L., Novak, J.: Mathematical issues in a fully-constrained formulation of Einstein equations. Phys. Rev. D **77**, 084007 (2008)
101. Cordero-Carrión, I., Cerdá-Durán. P., Ibánez, J.M.: Gravitational waves in dynamical spacetimes with matter content in the Fully Constrained Formulation. preprint arXiv:1108.0571.
102. Shibata, M., Uryu, K., Friedman, J.L.: Deriving formulations for numerical computation of binary neutron stars in quasicircular orbits: Phys. Rev. D 70. 044044 (2004); errata in Phys. Rev. D 70:129901(E) (2004)
103. Uryu, K., Limousin, F., Friedman, J.L., Gourgoulhon, E., Shibata, M.: Binary neutron stars: equilibrium models beyond spatial conformal flatness. Phys. Rev. Lett. **97**, 171101 (2006)

104. Uryu, K., Limousin, F., Friedman, J.L., Gourgoulhon, E., Shibata, M.: Nonconformally flat initial data for binary compact objects. Phys. Rev. D **80**, 124004 (2009)
105. Lin, L.-.M., Novak, J.: Rotating star initial data for a constrained scheme in numerical relativity. Class. Quantum Grav. **23**, 4545 (2006)

Chapter 11
Evolution Schemes

Abstract Various approaches to evolve forward in time the 3+1 Einstein equations are discussed. After a review of constrained schemes, we focus on free evolution schemes, giving some details about the propagation of the constraints. Among the free evolution schemes, a particular important one is the BSSN scheme, which is presented here in details.

11.1 Introduction

Even after having selected the foliation and the propagation of spatial coordinates (Chap. 10), there remains various strategies to integrate the 3+1 Einstein equations, either in their original form (5.68)–(5.71), or in the conformal form (7.88)–(7.93). In particular, the constraint equations (5.70)–(5.71) or (7.92)–(7.93) may be solved or not during the evolution, giving rise to respectively the so-called *free evolution schemes* and the *constrained schemes*. We discuss here the two types of schemes (Sects. 11.2 and 11.3), and present afterwards a widely used free evolution scheme: the BSSN one (Sect. 11.4).

Some review articles on the subject are those by Stewart [1], Friedrich and Rendall [2], Lehner [3], Shinkai and Yoneda [4, 5], Baumgarte and Shapiro [6], and Lehner and Reula [7]. We also recommend the textbooks by Alcubierre [8] and Baumgarte and Shapiro [9].

11.2 Constrained Schemes

A ***constrained scheme*** is a time scheme for integrating the 3+1 Einstein system in which some (***partially constrained scheme***) or all (***fully constrained scheme***) of the four constraints are used to compute some metric coefficients at each step of the numerical evolution.

É. Gourgoulhon, *3+1 Formalism in General Relativity*, Lecture Notes in Physics 846,
DOI: 10.1007/978-3-642-24525-1_11,

In the eighties, partially constrained schemes, with only the Hamiltonian constraint enforced, have been widely used in 2-D (axisymmetric) computations (e.g. Bardeen and Piran [10], Stark and Piran [11], Evans [12]). Still in the 2-D axisymmetric case, fully constrained schemes have been used by Evans [13] and Shapiro and Teukolsky [14] for non-rotating spacetimes, and by Abrahams, Cook, Shapiro and Teukolsky [15] for rotating ones. More recently the axisymmetric codes of Choptuik, Hirschmann, Liebling and Pretorius [16] and Rinne [17] are based on a constrained scheme too.

Regarding 3D numerical relativity, a fully constrained scheme based on the original 3+1 Einstein system (5.68)–(5.71) has been used to evolve a single black hole by Anderson and Matzner [18]. Another fully constrained scheme has been devised by Bonazzola, Gourgoulhon, Grandclément and Novak [19], but this time for the conformal 3+1 Einstein system (7.88)–(7.93). It makes use of maximal slicing and Dirac gauge (Sect. 10.4.2). This scheme has been improved by Cordero-Carrión et al. [20] to fix a non-uniqueness issue similar to the XCTS one discussed in Sect. 9.3.4. A mathematical analysis of this scheme has been performed in Ref. [21] and some applications are presented in Refs. [19–22, 23].

11.3 Free Evolution Schemes

11.3.1 Definition and Framework

A ***free evolution scheme*** is a time scheme for integrating the 3+1 Einstein system in which the constraint equations are solved only to get the initial data, e.g. by following one of the prescriptions discussed in Chap. 9. The subsequent evolution is performed via the dynamical equations only, without enforcing the constraints. Actually, facing the 3+1 Einstein system (5.68)–(5.71), we realize that the dynamical equation (5.69), coupled with the kinematic relation (5.68) and some choices for the lapse function and shift vector (as discussed in Chap. 10), is sufficient to get the values of $\boldsymbol{\gamma}$, $\boldsymbol{K}$, N and $\boldsymbol{\beta}$ at all times t, from which we can reconstruct the full spacetime metric $\boldsymbol{g}$.

A natural question which arises then is: to which extent does the metric $\boldsymbol{g}$ hence obtained fulfill the Einstein equation (5.1)? The dynamical part, Eq. (5.69), is fulfilled by construction, but what about the constraints (5.70) and (5.71)? If they were violated by the solution $(\boldsymbol{\gamma}, \boldsymbol{K})$ of the dynamical equation, then the obtained metric $\boldsymbol{g}$ would not satisfy Einstein equation. The key point is that, as we shall see in Sect. 11.3.2, provided that the constraints are satisfied at $t=0$, the dynamical equation (5.69) ensures that they are satisfied for all $t > 0$.

11.3.2 Propagation of the Constraints

Let us derive evolution equations for the constraints, or more precisely, for the constraint violations. These evolution equations will be consequences of the Bianchi identities.[1] We denote by $\boldsymbol{G}$ the Einstein tensor [cf. Eq. (2.80)]

$$\boldsymbol{G} := {}^4\boldsymbol{R} - \frac{1}{2}{}^4R\boldsymbol{g}, \tag{11.1}$$

so that the Einstein equation (5.1) is written

$$\boldsymbol{G} = 8\pi\boldsymbol{T}. \tag{11.2}$$

The ***Hamiltonian constraint violation*** is the scalar field defined by

$$\boxed{H := \boldsymbol{G}(\boldsymbol{n},\boldsymbol{n}) - 8\pi\boldsymbol{T}(\boldsymbol{n},\boldsymbol{n})}, \tag{11.3}$$

i.e.

$$H = {}^4\boldsymbol{R}(\boldsymbol{n},\boldsymbol{n}) + \frac{1}{2}{}^4R - 8\pi E, \tag{11.4}$$

where we have used the relations $\boldsymbol{g}(\boldsymbol{n},\boldsymbol{n}) = -1$ and $\boldsymbol{T}(\boldsymbol{n},\boldsymbol{n}) = E$ [Eq. (5.4)]. Thanks to the scalar Gauss equation (3.75) we may write

$$\boxed{H = \frac{1}{2}\left(R + K^2 - K_{ij}K^{ij}\right) - 8\pi E}. \tag{11.5}$$

Similarly we define the ***momentum constraint violation*** as the 1-form field

$$\boxed{\boldsymbol{M} := -\boldsymbol{G}(\boldsymbol{n}, \vec{\boldsymbol{\gamma}}(.)) + 8\pi\boldsymbol{T}(\boldsymbol{n}, \vec{\boldsymbol{\gamma}}(.))}. \tag{11.6}$$

By means of the contracted Codazzi equation (3.82) and the relation $\boldsymbol{T}(\boldsymbol{n}, \vec{\boldsymbol{\gamma}}(.)) = -\boldsymbol{p}$ [Eq. (5.5)], we get

$$\boxed{M_i = D_j K^j{}_i - D_i K - 8\pi p_i}, \tag{11.7}$$

From the above expressions, we see that the Hamiltonian constraint (5.70) and the momentum constraint (5.71) are equivalent to respectively

$$H = 0 \tag{11.8}$$

$$M_i = 0. \tag{11.9}$$

[1] The following computation is inspired from Frittelli's article [24].

Finally we define the ***dynamical equation violation*** as the spatial tensor field

$$\boxed{\boldsymbol{F} := \vec{\boldsymbol{\gamma}}^{*4}\boldsymbol{R} - 8\pi\vec{\boldsymbol{\gamma}}^{*}\left(\boldsymbol{T} - \frac{1}{2}T\boldsymbol{g}\right)}. \tag{11.10}$$

Indeed, let us recall that the dynamical part of the 3+1 Einstein system, Eq. (5.69) is nothing but the spatial projection of the Einstein equation written in terms of the Ricci tensor ${}^4\boldsymbol{R}$, i.e. Eq. (5.2), instead of the Einstein tensor, i.e. Eq. (11.2) (cf. Sect. 5.1.3). Introducing the stress tensor $\boldsymbol{S} = \vec{\boldsymbol{\gamma}}^{*}\boldsymbol{T}$ [Eq. Sect. 5.11] and using the relations $T = S - E$ [Eq. (5.15)] and $\vec{\boldsymbol{\gamma}}^{*}\boldsymbol{g} = \boldsymbol{\gamma}$, we can write $\boldsymbol{F}$ as

$$\boldsymbol{F} = \vec{\boldsymbol{\gamma}}^{*4}\boldsymbol{R} - 8\pi\left[\boldsymbol{S} + \frac{1}{2}(E - S)\boldsymbol{\gamma}\right]. \tag{11.11}$$

From Eq. (5.16), we see that the dynamical part of Einstein equation is equivalent to

$$\boldsymbol{F} = 0. \tag{11.12}$$

This is also clear if we replace $\vec{\boldsymbol{\gamma}}^{*4}\boldsymbol{R}$ in Eq. (11.11) by the expression (4.43): we immediately get Eq. (5.69).

Let us express $\vec{\boldsymbol{\gamma}}^{*}(\boldsymbol{G} - 8\pi\boldsymbol{T})$ in terms of $\boldsymbol{F}$. Using Eq. (11.1), we have

$$\vec{\boldsymbol{\gamma}}^{*}(\boldsymbol{G} - 8\pi\boldsymbol{T}) = \vec{\boldsymbol{\gamma}}^{*4}\boldsymbol{R} - \frac{1}{2}{}^4R\boldsymbol{\gamma} - 8\pi\boldsymbol{S}.$$

Comparing with Eq. (11.11), we get

$$\vec{\boldsymbol{\gamma}}^{*}(\boldsymbol{G} - 8\pi\boldsymbol{T}) = \boldsymbol{F} - \frac{1}{2}\left[{}^4R + 8\pi(S - E)\right]\boldsymbol{\gamma}. \tag{11.13}$$

Besides, the trace of Eq. (11.11) is

$$\begin{aligned}
F &= \mathrm{tr}_{\boldsymbol{\gamma}}\boldsymbol{F} = \gamma^{ij}F_{ij} = \gamma^{\mu\nu}F_{\mu\nu} \\
&= \underbrace{\gamma^{\mu\nu}\gamma^{\rho}{}_{\mu}\gamma^{\sigma}{}_{\nu}}_{\gamma^{\rho\nu}}{}^4R_{\rho\sigma} - 8\pi\left[S + \frac{1}{2}(E - S)\times 3\right] \\
&= \gamma^{\rho\sigma\,4}R_{\rho\sigma} + 4\pi(S - 3E) = {}^4R + {}^4R_{\rho\sigma}n^{\rho}n^{\sigma} + 4\pi(S - 3E).
\end{aligned}$$

Now, from Eq. (11.4), ${}^4R_{\rho\sigma}n^{\rho}n^{\sigma} = H - {}^4R/2 + 8\pi E$, so that the above relation becomes

$$F = {}^4R + H - \frac{1}{2}{}^4R + 8\pi E + 4\pi(S - 3E) = H + \frac{1}{2}\left[{}^4R + 8\pi(S - E)\right].$$

This enables us to write Eq. (11.13) as

$$\vec{\boldsymbol{\gamma}}^{*}(\boldsymbol{G} - 8\pi\boldsymbol{T}) = \boldsymbol{F} + (H - F)\boldsymbol{\gamma}. \tag{11.14}$$

Similarly to the 3+1 decomposition (5.14) of the stress–energy tensor, the 3+1 decomposition of $\boldsymbol{G} - 8\pi\boldsymbol{T}$ is

$$\boldsymbol{G} - 8\pi\boldsymbol{T} = \vec{\boldsymbol{\gamma}}^*(\boldsymbol{G} - 8\pi\boldsymbol{T}) + \underline{\boldsymbol{n}} \otimes \boldsymbol{M} + \boldsymbol{M} \otimes \underline{\boldsymbol{n}} + H\underline{\boldsymbol{n}} \otimes \underline{\boldsymbol{n}},$$

$\vec{\boldsymbol{\gamma}}^*(\boldsymbol{G} - 8\pi\boldsymbol{T})$ playing the role of $\boldsymbol{S}$, $\boldsymbol{M}$ that of $\boldsymbol{p}$ and H that of E. Thanks to Eq. (11.14), we may write

$$\boxed{\boldsymbol{G} - 8\pi\boldsymbol{T} = \boldsymbol{F} + (H - F)\boldsymbol{\gamma} + \underline{\boldsymbol{n}} \otimes \boldsymbol{M} + \boldsymbol{M} \otimes \underline{\boldsymbol{n}} + H\underline{\boldsymbol{n}} \otimes \underline{\boldsymbol{n}}}, \tag{11.15}$$

or, in index notation,

$$G_{\alpha\beta} - 8\pi T_{\alpha\beta} = F_{\alpha\beta} + (H - F)\gamma_{\alpha\beta} + n_\alpha M_\beta + M_\alpha n_\beta + H n_\alpha n_\beta. \tag{11.16}$$

This identity can be viewed as the 3+1 decomposition of Einstein equation (11.2) in terms of the dynamical equation violation $\boldsymbol{F}$, the Hamiltonian constraint violation H and the momentum constraint violation $\boldsymbol{M}$.

The next step consists in invoking the contracted Bianchi identity (2.79):

$$\boxed{\boldsymbol{\nabla} \cdot \vec{\boldsymbol{G}} = 0}, \tag{11.17}$$

i.e. $\nabla_\mu G^\mu{}_\alpha = 0$ [cf. Eqs. (2.39) and (2.58)]. Let us recall that this identity is purely geometrical and holds independently of Einstein equation. In addition, we assume that the matter obeys the energy-momentum conservation law (6.1):

$$\boxed{\boldsymbol{\nabla} \cdot \vec{\boldsymbol{T}} = 0}. \tag{11.18}$$

In view of the Bianchi identity (11.17), Eq. (11.18) is a necessary condition for the Einstein equation (11.2) to hold.

Remark 11.1 We assume here specifically that Eq. (11.18) holds, because in the following we do not demand that the whole Einstein equation is satisfied, but only its dynamical part, i.e. Eq. (11.12).

As we have seen in Chap. 6, in order for Eq. (11.18) to be satisfied, the matter energy density E and momentum density $\boldsymbol{p}$ (both relative to the Eulerian observer) must obey to the evolution equations (6.10) and (6.20).

Thanks to the Bianchi identity (11.17) and to the energy-momentum conservation law (11.18), the divergence of Eq. (11.15) leads to, successively,

$$\begin{aligned}
&\nabla_\mu \left(G^\mu{}_\alpha - 8\pi T^\mu{}_\alpha\right) = 0\\
&\nabla_\mu \left[F^\mu{}_\alpha + (H - F)\gamma^\mu{}_\alpha + n^\mu M_\alpha + M^\mu n_\alpha + H n^\mu n_\alpha\right] = 0,\\
&\nabla_\mu F^\mu{}_\alpha + D_\alpha(H - F) + (H - F)\left(\nabla_\mu n^\mu n_\alpha + n^\mu \nabla_\mu n_\alpha\right) - K M_\alpha + n^\mu \nabla_\mu M_\alpha\\
&\qquad + \nabla_\mu M^\mu n_\alpha - M^\mu K_{\mu\alpha} + n^\mu \nabla_\mu H n_\alpha - H K n_\alpha + H D_\alpha \ln N = 0,\\
&\nabla_\mu F^\mu_\alpha + D_\alpha(H - F) + (2H - F)(D_\alpha \ln N - K n_\alpha) - K M_\alpha + n^\mu \nabla_\mu M_\alpha,\\
&\qquad + \nabla_\mu M^\mu n_\alpha - K_{\alpha\mu} M^\mu + n^\mu \nabla_\mu H n_\alpha = 0,
\end{aligned} \tag{11.19}$$

where we have used Eq. (4.24) to express the $\boldsymbol{\nabla}\underline{\boldsymbol{n}}$ in terms of $\boldsymbol{K}$ and $\boldsymbol{D}\ln N$ (in particular $\nabla_\mu n^\mu = -K$). Let us contract Eq. (11.19) with $\boldsymbol{n}$: we get, successively,

$$\begin{aligned}
&n^\nu \nabla_\mu F^\mu{}_\nu + (2H - F)K + n^\nu n^\mu \nabla_\mu M_\nu - \nabla_\mu M^\mu - n^\mu \nabla_\mu H = 0, \\
&- F^\mu{}_\nu \nabla_\mu n^\nu + (2H - F)K - M_\nu n^\mu \nabla_\mu n^\nu - \nabla_\mu M^\mu - n^\mu \nabla_\mu H = 0, \\
&K^{\mu\nu} F_{\mu\nu} + (2H - F)K - M^\nu D_\nu \ln N - \nabla_\mu M^\mu - n^\mu \nabla_\mu H = 0.
\end{aligned} \tag{11.20}$$

Now the $\boldsymbol{\nabla}$-divergence of $\boldsymbol{M}$ is related to the $\boldsymbol{D}$-one by Eq. (6.6): $\nabla_\mu M^\mu = D_\mu M^\mu + M^\mu D_\mu \ln N$. Hence Eq. (11.20) can be written

$$n^\mu \nabla_\mu H = -D_\mu M^\mu - 2M^\mu D_\mu \ln N + K(2H - F) + K^{\mu\nu} F_{\mu\nu}.$$

Noticing that

$$n^\mu \nabla_\mu H = \frac{1}{N} m^\mu \nabla_\mu H = \frac{1}{N} \mathcal{L}_{\boldsymbol{m}} H = \frac{1}{N} \left(\frac{\partial}{\partial t} - \mathcal{L}_{\boldsymbol{\beta}} \right) H, \tag{11.21}$$

where $\boldsymbol{m}$ is the normal evolution vector (cf. Sect. 4.3.2), we get the following evolution equation for the Hamiltonian constraint violation

$$\boxed{\left(\frac{\partial}{\partial t} - \mathcal{L}_{\boldsymbol{\beta}} \right) H = -D_i(NM^i) - M^i D_i N + NK(2H - F) + NK^{ij} F_{ij}}. \tag{11.22}$$

Let us now project Eq. (11.19) onto Σ_t :

$$\gamma^{\nu\alpha} \nabla_\mu F^\mu{}_\nu + D^\alpha (H - F) + (2H - F) D^\alpha \ln N - K M^\alpha + \gamma^\alpha{}_\nu n^\mu \nabla_\mu M^\nu - K^\alpha{}_\mu M^\mu = 0. \tag{11.23}$$

Now the $\boldsymbol{\nabla}$-divergence of $\boldsymbol{F}$ is related to the $\boldsymbol{D}$-one by

$$\begin{aligned}
D_\mu F^{\mu\alpha} &= \gamma^\rho{}_\mu \gamma^\mu{}_\sigma \gamma^{\nu\alpha} \nabla_\rho F^\sigma{}_\nu = \gamma^\rho{}_\sigma \gamma^{\nu\alpha} \nabla_\rho F^\sigma{}_\nu = \gamma^{\nu\alpha} \left(\nabla_\rho F^\rho{}_\nu + n^\rho n_\sigma \nabla_\rho F^\sigma{}_\nu \right) \\
&= \gamma^{\nu\alpha} \left(\nabla_\rho F^\rho{}_\nu - F^\sigma{}_\nu n^\rho \nabla_\rho n_\sigma \right) \\
&= \gamma^{\nu\alpha} \nabla_\mu F^\mu{}_\nu - F^{\alpha\mu} D_\mu \ln N.
\end{aligned} \tag{11.24}$$

Besides, we have

$$\begin{aligned}
\gamma^\alpha{}_\nu n^\mu \nabla_\mu M^\nu &= \frac{1}{N} \gamma^\alpha{}_\nu m^\mu \nabla_\mu M^\nu = \frac{1}{N} \gamma^\alpha{}_\nu \left(\mathcal{L}_{\boldsymbol{m}} M^\nu + M^\mu \nabla_\mu m^\nu \right) \\
&= \frac{1}{N} \left[\mathcal{L}_{\boldsymbol{m}} M^\alpha + \gamma^\alpha{}_\nu M^\mu (\nabla_\mu N n^\nu + N \nabla_\mu n^\nu) \right] \\
&= \frac{1}{N} \mathcal{L}_{\boldsymbol{m}} M^\alpha - K^\alpha{}_\mu M^\mu,
\end{aligned} \tag{11.25}$$

where property (4.36) has been used to write $\gamma^\alpha{}_\nu \mathcal{L}_{\boldsymbol{m}} M^\nu = \mathcal{L}_{\boldsymbol{m}} M^\alpha$.

Thanks to Eqs. (11.24) and (11.25), and to the relation $\mathcal{L}_{\boldsymbol{m}} = \partial/\partial t - \mathcal{L}_{\boldsymbol{\beta}}$, Eq. (11.23) yields an evolution equation for the momentum constraint violation:

$$\boxed{\left(\frac{\partial}{\partial t} - \mathcal{L}_{\beta}\right) M^i = -D_j(NF^{ij}) + 2NK^i{}_j M^j + NKM^i + ND^i(F - H) + (F - 2H)D^i N}. \tag{11.26}$$

Let us now assume that the dynamical Einstein equation is satisfied, then $\boldsymbol{F} = 0$ [Eq. (11.12)] and Eqs. (11.22) and (11.26) reduce to

$$\left(\frac{\partial}{\partial t} - \mathcal{L}_{\beta}\right) H = -D_i(NM^i) + 2NKH - M^i D_i N \tag{11.27}$$

$$\left(\frac{\partial}{\partial t} - \mathcal{L}_{\beta}\right) M^i = -D^i(NH) + 2NK^i{}_j M^j + NKM^i - HD^i N. \tag{11.28}$$

If the constraints are satisfied at $t=0$, then $H|_{t=0} = 0$ and $M^i|_{t=0} = 0$. The above system gives then

$$\left.\frac{\partial H}{\partial t}\right|_{t=0} = 0 \tag{11.29}$$

$$\left.\frac{\partial M^i}{\partial t}\right|_{t=0} = 0. \tag{11.30}$$

We conclude that, at least in the case where all the fields are analytical (in order to invoke the Cauchy–Kovalevskaya theorem),

$$\forall t \geq 0, \quad H = 0 \quad \text{and} \quad M^i = 0, \tag{11.31}$$

i.e. the constraints are preserved by the dynamical evolution equation (5.69). Even if the hypothesis of analyticity is relaxed, the result still holds because the system (11.27)–(11.28) is symmetric hyperbolic [24].

Remark 11.2 The above result on the preservation of the constraints in a free evolution scheme holds only if the matter source obeys the energy-momentum conservation law (11.18).

11.3.3 Constraint-Violating Modes

The constraint preservation property established in the preceding section adds some substantial support to the concept of free evolution scheme. However this is a mathematical result and it does not guarantee that numerical solutions will not violate the constraints. Indeed numerical codes based on free evolution schemes have been plagued for a long time by the so-called ***constraint violating modes***. The latter are solutions $(\boldsymbol{\gamma}, \boldsymbol{K}, N, \boldsymbol{\beta})$ which satisfy $\boldsymbol{F} = 0$ up to numerical accuracy but with $H \neq 0$ and $\boldsymbol{M} \neq 0$, even though initially $H = 0$ and $\boldsymbol{M} = 0$ (up to numerical accuracy).

The reasons for the appearance of these constraint-violating modes are twofold: (i) due to numerical errors, the conditions $H = 0$ and $\boldsymbol{M} = 0$ are slightly violated in the initial data, and the evolution equations amplify this violation (in most cases exponentially !) and (ii) constraint violations may flow into the computational domain from boundary conditions imposed at timelike boundaries. Notice that the demonstration in Sect. 11.3.2 did not take into account any boundary and could not rule out (ii).

An impressive amount of works have then been devoted to this issue (see [4] for a review and Ref. [25, 26] for recent solutions to problem (ii)). We mention hereafter shortly the symmetric hyperbolic formulations, before discussing the most successful approach to date: the BSSN scheme.

11.3.4 Symmetric Hyperbolic Formulations

The idea is to introduce auxiliary variables so that the dynamical equations become a first-order symmetric hyperbolic system, because these systems are known to be well posed (see e.g. [1, 27]). This comprises the formulation developed in 2001 by Kidder et al. [28] (***KST formulation***), which constitutes some generalization of previous formulations developed by Frittelli and Reula [29] and by Andersson and York [30], the latter being known as the ***Einstein–Christoffel system***.

11.4 BSSN Scheme

11.4.1 Introduction

The ***BSSN scheme*** is a free evolution scheme for the conformal 3+1 Einstein system (7.88)–(7.93) which has been devised by Shibata and Nakamura in 1995 [31]. It has been re-analyzed by Baumgarte and Shapiro in 1999 [32], with a slight modification, and bears since then the name *BSSN* for *Baumgarte–Shapiro–Shibata–Nakamura*.

11.4.2 Expression of the Ricci Tensor of the Conformal Metric

The starting point of the BSSN formulation is the conformal 3+1 Einstein system (7.88)–(7.93). One then proceeds by expressing the Ricci tensor $\tilde{\boldsymbol{R}}$ of the conformal metric $\tilde{\boldsymbol{\gamma}}$, which appears in Eq. (7.91), in terms of the derivatives of $\tilde{\boldsymbol{\gamma}}$. To this aim, we consider the expression of the Ricci tensor in terms of the Christoffel symbols $\tilde{\Gamma}^k{}_{ij}$ of the metric $\tilde{\boldsymbol{\gamma}}$ with respect to the coordinates (x^i) [cf. Eqs. (2.69) and (2.76)]:

$$\tilde{R}_{ij} = \frac{\partial}{\partial x^k}\tilde{\Gamma}^k{}_{ij} - \frac{\partial}{\partial x^j}\tilde{\Gamma}^k{}_{ik} + \tilde{\Gamma}^k{}_{ij}\tilde{\Gamma}^l{}_{kl} - \tilde{\Gamma}^k{}_{il}\tilde{\Gamma}^l{}_{kj}. \tag{11.32}$$

Let us introduce the type (1,2) tensor field $\boldsymbol{\Delta}$ defined by

$$\boxed{\Delta^k{}_{ij} := \tilde{\Gamma}^k{}_{ij} - \bar{\Gamma}^k{}_{ij}}, \tag{11.33}$$

where the $\bar{\Gamma}^k{}_{ij}$'s denote the Christoffel symbols of the flat metric $\boldsymbol{f}$ with respect to the coordinates (x^i). As already noticed in Sect. 10.3.4, the identity (11.33) does define a tensor field, although each set of Christoffel symbols, $\tilde{\Gamma}^k{}_{ij}$ or $\bar{\Gamma}^k{}_{ij}$, is by no means the set of components of any tensor field. Actually an alternative expression of $\Delta^k{}_{ij}$, which is manifestly covariant, is

$$\boxed{\Delta^k{}_{ij} = \tfrac{1}{2}\tilde{\gamma}^{kl}\left(\bar{D}_i\tilde{\gamma}_{lj} + \bar{D}_j\tilde{\gamma}_{il} - \bar{D}_l\tilde{\gamma}_{ij}\right)}, \tag{11.34}$$

where $\bar{D}_i$ stands for the covariant derivative associated with the flat metric $\boldsymbol{f}$. It is not difficult to establish the equivalence of Eqs. (11.33) and (11.34): starting from the latter, we have

$$\begin{aligned}
\Delta^k{}_{ij} &= \frac{1}{2}\tilde{\gamma}^{kl}\Bigg(\frac{\partial\tilde{\gamma}_{lj}}{\partial x^i} - \bar{\Gamma}^m{}_{il}\tilde{\gamma}_{mj} - \bar{\Gamma}^m{}_{ij}\tilde{\gamma}_{lm} + \frac{\partial\tilde{\gamma}_{il}}{\partial x^j} - \bar{\Gamma}^m{}_{ji}\tilde{\gamma}_{ml} - \bar{\Gamma}^m{}_{jl}\tilde{\gamma}_{im} \\
&\qquad - \frac{\partial\tilde{\gamma}_{ij}}{\partial x^l} + \bar{\Gamma}^m{}_{li}\tilde{\gamma}_{mj} + \bar{\Gamma}^m{}_{lj}\tilde{\gamma}_{im}\Bigg) \\
&= \tilde{\Gamma}^k{}_{ij} + \frac{1}{2}\tilde{\gamma}^{kl}\left(-2\bar{\Gamma}^m{}_{ij}\tilde{\gamma}_{lm}\right) = \tilde{\Gamma}^k{}_{ij} - \underbrace{\tilde{\gamma}^{kl}\tilde{\gamma}_{lm}}_{\delta^k{}_m}\bar{\Gamma}^m{}_{ij} \\
&= \tilde{\Gamma}^k{}_{ij} - \bar{\Gamma}^k{}_{ij}.
\end{aligned}$$

Remark 11.3 While it is a well defined tensor field, $\boldsymbol{\Delta}$ depends upon the background flat metric $\boldsymbol{f}$, which is not unique on the hypersurface Σ_t.

A useful property is obtained by contracting Eq. (11.33) on the indices k and j:

$$\Delta^k{}_{ik} = \tilde{\Gamma}^k{}_{ik} - \bar{\Gamma}^k{}_{ik} = \frac{1}{2}\frac{\partial}{\partial x^i}\ln\tilde{\gamma} - \frac{1}{2}\frac{\partial}{\partial x^i}\ln f,$$

where $\tilde{\gamma} := \det\tilde{\gamma}_{ij}$ and $f := \det f_{ij}$. Since by construction $\tilde{\gamma} = f$ [Eq. (7.18)], we get

$$\boxed{\Delta^k{}_{ik} = 0}. \tag{11.35}$$

Remark 11.4 If the coordinates (x^i) are of Cartesian type, then $\bar{\Gamma}^k_{ij} = 0$, $\Delta^k{}_{ij} = \tilde{\Gamma}^k_{ij}$ and $\bar{D}_i = \partial/\partial x^i$. This is actually the case considered in the original articles of the BSSN formalism [31, 32]. We follow here the method of Ref. [19] to allow for non Cartesian coordinates, e.g. spherical ones.

Replacing $\tilde{\Gamma}^k_{ij}$ by $\Delta^k{}_{ij} + \bar{\Gamma}^k_{ij}$ [Eq. (11.33)] in the expression (11.32) of the Ricci tensor yields

$$\begin{aligned}
\tilde{R}_{ij} &= \frac{\partial}{\partial x^k}(\Delta^k{}_{ij} + \bar{\Gamma}^k{}_{ij}) - \frac{\partial}{\partial x^j}(\Delta^k{}_{ik} + \bar{\Gamma}^k{}_{ik}) + (\Delta^k{}_{ij} + \bar{\Gamma}^k{}_{ij})(\Delta^l{}_{kl} + \bar{\Gamma}^l{}_{kl}) \\
&\quad - (\Delta^k{}_{il} + \bar{\Gamma}^k{}_{il})(\Delta^l{}_{kj} + \bar{\Gamma}^l{}_{kj}) \\
&= \frac{\partial}{\partial x^k}\Delta^k{}_{ij} + \frac{\partial}{\partial x^k}\bar{\Gamma}^k{}_{ij} - \frac{\partial}{\partial x^j}\Delta^k{}_{ik} - \frac{\partial}{\partial x^j}\bar{\Gamma}^k{}_{ik} + \Delta^k{}_{ij}\Delta^l{}_{kl} + \bar{\Gamma}^l{}_{kl}\Delta^k{}_{ij} \\
&\quad + \bar{\Gamma}^k{}_{ij}\Delta^l{}_{kl} + \bar{\Gamma}^k{}_{ij}\bar{\Gamma}^l{}_{kl} - \Delta^k{}_{il}\Delta^l{}_{kj} - \bar{\Gamma}^l{}_{kj}\Delta^k{}_{il} - \bar{\Gamma}^k{}_{il}\Delta^l{}_{kj} \\
&\quad - \bar{\Gamma}^k{}_{il}\bar{\Gamma}^l{}_{kj}. \qquad (11.36)
\end{aligned}$$

Now since the metric $\boldsymbol{f}$ is flat, its Ricci tensor vanishes identically, so that

$$\frac{\partial}{\partial x^k}\bar{\Gamma}^k{}_{ij} - \frac{\partial}{\partial x^j}\bar{\Gamma}^k{}_{ik} + \bar{\Gamma}^k{}_{ij}\bar{\Gamma}^l{}_{kl} - \bar{\Gamma}^k{}_{il}\bar{\Gamma}^l{}_{kj} = 0.$$

Hence Eq. (11.36) reduces to

$$\begin{aligned}
\tilde{R}_{ij} &= \frac{\partial}{\partial x^k}\Delta^k{}_{ij} - \frac{\partial}{\partial x^j}\Delta^k{}_{ik} + \Delta^k{}_{ij}\Delta^l{}_{kl} + \bar{\Gamma}^l{}_{kl}\Delta^k{}_{ij} + \bar{\Gamma}^k{}_{ij}\Delta^l{}_{kl} - \Delta^k{}_{il}\Delta^l{}_{kj} \\
&\quad - \bar{\Gamma}^l{}_{kj}\Delta^k{}_{il} - \bar{\Gamma}^k{}_{il}\Delta^l{}_{kj}.
\end{aligned}$$

Property (11.35) enables us to simplify this expression further:

$$\begin{aligned}
\tilde{R}_{ij} &= \frac{\partial}{\partial x^k}\Delta^k{}_{ij} + \bar{\Gamma}^l{}_{kl}\Delta^k{}_{ij} - \bar{\Gamma}^l{}_{kj}\Delta^k{}_{il} - \bar{\Gamma}^k{}_{il}\Delta^l{}_{kj} - \Delta^k{}_{il}\Delta^l{}_{kj} \\
&= \frac{\partial}{\partial x^k}\Delta^k{}_{ij} + \bar{\Gamma}^k{}_{kl}\Delta^l{}_{ij} - \bar{\Gamma}^l{}_{ki}\Delta^k{}_{lj} - \bar{\Gamma}^l{}_{kj}\Delta^k{}_{il} - \Delta^k{}_{il}\Delta^l{}_{kj}.
\end{aligned}$$

We recognize in the first four terms of the right-hand side the covariant derivative $\bar{D}_k\Delta^k{}_{ij}$, hence

$$\tilde{R}_{ij} = \bar{D}_k\Delta^k{}_{ij} - \Delta^k{}_{il}\Delta^l{}_{kj}. \qquad (11.37)$$

Remark 11.5 Even if $\Delta^k{}_{ik}$ would not vanish, we would have obtained an expression of the Ricci tensor with exactly the same structure as Eq. (11.32), with the partial derivatives $\partial/\partial x^i$ replaced by the covariant derivatives $\bar{D}_i$ and the Christoffel symbols $\tilde{\Gamma}^k{}_{ij}$ replaced by the tensor components $\Delta^k{}_{ij}$. Indeed Eq. (11.37) can be seen as being nothing but a particular case of the more general formula obtained in Sect. 7.3.1 and relating the Ricci tensors associated with two different metrics, namely Eq. (7.40). Performing in the latter the substitutions $\boldsymbol{\gamma} \to \tilde{\boldsymbol{\gamma}}$, $\tilde{\boldsymbol{\gamma}} \to \boldsymbol{f}$, $R_{ij} \to \tilde{R}_{ij}$, $\tilde{R}_{ij} \to 0$ (for $\boldsymbol{f}$ is flat), $\tilde{D}_i \to \bar{D}_i$ and $C^k{}_{ij} \to \Delta^k{}_{ij}$ [compare Eqs. (7.29) and (11.33)] and using property (11.35), we get immediately Eq. (11.37).

If we substitute expression (11.34) for $\Delta^k{}_{ij}$ into Eq. (11.37), we get

$$
\begin{aligned}
\tilde{R}_{ij} &= \frac{1}{2}\bar{D}_k\left[\tilde{\gamma}^{kl}\left(\bar{D}_i\tilde{\gamma}_{lj}+\bar{D}_j\tilde{\gamma}_{il}-\bar{D}_l\tilde{\gamma}_{ij}\right)\right]-\Delta^k{}_{il}\Delta^l{}_{kj}\\
&= \frac{1}{2}\Bigg\{\bar{D}_k\Big[\bar{D}_i(\underbrace{\tilde{\gamma}^{kl}\tilde{\gamma}_{lj}}_{\delta^k{}_j})-\tilde{\gamma}_{lj}\bar{D}_i\tilde{\gamma}^{kl}+\bar{D}_j(\underbrace{\tilde{\gamma}^{kl}\tilde{\gamma}_{il}}_{\delta^k{}_i})-\tilde{\gamma}_{il}\bar{D}_j\tilde{\gamma}^{kl}\Big]-\bar{D}_k\tilde{\gamma}^{kl}\bar{D}_l\tilde{\gamma}_{ij}\\
&\quad -\tilde{\gamma}^{kl}\bar{D}_k\bar{D}_l\tilde{\gamma}_{ij}\Bigg\}-\Delta^k{}_{il}\Delta^l{}_{kj}\\
&= \frac{1}{2}\Big(-\bar{D}_k\tilde{\gamma}_{lj}\bar{D}_i\tilde{\gamma}^{kl}-\tilde{\gamma}_{lj}\bar{D}_k\bar{D}_i\tilde{\gamma}^{kl}-\bar{D}_k\tilde{\gamma}_{il}\bar{D}_j\tilde{\gamma}^{kl}-\tilde{\gamma}_{il}\bar{D}_k\bar{D}_j\tilde{\gamma}^{kl}-\bar{D}_k\tilde{\gamma}^{kl}\bar{D}_l\tilde{\gamma}_{ij}\\
&\quad -\tilde{\gamma}^{kl}\bar{D}_k\bar{D}_l\tilde{\gamma}_{ij}\Big)-\Delta^k{}_{il}\Delta^l{}_{kj}.
\end{aligned}
$$

Hence we can write, using $\bar{D}_k\bar{D}_i = \bar{D}_i\bar{D}_k$ (since $\boldsymbol{f}$ is flat) and exchanging some indices k and l,

$$
\boxed{\tilde{R}_{ij} = -\tfrac{1}{2}\left(\tilde{\gamma}^{kl}\bar{D}_k\bar{D}_l\tilde{\gamma}_{ij}+\tilde{\gamma}_{ik}\bar{D}_j\bar{D}_l\tilde{\gamma}^{kl}+\tilde{\gamma}_{jk}\bar{D}_i\bar{D}_l\tilde{\gamma}^{kl}\right)+\mathcal{Q}_{ij}(\tilde{\boldsymbol{\gamma}},\bar{\boldsymbol{D}}\tilde{\boldsymbol{\gamma}})}, \qquad (11.38)
$$

where

$$
\mathcal{Q}_{ij}(\tilde{\boldsymbol{\gamma}},\bar{\boldsymbol{D}}\tilde{\boldsymbol{\gamma}}) := -\frac{1}{2}\left(\bar{D}_k\tilde{\gamma}_{lj}\bar{D}_i\tilde{\gamma}^{kl}+\bar{D}_k\tilde{\gamma}_{il}\bar{D}_j\tilde{\gamma}^{kl}+\bar{D}_k\tilde{\gamma}^{kl}\bar{D}_l\tilde{\gamma}_{ij}\right)-\Delta^k{}_{il}\Delta^l{}_{kj} \qquad (11.39)
$$

is a term which does not contain any second derivative of $\tilde{\boldsymbol{\gamma}}$ and which is quadratic in the first derivatives.

11.4.3 Reducing the Ricci Tensor to a Laplace Operator

If we consider the Ricci tensor as a differential operator acting on the conformal metric $\tilde{\boldsymbol{\gamma}}$, its principal part (or *principal symbol*, cf. Sec. A.2.2) is given by the three terms involving second derivatives in the right-hand side of Eq. (11.38). We recognize in the first term, $\tilde{\gamma}^{kl}\bar{D}_k\bar{D}_l\tilde{\gamma}_{ij}$, a kind of Laplace operator acting on $\tilde{\gamma}_{ij}$. Actually, for a weak gravitational field, i.e. for $\tilde{\gamma}^{ij} = f^{ij} + h^{ij}$ with $f_{ik}f_{jl}h^{kl}h^{ij} \ll 1$, we have, at the linear order in $\boldsymbol{h}$, $\tilde{\gamma}^{kl}\bar{D}_k\bar{D}_l\tilde{\gamma}_{ij} \simeq \Delta_f\tilde{\gamma}_{ij}$, where $\Delta_f = f^{kl}\bar{D}_k\bar{D}_l$ is the Laplace operator associated with the metric $\boldsymbol{f}$. If we combine Eqs. (7.89) and (7.91), the Laplace operator in $\tilde{R}_{ij}$ gives rise to a *wave operator* for $\tilde{\gamma}_{ij}$, namely

$$
\left[\left(\frac{\partial}{\partial t}-\mathcal{L}_{\boldsymbol{\beta}}\right)^2-\frac{N^2}{\Psi^4}\tilde{\gamma}^{kl}\bar{D}_k\bar{D}_l\right]\tilde{\gamma}_{ij} = \cdots
$$

Unfortunately the other two terms that involve second derivatives in Eq. (11.38), namely $\tilde{\gamma}_{ik}\bar{D}_j\bar{D}_l\tilde{\gamma}^{kl}$ and $\tilde{\gamma}_{jk}\bar{D}_i\bar{D}_l\tilde{\gamma}^{kl}$, spoil the elliptic character of the operator acting on $\tilde{\gamma}_{ij}$ in $\tilde{R}_{ij}$, so that the combination of Eqs. (7.89) and (7.91) does no longer lead to a wave operator.

To restore the Laplace operator, Shibata and Nakamura [31] have considered the term $\bar{D}_l\tilde{\gamma}^{kl}$ which appears in the second and third terms of Eq. (11.38) as a variable independent from $\tilde{\gamma}_{ij}$. We recognize in this term the opposite of the vector $\tilde{\boldsymbol{\Gamma}}$ that has been introduced in Sect. 10.3.4 [cf. Eq. (10.54)]:

$$\boxed{\tilde{\Gamma}^i = -\bar{D}_j\tilde{\gamma}^{ij}}. \tag{11.40}$$

Equation (11.38) then becomes

$$\boxed{\tilde{R}_{ij} = \frac{1}{2}\left(-\tilde{\gamma}^{kl}\bar{D}_k\bar{D}_l\tilde{\gamma}_{ij} + \tilde{\gamma}_{ik}\bar{D}_j\tilde{\Gamma}^k + \tilde{\gamma}_{jk}\bar{D}_i\tilde{\Gamma}^k\right) + \mathcal{Q}_{ij}(\tilde{\boldsymbol{\gamma}}, \bar{\boldsymbol{D}}\tilde{\boldsymbol{\gamma}})}. \tag{11.41}$$

Remark 11.6 Actually, Shibata and Nakamura [31] have introduced the covector $F_i := \bar{D}^j\tilde{\gamma}_{ij}$ instead of $\tilde{\Gamma}^i$. As Eq. (10.67) shows, the two quantities are closely related. They are even equivalent in the linear regime. The quantity $\tilde{\Gamma}^i$ has been introduced by Baumgarte and Shapiro [32]. It has the advantage over F_i to encompass all the second derivatives of $\tilde{\gamma}_{ij}$ that are not part of the Laplacian. If one use F_i, this is true only at the linear order (weak field region). Indeed, by means of Eq. (10.67), we can write

$$\tilde{R}_{ij} = \frac{1}{2}\left(-\tilde{\gamma}^{kl}\bar{D}_k\bar{D}_l\tilde{\gamma}_{ij} + \bar{D}_jF_i + \bar{D}_iF_j + h^{kl}\bar{D}_i\bar{D}_k\tilde{\gamma}_{jl} + h^{kl}\bar{D}_i\bar{D}_k\tilde{\gamma}_{jl}\right) + \mathcal{Q}'_{ij}(\tilde{\boldsymbol{\gamma}}, \bar{\boldsymbol{D}}\tilde{\boldsymbol{\gamma}}),$$

where $h^{kl} := \tilde{\gamma}^{kl} - f^{kl}$. When compared with (11.41), the above expression contains the additional terms $h^{kl}\bar{D}_i\bar{D}_k\tilde{\gamma}_{jl}$ and $h^{kl}\bar{D}_i\bar{D}_k\tilde{\gamma}_{jl}$, which are quadratic in the deviation of $\tilde{\boldsymbol{\gamma}}$ from the flat metric.

The Ricci scalar $\tilde{R}$, which appears along $\tilde{R}_{ij}$ in Eq. (7.91), is deduced from the trace of Eq. (11.41):

$$\begin{aligned}\tilde{R} = \tilde{\gamma}^{ij}\tilde{R}_{ij} &= \frac{1}{2}\Big(-\tilde{\gamma}^{kl}\tilde{\gamma}^{ij}\bar{D}_k\bar{D}_l\tilde{\gamma}_{ij} + \underbrace{\tilde{\gamma}^{ij}\tilde{\gamma}_{ik}}_{\delta^j{}_k}\bar{D}_j\tilde{\Gamma}^k + \underbrace{\tilde{\gamma}^{ij}\tilde{\gamma}_{jk}}_{\delta^i{}_k}\bar{D}_i\tilde{\Gamma}^k\Big) + \tilde{\gamma}^{ij}\mathcal{Q}_{ij}(\tilde{\boldsymbol{\gamma}}, \bar{\boldsymbol{D}}\tilde{\boldsymbol{\gamma}}) \\ &= \frac{1}{2}\left[-\tilde{\gamma}^{kl}\bar{D}_k\left(\tilde{\gamma}^{ij}\bar{D}_l\tilde{\gamma}_{ij}\right) + \tilde{\gamma}^{kl}\bar{D}_k\tilde{\gamma}^{ij}\bar{D}_l\tilde{\gamma}_{ij} + 2\bar{D}_k\tilde{\Gamma}^k\right] + \tilde{\gamma}^{ij}\mathcal{Q}_{ij}(\tilde{\boldsymbol{\gamma}}, \bar{\boldsymbol{D}}\tilde{\boldsymbol{\gamma}}).\end{aligned}$$

Now, from Eq. (11.34), $\tilde{\gamma}^{ij}\bar{D}_l\tilde{\gamma}_{ij} = 2\Delta^k{}_{lk}$, and from Eq. (11.35), $\Delta^k{}_{lk} = 0$. Thus the first term in the right-hand side of the above equation vanishes and we get

$$\boxed{\tilde{R} = \bar{D}_k\tilde{\Gamma}^k + \mathcal{Q}(\tilde{\boldsymbol{\gamma}}, \bar{\boldsymbol{D}}\tilde{\boldsymbol{\gamma}})}, \tag{11.42}$$

where

$$\mathcal{Q}(\tilde{\boldsymbol{\gamma}}, \bar{\boldsymbol{D}}\tilde{\boldsymbol{\gamma}}) := \frac{1}{2}\tilde{\gamma}^{kl}\bar{D}_k\tilde{\gamma}^{ij}\bar{D}_l\tilde{\gamma}_{ij} + \tilde{\gamma}^{ij}\mathcal{Q}_{ij}(\tilde{\boldsymbol{\gamma}}, \bar{\boldsymbol{D}}\tilde{\boldsymbol{\gamma}}) \tag{11.43}$$

is a term that does not contain any second derivative of $\tilde{\boldsymbol{\gamma}}$ and is quadratic in the first derivatives.

The idea of introducing auxiliary variables, such as $\tilde{\Gamma}^i$ or F_i, to reduce the Ricci tensor to a Laplace-like operator traces back to Nakamura, Oohara and Kojima [33]. In that work, such a treatment was performed for the Ricci tensor $\boldsymbol{R}$ of the physical metric $\boldsymbol{\gamma}$, whereas in Shibata and Nakamura's study [31], it was done for the Ricci tensor $\tilde{\boldsymbol{R}}$ of the conformal metric $\tilde{\boldsymbol{\gamma}}$. The same considerations had been put forward much earlier for the four-dimensional Ricci tensor ${}^4\boldsymbol{R}$. Indeed, this is the main motivation for the *harmonic coordinates* mentioned in Sect. 10.2.3: in 1921 de Donder [34] introduced these coordinates in order to write the principal part of the Ricci tensor as a wave operator acting on the metric coefficients $g_{\alpha\beta}$:

$$ {}^4R_{\alpha\beta} = -\frac{1}{2} g^{\mu\nu} \frac{\partial}{\partial x^\mu} \frac{\partial}{\partial x^\nu} g_{\alpha\beta} + \mathscr{Q}_{\alpha\beta}(\boldsymbol{g}, \boldsymbol{\partial g}), \tag{11.44} $$

where $\mathscr{Q}_{\alpha\beta}(\boldsymbol{g}, \boldsymbol{\partial g})$ is a term which does not contain any second derivative of $\boldsymbol{g}$ and which is quadratic in the first derivatives. In the current context, the analogue of harmonic coordinates would be to set $\tilde{\Gamma}^i = 0$, for then Eq. (11.41) would resemble Eq. (11.44). The choice $\tilde{\Gamma}^i = 0$ corresponds to the *Dirac gauge* discussed in Sect. 10.4.2. However the philosophy of the BSSN formulation is to leave free the coordinate choice, allowing for any value of $\tilde{\Gamma}^i$. In this respect, a closer 4-dimensional analogue of BSSN is the *generalized harmonic decomposition* introduced by Friedrich [35] and Garfinkle [36] (see also Ref. [37, 38]) and implemented by Pretorius for the binary black hole problem [39–42].

The allowance for any coordinate system means that $\tilde{\Gamma}^i$ becomes a new variable, in addition to $\tilde{\gamma}_{ij}$, $\tilde{A}_{ij}$, Ψ, K, N and β^i. One then needs an evolution equation for it. But we have already derived such an equation in Sect. 10.3.4, namely Eq. (10.57). Equation (11.40) is then a constraint on the system, in addition to the Hamiltonian and momentum constraints.

11.4.4 The Full Scheme

By collecting together Eqs. (7.88)–(7.91), (11.41), (11.42) and (10.57), we can write the complete system of evolution equations for the BSSN scheme:

$$ \boxed{\left(\frac{\partial}{\partial t} - \mathscr{L}_{\boldsymbol{\beta}}\right) \Psi = \frac{\Psi}{6} \left(\tilde{D}_i \beta^i - NK\right)} \tag{11.45} $$

$$ \boxed{\left(\frac{\partial}{\partial t} - \mathscr{L}_{\boldsymbol{\beta}}\right) \tilde{\gamma}_{ij} = -2N\tilde{A}_{ij} - \frac{2}{3}\tilde{D}_k \beta^k \tilde{\gamma}_{ij}} \tag{11.46} $$

$$ \boxed{\left(\frac{\partial}{\partial t} - \mathscr{L}_{\boldsymbol{\beta}}\right) K = -\Psi^{-4}\left(\tilde{D}_i \tilde{D}^i N + 2\tilde{D}_i \ln \Psi \tilde{D}^i N\right) + N\left[4\pi(E+S) + \tilde{A}_{ij}\tilde{A}^{ij} + \frac{K^2}{3}\right]} \tag{11.47} $$

$$\boxed{\begin{aligned}\left(\frac{\partial}{\partial t}-\mathscr{L}_{\boldsymbol{\beta}}\right)\tilde{A}_{ij} &= -\frac{2}{3}\bar{D}_k\beta^k\tilde{A}_{ij}+N\left[K\tilde{A}_{ij}-2\tilde{\gamma}^{kl}\tilde{A}_{ik}\tilde{A}_{jl}-8\pi\left(\Psi^{-4}S_{ij}-\frac{1}{3}S\tilde{\gamma}_{ij}\right)\right]\\ &+\Psi^{-4}\Big\{-\tilde{D}_i\tilde{D}_jN+2\tilde{D}_i\ln\Psi\,\tilde{D}_jN+2\tilde{D}_j\ln\Psi\,\tilde{D}_iN\\ &+\frac{1}{3}\left(\tilde{D}_k\tilde{D}^kN-4\tilde{D}_k\ln\Psi\,\tilde{D}^kN\right)\tilde{\gamma}_{ij}\\ &+N\Big[\frac{1}{2}\left(-\tilde{\gamma}^{kl}\bar{D}_k\bar{D}_l\tilde{\gamma}_{ij}+\tilde{\gamma}_{ik}\bar{D}_j\tilde{\Gamma}^k+\tilde{\gamma}_{jk}\bar{D}_i\tilde{\Gamma}^k\right)+\mathscr{Q}_{ij}(\tilde{\boldsymbol{\gamma}},\bar{\boldsymbol{D}}\tilde{\boldsymbol{\gamma}})\\ &-\frac{1}{3}\left(\bar{D}_k\tilde{\Gamma}^k+\mathscr{Q}(\tilde{\boldsymbol{\gamma}},\bar{\boldsymbol{D}}\tilde{\boldsymbol{\gamma}})\right)\tilde{\gamma}_{ij}-2\tilde{D}_i\tilde{D}_j\ln\Psi+4\tilde{D}_i\ln\Psi\,\tilde{D}_j\ln\Psi\\ &+\frac{2}{3}\left(\tilde{D}_k\tilde{D}^k\ln\Psi-2\tilde{D}_k\ln\Psi\,\tilde{D}^k\ln\Psi\right)\tilde{\gamma}_{ij}\Big]\Big\}.\end{aligned}}\tag{11.48}$$

$$\boxed{\begin{aligned}\left(\frac{\partial}{\partial t}-\mathscr{L}_{\boldsymbol{\beta}}\right)\tilde{\Gamma}^i &= \frac{2}{3}\bar{D}_k\beta^k\tilde{\Gamma}^i+\tilde{\gamma}^{jk}\bar{D}_j\bar{D}_k\beta^i+\frac{1}{3}\tilde{\gamma}^{ij}\bar{D}_j\bar{D}_k\beta^k-2\tilde{A}^{ij}\bar{D}_jN\\ &-2N\left[8\pi\Psi^4p^i-\tilde{A}^{jk}\Delta^i{}_{jk}-6\tilde{A}^{ij}\bar{D}_j\ln\Psi+\frac{2}{3}\tilde{\gamma}^{ij}\bar{D}_jK\right]\end{aligned}},\tag{11.49}$$

where $\mathscr{Q}_{ij}(\tilde{\boldsymbol{\gamma}},\bar{\boldsymbol{D}}\tilde{\boldsymbol{\gamma}})$ and $\mathscr{Q}(\tilde{\boldsymbol{\gamma}},\bar{\boldsymbol{D}}\tilde{\boldsymbol{\gamma}})$ are defined by Eqs. (11.39) and (11.43) and we have used $\mathscr{L}_{\boldsymbol{\beta}}\tilde{\Gamma}^i = \beta^k\bar{D}_k\tilde{\Gamma}^i - \tilde{\Gamma}^k\bar{D}_k\beta^i$ to rewrite Eq. (10.57). These equations must be supplemented with the constraints (7.92) (Hamiltonian constraint), (7.93) (momentum constraint), (7.18) ("unit" determinant of $\tilde{\gamma}_{ij}$), (7.61) ($\tilde{\boldsymbol{A}}$ traceless) and (11.40) (definition of $\tilde{\boldsymbol{\Gamma}}$) :

$$\boxed{\tilde{D}_i\tilde{D}^i\Psi-\tfrac{1}{8}\tilde{R}\Psi+\left(\tfrac{1}{8}\tilde{A}_{ij}\tilde{A}^{ij}-\tfrac{1}{12}K^2+2\pi E\right)\Psi^5=0}\tag{11.50}$$

$$\boxed{\tilde{D}^j\tilde{A}_{ij}+6\tilde{A}_{ij}\tilde{D}^j\ln\Psi-\tfrac{2}{3}\tilde{D}_iK=8\pi p_i}\tag{11.51}$$

$$\boxed{\det(\tilde{\gamma}_{ij})=f}\tag{11.52}$$

$$\boxed{\tilde{\gamma}^{ij}\tilde{A}_{ij}=0}\tag{11.53}$$

$$\boxed{\tilde{\Gamma}^i+\bar{D}_j\tilde{\gamma}^{ij}=0}.\tag{11.54}$$

The unknowns for the BSSN system are Ψ, $\tilde{\gamma}_{ij}$, K, $\tilde{A}_{ij}$ and $\tilde{\Gamma}^i$. They involve 1+6+1+6+3 = 17 components, which are evolved via the 17-component equations (11.45)–(11.49). The constraints (11.50)–(11.54) involve 1+3+1+1+3 = 9 components, reducing the number of degrees of freedom to $17-9=8$. The coordinate choice, via the lapse function N and the shift vector β^i, reduces this number to $8-4=4=2\times 2$, which corresponds to the 2 degrees of freedom of the gravitational field expressed in terms of the couple $(\tilde{\gamma}_{ij},\tilde{A}_{ij})$.

The complete system to be solved must involve some additional equations resulting from the choice of lapse N and shift vector $\boldsymbol{\beta}$, as discussed in Chap. 10. The well-posedness of the whole system is discussed in Refs. [43] and [44], for some usual coordinate choices, like harmonic slicing (Sect. 10.2.3) with hyperbolic gamma driver (Sect. 10.3.5).

11.4.5 Applications

The BSSN scheme is by far the most widely used evolution scheme in contemporary numerical relativity. It has notably been used for computing gravitational collapses [45–52], mergers of binary neutron stars [53–61], mergers of binary black holes [62–74] and mergers of neutron star—black holes binaries [75] (see [76–78] for a review). In addition, most recent codes for general relativistic MHD employ the BSSN formulation [79– 84].

References

1. Stewart, J.M.: The Cauchy problem and the initial boundary value problem in numerical relativity. Class Quantum Grav. **15**, 2865 (1998)
2. Friedrich, H., Rendall, A.: The Cauchy problem for the Einstein equations, in Einstein's field equations and their physical implications: selected essays in honour of Jürgen Ehlers. In: Schmidt, B.G. (eds) Lecture Notes in Physics 540, pp. 127. Springer, Berlin (2000)
3. Lehner, L.: Numerical relativity: a review. Class. Quantum Grav. **18**, R25 (2001)
4. Shinkai, H., Yoneda, G.: Re-formulating the Einstein equations for stable numerical simulations: Formulation problem in numerical relativity, to appear (?) in Progress in Astronomy and Astrophysics (Nova Science Publ.), preprint gr-qc/0209111
5. Shinkai, H.: Introduction to numerical relativity. Lecture Notes for APCTP Winter School on Gravitation and Cosmology, Jan 17–18 Seoul, Korea (2003) available at http://www.einstein1905.info/winterAPCTP/
6. Baumgarte, T.W., Shapiro, S.L.: Numerical relativity and compact binaries. Phys. Rep. **376**, 41 (2003)
7. Lehner, L. and Reula, O.: Status quo and open problems in the numerical construction of spacetimes, in Ref. [85] , p. 205.
8. Alcubierre, M.: Introduction to 3+1 Numerical Relativity. Oxford University Press, Oxford (2008)
9. Baumgarte, T.W., Shapiro, S.L.: Numerical Relativity. Solving Einstein's Equations on the Computer. Cambridge University Press, Cambridge (2010)
10. Bardeen, J.M., Piran, T.: General relativistic axisymmetric rotating systems: coordinates and equations. Phys. Rep. **96**, 206 (1983)
11. Stark, R.F., Piran, T.: Gravitational-wave emission from rotating gravitational collapse. Phys. Rev. Lett. **55**, 891 (1985)
12. Evans, C.R.: An approach for calculating axisymmetric gravitational collapse. In: Centrella, J. (eds) Dynamical Spacetimes and Numerical Relativity, pp. 3. Cambridge University Press, Cambridge (1986)

13. Evans, C.R.: Enforcing the momentum constraints during axisymmetric spacelike simulations. In: Evans, C.R., Finn, L.S., Hobill, D.W. (eds) Frontiers in Numerical Relativity., pp. 194. Cambridge University Press, Cambridge (1989)
14. Shapiro, S.L., Teukolsky, S.A.: Collisions of relativistic clusters and the formation of black holes. Phys. Rev. D **45**, 2739 (1992)
15. Abrahams, A.M., Cook, G.B., Shapiro, S.L., Teukolsky, S.A.: Solving Einstein's equations for rotating spacetimes: Evolution of relativistic star clusters. Phys. Rev. D **49**, 5153 (1994)
16. Choptuik, M.W., Hirschmann, E.W., Liebling, S.L., Pretorius, F.: An axisymmetric gravitational collapse code. Class. Quantum Grav. **20**, 1857 (2003)
17. Rinne, O.: Constrained evolution in axisymmetry and the gravitational collapse of prolate Brill waves. Class. Quantum Grav. **25**, 135009 (2008)
18. Anderson, M., Matzner, R.A.: Extended lifetime in computational evolution of isolated black holes. Found. Phys. **35**, 1477 (2005)
19. Bonazzola, S., Gourgoulhon, E., Grandclément, P., Novak, J.: Constrained scheme for the Einstein equations based on the Dirac gauge and spherical coordinates. Phys. Rev. D **70**, 104007 (2004)
20. Cordero-Carrión, I., Cerdá-Durán, P., Dimmelmeier, H., Jaramillo, J.L., Novak, J., Gourgoulhon, E.: Improved constrained scheme for the Einstein equations: an approach to the uniqueness issue. Phys. Rev. D **79**, 024017 (2009)
21. Cordero-Carrión, I., Ibáñez, J.M., Gourgoulhon, E., Jaramillo, J.L., Novak, J.: Mathematical issues in a fully-constrained formulation of Einstein equations. Phys. Rev. D **77**, 084007 (2008)
22. Cordero-Carrión, I., Cerdá-Durán, P., Ibáñez, J.M.: Dynamical spacetimes and gravitational radiation in a fully constrained formulation. J. Phys.: Conf. Ser. **228**, 012055 (2010)
23. Cordero-Carrión I., Cerdá-Durán P., and Ibáñez J.M.: Gravitational waves in dynamical spacetimes with matter content in the Fully Constrained Formulation, preprint arXiv:1108.0571
24. Frittelli, S.: Note on the propagation of the constraints in standard 3+1 general relativity. Phys. Rev. D **55**, 5992 (1997)
25. Kidder, L.E., Lindblom, L., Scheel, M.A., Buchman, L.T., Pfeiffer, H.P.: Boundary conditions for the Einstein evolution system. Phys. Rev. D **71**, 064020 (2005)
26. Sarbach, O., Tiglio, M.: Boundary conditions for Einstein's field equations: mathematical and numerical analysis. J. Hyper. Diff. Equat. **2**, 839 (2005)
27. Reula O.: Strong Hyperbolicity, lecture at the VII Mexican School on Gravitation and Mathematical Physics (Playa del Carmen, Nov 26–Dec 2, Mexico, 2006); http://www.smf.mx/~dgfm-smf/EscuelaVII/
28. Kidder, L.E., Scheel, M.A., Teukolsky, S.A.: Extending the lifetime of 3D black hole computations with a new hyperbolic system of evolution equations. Phys. Rev. D **64**, 064017 (2001)
29. Frittelli, S., Reula, O.A.: First-order symmetric hyperbolic Einstein equations with arbitrary fixed gauge. Phys. Rev. Lett. **76**, 4667 (1996)
30. Anderson, A., York, J.W.: Fixing Einstein's equations. Phys. Rev. Lett. **82**, 4384 (1999)
31. Shibata, M., Nakamura, T.: Evolution of three-dimensional gravitational waves: Harmonic slicing case. Phys. Rev. D **52**, 5428 (1995)
32. Baumgarte, T.W., Shapiro, S.L.: Numerical integration of Einstein's field equations. Phys. Rev. D **59**, 024007 (1999)
33. Nakamura, T., Oohara, K., Kojima, Y.: General relativistic collapse to black holes and gravitational waves from black holes. Prog. Theor. Phys. Suppl. **90**, 1 (1987)
34. De Donder, T.: La Gravifique einsteinienne, Gauthier-Villars, Paris (1921). http://www.numdam.org/item?id=AIHP_1930__1_2_77_0
35. Friedrich, H.: On the hyperbolicity of Einstein's and other gauge field equations. Commun. Math. Phys. **100**, 525 (1985)
36. Garfinkle, D.: Harmonic coordinate method for simulating generic singularities. Phys. Rev. D **65**, 044029 (2002)
37. Gundlach, C., Calabrese, G., Hinder, I., Martí n-García, J.M.: Constraint damping in the Z4 formulation and harmonic gauge. Class. Quantum Grav. **22**, 3767 (2005)

38. Lindblom, L., Scheel, M.A., Kidder, L.E., Owen, R., Rinne, O.: A new generalized harmonic evolution system. Class. Quantum Grav. **23**, S447 (2006)
39. Pretorius, F.: Numerical relativity using a generalized harmonic decomposition. Class. Quantum Grav. **22**, 425 (2005)
40. Pretorius, F.: Evolution of binary black-hole spacetimes. Phys. Rev. Lett. **95**, 121101 (2005)
41. Pretorius, F.: Simulation of binary black hole spacetimes with a harmonic evolution scheme. Class. Quantum Grav. **23**, S529 (2006)
42. Pretorius, F.: Binary black hole coalescence. In: Colpi, M., Casella, P., Gorini, V., Moschella, U., Possenti, A. (eds) Physics of Relativistic Objects in Compact Binaries: From Birth to Coalescence, pp. 305. Springer, Dordrecht/Canopus (2009)
43. Beyer, H., Sarbach, O.: Well-posedness of the Baumgarte-Shapiro-Shibata-Nakamura formulation of Einstein's field equations. Phys. Rev. D **70**, 104004 (2004)
44. Gundlach, C., Martín-García, J.M.: Well-posedness of formulations of the Einstein equations with dynamical lapse and shift conditions. Phys. Rev. D **74**, 024016 (2006)
45. Shibata, M., Baumgarte, T.W., Shapiro, S.L.: Stability and collapse of rapidly rotating, supramassive neutron stars: 3D simulations in general relativity. Phys. Rev. D **61**, 044012 (2000)
46. Shibata, M.: Axisymmetric general relativistic hydrodynamics: Long-term evolution of neutron stars and stellar collapse to neutron stars and black holes. Phys. Rev. D **67**, 024033 (2003)
47. Sekiguchi, Y., Shibata, M.: Axisymmetric collapse simulations of rotating massive stellar cores in full general relativity: Numerical study for prompt black hole formation. Phys. Rev. D **71**, 084013 (2005)
48. Shibata, M., Sekiguchi, Y.: Three-dimensional simulations of stellar core collapse in full general relativity: Nonaxisymmetric dynamical instabilities. Phys. Rev. D **71**, 024014 (2005)
49. Baiotti, L., Hawke, I., Montero, P.J., Löffler, F., Rezzolla, L., Stergioulas, N., Font, J.A., Seidel, E.: Three-dimensional relativistic simulations of rotating neutron-star collapse to a Kerr black hole. Phys. Rev. D **71**, 024035 (2005)
50. Baiotti, L., Hawke, I., Rezzolla, L., Schnetter, E.: Gravitational-wave emission from rotating gravitational collapse in three dimensions. Phys. Rev. Lett. **94**, 131101 (2005)
51. Baiotti, L., Rezzolla, L.: Challenging the paradigm of singularity excision in gravitational collapse. Phys. Rev. Lett. **97**, 141101 (2006)
52. Sekiguchi Y., Shibata M.: Formation of black hole and accretion disk in a massive high-entropy stellar core collapse, Astrophys. J. (in press); preprint arXiv:1009.5303
53. Shibata, M.: Fully general relativistic simulation of coalescing binary neutron stars: Preparatory tests. Phys. Rev. D **60**, 104052 (1999)
54. Shibata, M., Uryu, K.: Simulation of merging binary neutron stars in full general relativity: $\Gamma = 2$ case. Phys. Rev. D **61**, 064001 (2000)
55. Shibata, M., Uryu, K.: Gravitational waves from the merger of binary neutron stars in a fully general relativistic simulation. Prog. Theor. Phys. **107**, 265 (2002)
56. Shibata, M., Taniguchi, K., Uryu, K.: Merger of binary neutron stars of unequal mass in full general relativity. Phys. Rev. D **68**, 084020 (2003)
57. Shibata, M., Taniguchi, K., Uryu, K.: Merger of binary neutron stars with realistic equations of state in full general relativity. Phys. Rev. D **71**, 084021 (2005)
58. Shibata, M., Taniguchi, K.: Merger of binary neutron stars to a black hole: Disk mass, short gamma-ray bursts, and quasinormal mode ringing. Phys. Rev. D **73**, 064027 (2006)
59. Baiotti, L., Giacomazzo, B., Rezzolla, L.: Accurate evolutions of inspiralling neutron-star binaries: Prompt and delayed collapse to a black hole. Phys. Rev. D **78**, 084033 (2008)
60. Hotokezaka, K., Kyutoku, K., Okawa, H., Shibata, M., Kiuchi, K.: Binary neutron star mergers: Dependence on the nuclear equation of state. Phys. Rev. D **83**, 124008 (2011)
61. Sekiguchi Y., Kiuchi K., Kyutoku K., Shibata M.: Gravitational waves and neutrino emission from the merger of binary neutron stars, Phys. Rev. Lett. **107**, 051102 (2011)
62. Baker, J.G., Centrella, J., Choi, D.-I., Koppitz, M., van Meter, J.: Gravitational-wave extraction from an inspiraling configuration of merging black holes. Phys. Rev. Lett. **96**, 111102 (2006)

63. Baker, J.G., Centrella, J., Choi, D.-I., Koppitz, M., van Meter, J.: Binary black hole merger dynamics and waveforms. Phys. Rev. D **73**, 104002 (2006)
64. van Meter, J.R., Baker, J.G., Koppitz, M., Choi, D.I.: How to move a black hole without excision: gauge conditions for the numerical evolution of a moving puncture. Phys. Rev. D **73**, 124011 (2006)
65. Campanelli, M., Lousto, C.O., Marronetti, P., Zlochower, Y.: Accurate evolutions of orbiting black-hole binaries without excision. Phys. Rev. Lett. **96**, 111101 (2006)
66. Campanelli, M., Lousto, C.O., Zlochower, Y.: Last orbit of binary black holes. Phys. Rev. D **73**, 061501(R) (2006)
67. Campanelli, M., Lousto, C.O., Zlochower, Y.: Spinning-black-hole binaries: The orbital hang-up. Phys. Rev. D **74**, 041501(R) (2006)
68. Campanelli, M., Lousto, C.O., Zlochower, Y.: Spin-orbit interactions in black-hole binaries. Phys. Rev. D **74**, 084023 (2006)
69. Sperhake, U.: Binary black-hole evolutions of excision and puncture data. Phys. Rev. D **76**, 104015 (2007)
70. Diener, P., Herrmann, F., Pollney, D., Schnetter, E., Seidel, E., Takahashi, R., Thornburg, J., Ventrella, J.: Accurate evolution of orbiting binary black holes. Phys. Rev. Lett. **96**, 121101 (2006)
71. Brügmann, B., González, J.A., Hannam, M., Husa, S., Sperhake, U., Tichy, W.: Calibration of moving puncture simulations. Phys. Rev. D **77**, 024027 (2008)
72. Marronetti, P., Tichy, W., Brügmann, B., González, J., Hannam, M., Husa, S., Sperhake, U.: Binary black holes on a budget: simulations using workstations. Class. Quantum Grav. **24**, S43 (2007)
73. Herrmann, F., Hinder, I., Shoemaker, D., Laguna, P.: Unequal mass binary black hole plunges and gravitational recoil. Class. Quantum Grav. **24**, S33 (2007)
74. Herrmann, F., Hinder, I., Shoemaker, D., Laguna, P., Matzner, R.A.: Gravitational recoil from spinning binary black hole mergers. Astrophys. J. **661**, 430 (2007)
75. Kyutoku, K., Shibata, M., Taniguchi, K.: Gravitational waves from nonspinning black hole-neutron star binaries: Dependence on equations of state. Phys. Rev. D **82**, 044049 (2010)
76. A. Buonanno : Binary black hole coalescence, in astrophysics of compact objects. AIP Conf. Proc. **968**, 307 (2008)
77. Centrella, J.M., Baker, J.G., Kelly, B.J., van Meter, J.R.: Black-hole binaries, gravitational waves, and numerical relativity. Rev. Mod. Phys. **82**, 3069 (2010)
78. Duez, M.D.: Numerical relativity confronts compact neutron star binaries: a review and status report. Class. Quantum Grav. **27**, 114002 (2010)
79. Duez, M.D., Liu, Y.T., Shapiro, S.L., Stephens, B.C.: Relativistic magnetohydrodynamics in dynamical spacetimes: Numerical methods and tests. Phys. Rev. D **72**, 024028 (2005)
80. Shibata, M., Sekiguchi, Y.: Magnetohydrodynamics in full general relativity: Formulation and tests. Phys. Rev. D **72**, 044014 (2005)
81. Shibata, M., Liu, Y.T., Shapiro, S.L., Stephens, B.C.: Magnetorotational collapse of massive stellar cores to neutron stars: Simulations in full general relativity. Phys. Rev. D **74**, 104026 (2006)
82. Giacomazzo, B., Rezzolla, L.: WhiskyMHD: a new numerical code for general relativistic magnetohydrodynamics. Class. Quantum Grav. **24**, S235 (2007)
83. Kiuchi, K., Shibata, M., Yoshida, S.: Evolution of neutron stars with toroidal magnetic fields: Axisymmetric simulation in full general relativity. Phys. Rev. D **78**, 024029 (2008)
84. Kiuchi, K., Yoshida, S., Shibata, M.: Non-axisymmetric instabilities of neutron star with toroidal magnetic fields, Astron. Astrophys. **532**, A30 (2011)
85. Chruściel, P.T., Friedrich, H. (eds): The Einstein equations and the large scale behavior of gravitational fields—50 years of the Cauchy problem in general relativity. Birkhäuser Verlag, Basel (2004)

Appendix A
Conformal Killing Operator and Conformal Vector Laplacian

In this Appendix, we investigate the main properties of two important vectorial operators on Riemannian manifolds: the *conformal Killing operator* and the associated *conformal vector Laplacian*. The framework is that of a single three-dimensional manifold Σ, endowed with a Riemannian metric (cf. Sect. 2.3.2). In practice, Σ is embedded in some spacetime $(\mathscr{M}, \boldsymbol{g})$, as being part of a 3+1 foliation $(\Sigma_t)_{t\in\mathbb{R}}$, but we shall not require such a feature here. For concreteness, we shall denote Riemannian metric on Σ by $\tilde{\gamma}$, because in most applications of the 3+1 formalism, the conformal Killing operator appears for the metric $\tilde{\gamma}$ conformally related to the physical metric γ and introduced in Chap. 7. But again, we shall not use the hypothesis that $\tilde{\gamma}$ is derived from some "physical" metric γ. So in all what follows, $\tilde{\gamma}$ can be replaced by the physical metric γ or any other Riemannian metric, as for instance the background metric f introduced in Chaps. 7 and 8.

A.1 Conformal Killing Operator

A.1.1 Definition

The ***conformal Killing operator*** $\tilde{\boldsymbol{L}}$ associated with the metric $\tilde{\gamma}$ is the linear mapping from the space $\mathscr{T}(\Sigma)$ of vector fields on Σ to the space of symmetric tensor fields of type $(2,0)$ defined by

$$\boxed{\forall \boldsymbol{v} \in \mathscr{T}(\Sigma), \quad (\tilde{L}v)^{ij} := \tilde{D}^i v^j + \tilde{D}^j v^i - \frac{2}{3}\tilde{D}_k v^k \tilde{\gamma}^{ij}}, \tag{A.1}$$

where $\tilde{\boldsymbol{D}}$ is the Levi-Civita connection associated with $\tilde{\gamma}$ and $\tilde{D}^i := \tilde{\gamma}^{ij}\tilde{D}_j$. An immediate property of $\tilde{\boldsymbol{L}}$ is to be traceless with respect to $\tilde{\gamma}$, thanks to the $-2/3$ factor in the above definition: for any vector $\boldsymbol{v}$,

$$\tilde{\gamma}_{ij}(\tilde{L}v)^{ij} = 0. \tag{A.2}$$

É. Gourgoulhon, *3+1 Formalism in General Relativity*, Lecture Notes in Physics 846, 273
DOI: 10.1007/978-3-642-24525-1, © Springer-Verlag Berlin Heidelberg 2012

A.1.2 Behavior Under Conformal Transformations

An important property of $\tilde{\boldsymbol{L}}$ is to be invariant, up to some scale factor, with respect to conformal transformations. Indeed let us consider a metric γ conformally related to $\tilde{\gamma}$:

$$\gamma = \Psi^4 \tilde{\gamma}. \tag{A.3}$$

In practice γ will be the metric induced on Σ by the spacetime metric $\boldsymbol{g}$ and Ψ the conformal factor defined in Chap. 7, but we shall not employ this here. So γ and $\tilde{\gamma}$ are any two Riemannian metrics on Σ that are conformally related (we could have called them γ_1 and γ_2) and Ψ is simply the conformal factor between them. We can employ the formulæ derived in Chap. 7 to relate the conformal Killing operator of $\tilde{\gamma}, \tilde{\boldsymbol{L}}$ with that of γ, $\boldsymbol{L}$ say. Formula (7.32) gives

$$\begin{aligned} D^j v^i = \gamma^{jk} D_k v^i &= \Psi^{-4} \tilde{\gamma}^{jk} \left[\tilde{D}_k v^i + 2 \left(v^l \tilde{D}_l \ln \Psi \delta^i_k + v^i \tilde{D}_k \ln \Psi - \tilde{D}^i \ln \Psi \tilde{\gamma}_{kl} v^l \right) \right] \\ &= \Psi^{-4} \left[\tilde{D}^j v^i + 2 \left(v^k \tilde{D}_k \ln \Psi \tilde{\gamma}^{ij} + v^i \tilde{D}^j \ln \Psi - v^j \tilde{D}^i \ln \Psi \right) \right]. \end{aligned}$$

Hence

$$D^i v^j + D^j v^i = \Psi^{-4} \left(\tilde{D}^i v^j + \tilde{D}^j v^i + 4 v^k \tilde{D}_k \ln \Psi \tilde{\gamma}^{ij} \right)$$

Besides, from Eq. (7.33),

$$-\frac{2}{3} D_k v^k \gamma^{ij} = -\frac{2}{3} \left(\tilde{D}_k v^k + 6 v^k \tilde{D}_k \ln \Psi \right) \Psi^{-4} \tilde{\gamma}^{ij}.$$

Adding the above two equations, we get the simple relation

$$\boxed{(Lv)^{ij} = \Psi^{-4} (\tilde{L}v)^{ij}}. \tag{A.4}$$

Hence the conformal Killing operator is invariant, up to the scale factor Ψ^{-4}, under a conformal transformation.

A.1.3 Conformal Killing Vectors

Let us examine the kernel of the conformal Killing operator, i.e. the subspace $\ker \tilde{\boldsymbol{L}}$ of $\mathscr{T}(\Sigma)$ constituted by vectors $\boldsymbol{v}$ satisfying

$$(\tilde{L}v)^{ij} = 0. \tag{A.5}$$

A vector field which obeys Eq. (A.5) is called a ***conformal Killing vector***. It is the generator of some conformal isometry of $(\Sigma, \tilde{\gamma})$. A ***conformal isometry*** of $(\Sigma, \tilde{\gamma})$ is a diffeomorphism $\Phi : \Sigma \to \Sigma$ for which there exists some scalar field Ω such that $\Phi_* \tilde{\gamma} = \Omega^2 \tilde{\gamma}$. Notice that any isometry is a conformal isometry (corresponding to $\Omega = 1$), which means that every Killing vector is a conformal Killing vector.

The latter property is obvious from the definition (A.1) of the conformal Killing operator. Notice also that any conformal isometry of $(\Sigma, \tilde{\gamma})$ is a conformal isometry of (Σ, γ), where γ is a metric conformally related to $\tilde{\gamma}$ [cf. Eq. (A.3)]. Of course, $(\Sigma, \tilde{\gamma})$ may not admit any conformal isometry at all, yielding $\ker \tilde{\boldsymbol{L}} = \{0\}$. The maximum dimension of $\ker \tilde{\boldsymbol{L}}$ is 10 (taking into account that Σ has dimension 3). If $(\Sigma, \tilde{\gamma})$ is the Euclidean space $(\mathbb{R}^3, \boldsymbol{f})$, the conformal isometries are constituted by the isometries (translations, rotations) augmented by the homotheties.

A.2 Conformal Vector Laplacian

A.2.1 Definition

The ***conformal vector Laplacian*** associated with the metric $\tilde{\gamma}$ is the endomorphism $\tilde{\Delta}_L$ of the space $\mathscr{T}(\Sigma)$ of vector fields on Σ defined by the divergence of the conformal Killing operator:

$$\boxed{\forall \boldsymbol{v} \in \mathscr{T}(\Sigma), \quad \tilde{\Delta}_L v^i := \tilde{D}_j (\tilde{L} v)^{ij}}. \tag{A.6}$$

From Eq. (A.1),

$$\begin{aligned} \tilde{\Delta}_L v^i &= \tilde{D}_j \tilde{D}^i v^j + \tilde{D}_j \tilde{D}^j v^i - \frac{2}{3} \tilde{D}^i \tilde{D}_k v^k = \tilde{D}^i \tilde{D}_j v^j + \tilde{R}^i_{\ j} v^j + \tilde{D}_j \tilde{D}^j v^i - \frac{2}{3} \tilde{D}^i \tilde{D}_j v^j \\ &= \tilde{D}_j \tilde{D}^j v^i + \frac{1}{3} \tilde{D}^i \tilde{D}_j v^j + \tilde{R}^i_{\ j} v^j, \end{aligned} \tag{A.7}$$

where we have used the contracted Ricci identity (7.39). Hence $\tilde{\Delta}_L v^i$ is a second order operator acting on the vector $\boldsymbol{v}$, which is the sum of (i) the vector Laplacian $\tilde{D}_j \tilde{D}^j v^i$, (ii) one third of the gradient of the divergence $\tilde{D}^i \tilde{D}_j v^j$ and (iii) the curvature term $\tilde{R}^i_{\ j} v^j$:

$$\boxed{\tilde{\Delta}_L v^i = \tilde{D}_j \tilde{D}^j v^i + \tfrac{1}{3} \tilde{D}^i \tilde{D}_j v^j + \tilde{R}^i_{\ j} v^j} \tag{A.8}$$

The conformal vector Laplacian plays an important role in 3+1 general relativity, for solving the constraint equations (Chap. 9), but also for the time evolution problem (Sect. 10.3.2). The main properties of $\tilde{\Delta}_L$ have been first investigated by York [1, 2].

A.2.2 Elliptic Character

Given a point $p \in \Sigma$ and a linear form $\boldsymbol{\xi} \in \mathscr{T}^*_p(\sigma)$, the ***principal symbol*** of $\tilde{\Delta}_L$ with respect to p and $\boldsymbol{\xi}$ is the linear map $\boldsymbol{P}_{(p,\xi)} : \mathscr{T}_p(\Sigma) \rightarrow \mathscr{T}_p(\Sigma)$ defined as follows (see e.g. [3]). First of all, keep only the terms involving the highest derivatives in $\tilde{\Delta}_L$ (i.e. the second order ones); in terms of components, the operator is then reduced to

$$v^i \longmapsto \tilde{\gamma}^{jk}\frac{\partial}{\partial x^j}\frac{\partial}{\partial x^k}v^i + \frac{1}{3}\tilde{\gamma}^{ik}\frac{\partial}{\partial x^k}\frac{\partial}{\partial x^j}v^j \tag{A.9}$$

Then replace each occurrence of $\partial/\partial x^j$ by the component ξ_j of the linear form $\boldsymbol{\xi}$, thereby obtaining a mapping which is no longer differential, i.e. that involves only values of the fields at the point p; this is the principal symbol of $\tilde{\Delta}_L$ at p with respect to $\boldsymbol{\xi}$:

$$\begin{aligned} \boldsymbol{P}_{(p,\xi)} : & \mathscr{T}_p(\Sigma) \longrightarrow \mathscr{T}_p(\Sigma) \\ & \boldsymbol{v} = (v^i) \longmapsto \boldsymbol{P}_{(p,\xi)}(\boldsymbol{v}) = \left(\tilde{\gamma}^{jk}(p)\xi_j\xi_k v^i + \frac{1}{3}\tilde{\gamma}^{ik}(p)\xi_k\xi_j v^j\right), \end{aligned} \tag{A.10}$$

A differential operator, such as $\tilde{\Delta}_L$, is said to be ***elliptic*** on Σ iff the principal symbol $\boldsymbol{P}_{(p,\xi)}$ is an isomorphism for every $p \in \Sigma$ and every non-vanishing linear form $\boldsymbol{\xi} \in \mathscr{T}^*_p(\Sigma)$. It is said to be ***strongly elliptic*** if all the eigenvalues of $\boldsymbol{P}_{(p,\xi)}$ are non-vanishing and have the same sign. To check whether this is actually the case for $\tilde{\Delta}_L$, let us consider the bilinear form $\tilde{\boldsymbol{P}}_{(p,\xi)}$ associated to the endomorphism $\boldsymbol{P}_{(p,\xi)}$ by the conformal metric:

$$\forall(\boldsymbol{v},\boldsymbol{w}) \in \mathscr{T}_p(\Sigma)^2, \quad \tilde{\boldsymbol{P}}_{(p,\xi)}(\boldsymbol{v},\boldsymbol{w}) = \tilde{\boldsymbol{\gamma}}\left(\boldsymbol{v},\boldsymbol{P}_{(p,\xi)}(\boldsymbol{w})\right). \tag{A.11}$$

Its matrix $\tilde{\boldsymbol{P}}_{ij}$ is deduced from the matrix $P^i_{\ j}$ of $\boldsymbol{P}_{(p,\xi)}$ by lowering the index i with $\tilde{\gamma}(p)$. We get

$$\tilde{P}_{ij} = \tilde{\gamma}^{kl}(p)\xi_k\xi_l\tilde{\gamma}_{ij}(p) + \frac{1}{3}\xi_i\xi_j. \tag{A.12}$$

Hence $\tilde{\boldsymbol{P}}_{(p,\xi)}$ is clearly a symmetric bilinear form. Moreover it is positive definite for $\boldsymbol{\xi} \neq 0$: for any vector $\boldsymbol{v} \in \mathscr{T}_p(\Sigma)$ such that $\boldsymbol{v} \neq 0$, we have

$$\tilde{\boldsymbol{P}}_{(p,\xi)}(\boldsymbol{v},\boldsymbol{v}) = \tilde{\gamma}^{kl}(p)\xi_k\xi_l\tilde{\gamma}_{ij}(p)v^iv^j + \frac{1}{3}(\xi_i v^i)^2 > 0,$$

where the inequality follows from the positive definite character of $\tilde{\boldsymbol{\gamma}}$. $\tilde{\boldsymbol{P}}_{(p,\xi)}$ being a positive definite symmetric bilinear form, we conclude that $\boldsymbol{P}_{(p,\xi)}$ is an isomorphism and that all its eigenvalues are real and strictly positive. Therefore $\tilde{\Delta}_L$ is a strongly elliptic operator.

A.2.3 Kernel

Let us now determine the kernel of $\tilde{\Delta}_L$. Clearly this kernel contains the kernel of the conformal Killing operator $\tilde{\boldsymbol{L}}$. Actually it is not larger than that kernel:

$$\boxed{\ker \tilde{\Delta}_L = \ker \tilde{\boldsymbol{L}}.} \tag{A.13}$$

Let us establish this property. For any vector field $\boldsymbol{v} \in \mathscr{T}(\Sigma)$, we have

$$\begin{aligned}\int_{\Sigma} \tilde{\gamma}_{ij} v^i \tilde{\Delta}_L v^j \sqrt{\tilde{\gamma}} \mathrm{d}^3 x &= \int_{\Sigma} \tilde{\gamma}_{ij} v^i \tilde{D}_l (\tilde{L}v)^{jl} \sqrt{\tilde{\gamma}}\, \mathrm{d}^3 x \\ &= \int_{\Sigma} \left\{ \tilde{D}_l \left[\tilde{\gamma}_{ij} v^i (\tilde{L}v)^{jl} \right] - \tilde{\gamma}_{ij} \tilde{D}_l v^i (\tilde{L}v)^{jl} \right\} \sqrt{\tilde{\gamma}}\, \mathrm{d}^3 x \\ &= \oint_{\partial\Sigma} \tilde{\gamma}_{ij} v^i (\tilde{L}v)^{jl} \tilde{s}_l \sqrt{\tilde{q}}\, \mathrm{d}^2 y - \int_{\Sigma} \tilde{\gamma}_{ij} \tilde{D}_l v^i (\tilde{L}v)^{jl} \sqrt{\tilde{\gamma}}\, \mathrm{d}^3 x, \qquad \text{(A.14)}\end{aligned}$$

where the Gauss–Ostrogradsky theorem has been used to get the last line. We shall consider two situations for $(\Sigma, \tilde{\gamma})$:

- Σ is a *closed manifold*, i.e. is compact without boundary;
- $(\Sigma, \tilde{\gamma})$ is an *asymptotically flat manifold*, in the sense made precise in Sect. 8.2.

In the former case the lack of boundary of Σ implies that the first integral in the right-hand side of Eq. (A.14) is zero. In the latter case, we will restrict our attention to vectors $\boldsymbol{v}$ which decay at spatial infinity according to (cf. Sect. 8.2)

$$v^i = O(r^{-1}) \qquad \text{(A.15)}$$

$$\frac{\partial v^i}{\partial x^j} = O(r^{-2}), \qquad \text{(A.16)}$$

where the components are to be taken with respect to the asymptotically Cartesian coordinate system (x^i) introduced in Sect. 8.2. The behavior (A.15, A.16) implies

$$v^i (\tilde{L}v)^{jl} = O(r^{-3}),$$

so that the surface integral in Eq. (A.14) vanishes. So for both cases (Σ closed or asymptotically flat) Eq. (A.14) reduces to

$$\int_{\Sigma} \tilde{\gamma}_{ij} v^i \tilde{\Delta}_L v^j \sqrt{\tilde{\gamma}}\, \mathrm{d}^3 x = - \int_{\Sigma} \tilde{\gamma}_{ij} \tilde{D}_l v^i (\tilde{L}v)^{jl} \sqrt{\tilde{\gamma}}\, \mathrm{d}^3 x. \qquad \text{(A.17)}$$

In view of the right-hand side integrand, let us evaluate

$$\begin{aligned}\tilde{\gamma}_{ij} \tilde{\gamma}_{kl} (\tilde{L}v)^{ik} (\tilde{L}v)^{jl} &= \tilde{\gamma}_{ij} \tilde{\gamma}_{kl} (\tilde{D}^i v^k + \tilde{D}^k v^i)(\tilde{L}v)^{jl} - \frac{2}{3} \tilde{D}_m v^m \underbrace{\tilde{\gamma}^{ik} \tilde{\gamma}_{ij}}_{\delta^k_j} \tilde{\gamma}_{kl} (\tilde{L}v)^{jl} \\ &= (\tilde{\gamma}_{kl} \tilde{D}_j v^k + \tilde{\gamma}_{ij} \tilde{D}_l v^i)(\tilde{L}v)^{jl} - \frac{2}{3} \tilde{D}_m v^m \underbrace{\tilde{\gamma}_{jl} (\tilde{L}v)^{jl}}_{0} \\ &= 2 \tilde{\gamma}_{ij} \tilde{D}_l v^i (\tilde{L}v)^{jl},\end{aligned}$$

where we have used the symmetry and the traceless property of $(\tilde{L}v)^{jl}$ to get the last line. Hence Eq. (A.17) becomes

$$\int_{\Sigma} \tilde{\gamma}_{ij} v^i \tilde{\Delta}_L v^j \sqrt{\tilde{\gamma}}\, \mathrm{d}^3 x = -\frac{1}{2} \int_{\Sigma} \tilde{\gamma}_{ij} \tilde{\gamma}_{kl} (\tilde{L}v)^{ik} (\tilde{L}v)^{jl} \sqrt{\tilde{\gamma}}\, \mathrm{d}^3 x.$$

Let us assume now that $\boldsymbol{v} \in \ker\tilde{\boldsymbol{\Delta}}_L$: $\tilde{\Delta}_L v^j = 0$. Then the left-hand side of the above equation vanishes, leaving

$$\int_\Sigma \tilde{\gamma}_{ij}\tilde{\gamma}_{kl}(\tilde{L}v)^{ik}(\tilde{L}v)^{jl}\sqrt{\tilde{\gamma}}\,\mathrm{d}^3x = 0. \tag{A.18}$$

Since $\tilde{\gamma}$ is a positive definite metric, we conclude that $(\tilde{L}v)^{ij} = 0$, i.e. that $\boldsymbol{v} \in \ker\tilde{\boldsymbol{L}}$. This demonstrates property (A.13). Hence the "harmonic functions" of the conformal vector Laplacian $\tilde{\Delta}_L$ are nothing but the conformal Killing vectors (one should add "which vanish at spatial infinity as (A.15) and (A.16)" in the case of an asymptotically flat space).

A.2.4 Solutions to the Conformal Vector Poisson Equation

Let now discuss the existence and uniqueness of solutions to the conformal vector Poisson equation

$$\boxed{\tilde{\Delta}_L v^i = S^i}, \tag{A.19}$$

where the vector field $\boldsymbol{S}$ is given (the source). Again, we shall distinguish two cases: the closed manifold case and the asymptotically flat one. When Σ is a closed manifold, we notice first that a necessary condition for the solution to exist is that the source must be orthogonal to any vector field in the kernel, in the sense that

$$\forall \boldsymbol{C} \in \ker\tilde{\boldsymbol{L}}, \quad \int_\Sigma \tilde{\gamma}_{ij}C^iS^j\sqrt{\tilde{\gamma}}\,\mathrm{d}^3x = 0. \tag{A.20}$$

This is easily established by replacing S^j by $\tilde{\Delta}_L v^i$ and performing the same integration by part as above to get

$$\int_\Sigma \tilde{\gamma}_{ij}C^iS^j\sqrt{\tilde{\gamma}}\,\mathrm{d}^3x = -\frac{1}{2}\int_\Sigma \tilde{\gamma}_{ij}\tilde{\gamma}_{kl}(\tilde{L}C)^{ik}(\tilde{L}v)^{jl}\sqrt{\tilde{\gamma}}\,\mathrm{d}^3x.$$

Since, by definition $(\tilde{L}C)^{ik} = 0$, Eq. (A.20) follows. If condition (A.20) is fulfilled, it can be shown that Eq. (A.19) admits a solution and that this solution is unique up to the addition of a conformal Killing vector.

Remark A.1 If the metric $\tilde{\gamma}$ does not admit any conformal Killing vector, condition (A.20) is trivially fulfilled.

In the asymptotically flat case, we assume that, in terms of the asymptotically Cartesian coordinates (x^i) introduced in Sect. 8.2

$$S^i = O(r^{-3}). \tag{A.21}$$

Moreover, because of the presence of the Ricci tensor in $\tilde{\Delta}_L$, one must add the decay condition

$$\frac{\partial^2 \tilde{\gamma}_{ij}}{\partial x^k \partial x^l} = O(r^{-3}) \tag{A.22}$$

to the asymptotic flatness conditions introduced in Sect. 8.2 [Eqs. (8.1)–(8.2)]. Indeed Eq. (A.22) along with Eqs. (8.1) guarantees that

$$\tilde{R}_{ij} = O(r^{-3}). \tag{A.23}$$

Then a general theorem by Cantor (1979) [4], regarding elliptic operators on asymptotically flat manifolds, can be invoked (see Appendix B of Ref. [5] as well as Ref. [6]) to conclude that the solution of Eq. (A.19) with the boundary condition

$$v^i = 0 \quad \text{when}\, r \to 0 \tag{A.24}$$

exists and is unique. The possibility to add a conformal Killing vector to the solution, as in the compact case, does no longer exist because there is no conformal Killing vector which vanishes at spatial infinity on asymptotically flat Riemannian manifolds.

Regarding the numerical techniques to solve the conformal vector Poisson equation (A.19), let us mention that a very accurate spectral method has been developed by Grandclément et al. [7] in the case of the Euclidean space: $(\Sigma, \tilde{\gamma}) = (\mathbb{R}^3, \boldsymbol{f})$. It is based on the use of Cartesian components of vector fields altogether with spherical coordinates. An alternative technique, using both spherical components and spherical coordinates is presented in Ref. [8].

References

1. York, J.W.: Conformally invariant orthogonal decomposition of symmetric tensors on Riemannian manifolds and the initial-value problem of general relativity. J. Math. Phys. **14**, 56 (1973)
2. York, J.W.: Covariant decompositions of symmetric tensors in the theory of gravitation. Ann. Inst. Henri Poincaré A 21:319 (1974). http://www.numdam.org/item?id=AIHPA_1974__21_4_319_0
3. Dain, S.: Elliptic systems. In: Frauendiener, J., Giulini, D.J.W., Perlick, V. (eds.) Analytical and Numerical Approaches to Mathematical Relativity, Lect. Notes Phys. **692**, Springer, Berlin (2006), p. 117
4. Cantor, M.: Some problems of global analysis on asymptotically simple manifolds. Compositio Mathematica **38**, 3 (1979). http://www.numdam.org/item?id=CM_1979__38_1_3_0
5. Smarr, L., York, J.W.: Radiation gauge in general relativity. Phys. Rev. D **17**, 1945 (1978)
6. Choquet-Bruhat, Y., Isenberg, J., York, J.W.: Einstein constraints on asymptotically Euclidean manifolds. Phys. Rev. D **61**, 084034 (2000)
7. Grandclément, P., Bonazzola, S., Gourgoulhon, E., Marck, J.-A.: A multi-domain spectral method for scalar and vectorial Poisson equations with non-compact sources. J. Comput. Phys. **170**, 231 (2001)
8. Bonazzola, S., Gourgoulhon, E., Grandclément, P., Novak, J.: Constrained scheme for the Einstein equations based on the Dirac gauge and spherical coordinates. Phys. Rev. D **70**, 104007 (2004)

Appendix B
Sage Codes

Sage is a modern and powerful computer algebra system; it is open-source and freely downloadable from

http://sagemath.org/

Various Sage codes are provided below to perform otherwise tedious computations. These codes can be downloaded as Sage worksheets from the book web page

http://relativite.obspm.fr/3p1

B.1 Riemann Tensor

Metric tensor:

```
# manifold dimension
n = 3
# coordinates
var('ro, th, ph, b') ; x = [ro, th, ph]
## var('r, th, ph') ; x = [r, th, ph]
## var('th, ph') ; x = [th, ph]
# H^3 metric :
g = [[b^2,0,0], [0,(b*sinh(ro))^ 2,0], \
   [0,0,(b*sinh(ro)*sin(th))^ 2]]
## R^3 metric :
## g = [[1,0,0],[0,r^2,0],[0,0,(r*sin(th))^2]]
## S^ 2 metric :
## g = [[r^2,0], [0, (r*sin(th))^ 2]]
g_mat = matrix(g)
ginv_mat = g_mat.inverse()
ginv = [[ginv_ mat[i,j].simplify_full() for j in range(n)] \
   for i in range(n)]
```


DOI: 10.1007/978-3-642-24525-1,

Christoffel symbols:

```
Chr0 = [[[ sum(ginv[i][l]/2 * (diff(g[l][k],x[j]) \
   + diff(g[j][l],x[k]) - diff(g[j][k],x[l])) \
   for l in range(n)) for k in range(n) ] for j in range(n) ] \
   for i in range(n) ]
Chr = [[[ Chr0[i][j][k].simplify_full() for k in range(n) ] \
   for j in range(n) ] for i in range(n) ]
```

Riemann tensor:

```
Riem0 = [[[[ diff(Chr[i][j][l],x[k]) - diff(Chr[i][j][k],x[l]) \
   + sum(Chr[i][m][k] * Chr[m][j][l] \
   - Chr[i][m][l] * Chr[m][j][k] for m in range(n))\
   for l in range(n) ] for k in range(n) ] for j in range(n) ] \
   for i in range(n) ]
Riem = [[[[ Riem0[i][j][k][l].simplify_full() \
   for l in range(n) ] for k in range(n) ] for j in range(n) ] \
   for i in range(n) ]
```

Ricci tensor and scalar curvature:

```
Ric0 = [[ sum(Riem[k][i][k][j] for k in range(n) ) \
   for j in range(n) ] for i in range(n) ]
Ric = [[ Ric0[i][j].simplify_full() for j in range(n) ] \
   for i in range(n) ]
Rscal0 = sum(sum(ginv[i][j] * Ric[i][j] \
   for j in range(n)) for i in range(n))
Rscal = Rscal0.simplify_full()
```

Check of maximally symmetric character:

```
max_sym0 = [[[[ Riem[i][j][k][l] - Rscal/(n*(n-1)) \
   * (identity_matrix(n)[i,k] * g[j][l] \
   - identity_matrix(n)[i,l] * g[j][k]) for l in range(n) ] \
   for k in range(n) ] for j in range(n) ] for i in range(n)]
max_sym = [[[[max_sym0[i][j][k][l].simplify_full ()\
   for l in range(n) ] for k in range(n) ] \
   for j in range(n) ] for i in range(n) ]
```

B.2 Hyperboloidal Slicing of Minkowski Spacetime

Here we present the Sage code used in Examples 4.1–4.4 and 5.1, devoted to the slicing of Minkowski spacetime $(\mathscr{M}, \boldsymbol{g})$ by a family of hyperbolic spaces $\Sigma_t \sim \mathbb{H}^3$. Hereafter (w, x, y, z) are Minkowskian coordinates of $\mathscr{M}$. Aiming to producing two-dimensional plots, we set $y = z = 0$ in the formulae of Examples 4.1–4.4 and 5.1. In other words, we work only in the (w, x) plane. In addition, we use units for which $b = 1$.

Preliminaries:

```
from matplotlib import rc
rc('text', usetex=True) # using TeX rendering in labels
var('w x') # declaring (w,x) as the basic coordinates
```

Scalar function f defining the foliation as $w = f(x, t)$ [Eq. (4.13)]:

```
def fw(x,t) :
...          return t + sqrt(1+x*x)
```

Lapse function [Eq. (4.14)], unit normal vector [Eq. (4.16)], 4-acceleration of Eulerian observers [vector metric-dual to the 1-form given by Eq. (4.21)] and shift vector [Eq. (5.41)]:

```
def nn(x) :
...          return sqrt(1+x*x)
def vect_n(w,x) :
...          return [nn(x), x]
def vect_a(w,x) :
...          return [x^ 2/nn(x), x]
def vect_beta(w,x) :
...          return [-x^ 2, -x*nn(x)]
```

Plotting selected slices:

```
ns = 9 # number of slices
tmin = −4 ; tmax = 4
# selected values of t :
ts = [ tmin + i*(tmax-tmin)/(ns-1.) for i in range(ns)]
graph = Graphics()
for i in range(ns) :
...          graph += plot(fw(x,ts[i]),(x,-3,3))
```

Adding the future light cone originating from (w,x) = (0,0):

```
cone = line([(−3,3),(0,0),(3,3)], color = ''green'', \
   linestyle='-', thickness=0.5)
graph += cone
```

Adding points A and B, as well as vectors $\boldsymbol{n}$ and $\boldsymbol{a}$ at these points:

```
x1 = 1 ; w1 = fw(x1, 0) # point B
w2 = w1 + vect_n(w1,x1)[0] ; x2 = x1 + vect_n(w1,x1)[1] # n at B
points = circle((x1,w1),0.05, fill=True, color=''black'') \
  + text(''$B$'', (x1-0.15,w1+0.18), fontsize=16, color=''black'')
graph + = line([(0,0), (x1,w1)], color=''black'',
   linestyle='--')
vectors = arrow((x1,w1), (x2,w2)) \
  + text('$\mathbf{n}$',(x2,w2+0.1), fontsize=16
w2 = w1 + vect_a(w1,x1)[0] ; x2 = x1 + vect_a(w1,x1)[1] # a at B
vectors += arrow((x1,w1), (x2,w2)) \
   + text ('$ \mathbf{a}$',(x2,w2+0.1), fontsize=16)
x1 = 0 ; w1 = fw(x1, 0) # point A
w2 = w1 + vect_n(w1,x1)[0] ; x2 = x1 + vect_n(w1,x1)[1] # n at A
points += circle((x1,w1),0.05, fill=True, color=''black'') \
  + text(''$A$'', (x1+0.15,w1+0.15), fontsize=16, color=''black'')
vectors += arrow((x1,w1), (x2,w2)) \
  + text('$\mathbf{n}$',(x2+0.15,w2-0.1), fontsize=16)
```

Producing Fig. 4.4:

```
label_slice = text('$\sigma_{-3}$', (3,0.35), fontsize=16) \
...    + text('$\sigma_{-2}$', (3,1.35), fontsize=16)\
...    + text('$\sigma}_{-1\}\$', (3,2.35), fontsize=16) \
...    + text('$\{sigma}_{0\}\$', (3,3.35), fontsize=16) \
...    + text('$\sigma_{1}\$', (3,4.15), fontsize=16) \
...    + text('$ \sigma}_{2}\$', (1.5,4.15), fontsize=16)
show(graph+label_slice+points+vectors, xmin=-3, xmax=3, \
  ymin=-1, ymax=4, aspect_ratio=1, \
  axes_labels=['$x/b$','$w/b$'], axes_pad=0, fontsize=12)
```

Producing Fig. 5.2:

```
x1 = 1 ; w1 = fw(x1, 0) # point B
w2 = w1 + vect_n(w1,x1)[0] ; x2 = x1 + vect_n(w1,x1)[1] # n at B
vectors = arrow((x1,w1), (x2,w2)) \
  + text(r'$\mathbf{n}$',(x2-0.1,w2+0.1), fontsize=16)
# Vector m = N n :
w3 = w1 + nn(x1)*vect_n(w1,x1)[0]
x3 = x1 + nn(x1)*vect_n(w1,x1)[1]
vectors += arrow((x1,w1), (x3,w3)) \
  + text(r'$N\mathbf{n}$',(x3+0.15,w3+0.1), fontsize=16)
# Vector d/dt :
```

```
w4 = w1 + 1 ; x4 = x1 ;
vectors += arrow((x1,w1), (x4,w4)) \
  + text(r'$\partial_t$',(x4+0.2,w4-0.2), fontsize=16)
# Shift vector :
w5 = w1 + vect_beta(w1,x1)[0] ; x5 = x1 + vect_beta(w1,x1)[1]
vectors += arrow((x1,w1), (x5,w5)) \
  + text(r'$\boldmath $\beta$',(x5,w5+0.2), fontsize=16)
joint_line = line([(x3,w3), (x4,w4), (x5,w5)], \
  color = ''black'', linestyle=':', thickness=2)
x_const_line = line([(x1,-1), (x1,4)], color = ''red'', \
  linestyle='--') + text(r'$\rho, \theta, \varphi) = \mathrm{const}$',\
     (x1+1, -0.8), color=''red'', fontsize=16)
show(graph+label_slice+points+vectors+joint_line+
   x_const_line, \
 xmin=-3, xmax=3, ymin=-1, ymax=4, aspect_ratio=1, \
 axes_labels=['$x/b$','$w/b$'], axes_pad=0, fontsize=12)
```

B.3 Other Sage Codes

Other Sage codes related to the 3+1 formalism can be found at the URL http://relativite.obspm.fr/3p1.

Index

É. Gourgoulhon, *3+1 Formalism in General Relativity*, Lecture Notes in Physics 846,
DOI: 10.1007/978-3-642-24525-1, © Springer-Verlag Berlin Heidelberg 2012